Complex Dynamics and Morphogenesis

Grenoble Sciences

The aim of Grenoble Sciences is twofold:

- to produce works corresponding to a clearly defined project, without the constraints of trends nor curriculum;

- to ensure the utmost scientific and pedagogic quality of the selected works: each project is selected by Grenoble Sciences with the help of anonymous referees. In order to optimize the work, the authors interact for a year (on average) with the members of a reading committee, whose names figure in the front pages of the work, which is then co-published with the most suitable publishing partner.

Contact
Tel.: (33) 4 76 51 46 95
E-mail: grenoble.sciences@univ-grenoble-alpes.fr
Website: https://grenoble-sciences.ujf-grenoble.fr

Scientific Director of Grenoble Sciences
Jean Bornarel, Emeritus Professor
at the Grenoble Alpes University, Grenoble, France

Grenoble Sciences is a department of the Grenoble Alpes University
supported by the **ministère de l'Éducation nationale,
de l'Enseignement supérieur et de la Recherche**
and the **région Auvergne-Rhône-Alpes**.

Complex Dynamics and Morphogenesis is an improved version of the original book *Dynamiques complexes et morphogenèse* by Chaouqi Misbah, Springer-Verlag France, 2011, ISBN 978 2 8178 0193 3.

The reading committee included the following members:

- **Bernard Billia**, CNRS Senior Researcher, IM2NP, Marseille;
- **François Charru**, Professor, Toulouse University;
- **Stéphane Douady**, CNRS Researcher, Laboratoire MSC, Paris;
- **Hamid Kellay**, Professor, Bordeaux University;
- **Paul Manneville**, CNRS Senior Researcher, LADHYX, Palaiseau;
- **Christian Wagner**, Professor, Saarland University.

Editorial coordination: Stéphanie Trine; translation from original French version: Joyce Lainé; copy-editing: Ali Woollatt and Stéphanie Trine; figures: Sylvie Bordage; cover illustration: Alice Giraud, from fig. 1.4 [© NOAA], 13.5a [© forcdan Fotolia], 15.2 [all rights reserved], 15.3d [from [78], reprinted with permission from Macmillan Publishers Ltd] and elements provided by the author.

Chaouqi Misbah

Complex Dynamics
and Morphogenesis

An Introduction to Nonlinear Science

Chaouqi Misbah
CNRS
Laboratoire Interdisciplinaire de Physique
Grenoble
France

Co-published by Springer International Publishing AG, Gewerbestrasse 11, 3330 Cham, Switzerland, and Editions Grenoble Sciences, Université Joseph Fourier, 230 rue de la Physique, BP 53 38041 Grenoble cedex 09, France.

ISBN 978-94-024-1467-7   ISBN 978-94-024-1020-4   (eBook)
DOI 10.1007/978-94-024-1020-4

Printed on acid-free paper

This Springer imprint is published by Springer Nature
The registered company is Springer Science+Business Media B.V.
The registered company address is: Van Godewijckstraat 30, 3311 GX Dordrecht, The Netherlands

To Florence, Hisham and Sammy
To my parents, sisters and brothers

Foreword

As stated by its subtitle, this book is an introduction to nonlinear sciences. This subtitle has the merit of emphasizing the multiplicity of this book which deals with biology, chemistry, and of course physics and mathematics.

If someone thought to divide science into linear and nonlinear science, no doubt that the latter would form the main part. However, the range of linear physics, i.e. physics described by first order equations, is surprisingly extended. For instance, linear elasticity is a good approximation in many cases. Maxwell's equations of electromagnetism are linear with respect to the electromagnetic field. Schrödinger's and Dirac's equations are linear with respect to the wavefunction. These partial differential equations allow the use of sophisticated mathematical techniques such as Green's functions and retarded potentials, making linear physics efficient, but also cumbersome.

Nonlinear science is quite different: analytic methods are of limited use. Computers may of course provide a workaround, but numerical results do not give the same insight as analytic arguments. Thus, a texbook on nonlinear science has to be less ambitious in its accuracy of description than a book on quantum mechanics, electromagnetism or elasticity – but it may be more pleasant to read. The present book by Chaouqi Misbah has this virtue. Since one cannot go more deeply into details, it presents the essential features, sometimes in a purely qualitative way. These features have an extraordinary diversity, in contrast to the rather monotonous landscape of linear physics.

In this book, the reader finds a variety of phenomena, many familiar but not understood – for example, a swing. It is unfortunate that children at school do not learn how to use a swing, because they would discover what parametric resonance is. As explained in the book, the greatest amplitude is reached if the child lowers their center of mass when the swing goes up, and raises it when the swing goes down, creating an excitation frequency at twice the natural frequency of the swing. This is typical of parametric resonance.

Swings are not the only entertainment the reader may find in this book. Another example is the transition to chaos by subharmonic cascade. Is it not engaging to play with a pocket computer and observe how the change of a parameter (a in formula (8.8)) changes the trivial fixed point into a discrete limit cycle of 2 points, then 4 points, 8 points, etc., before reaching complete chaos? This will be surprising for the newcomer to nonlinear science; another surprise awaits (section 8.5) with the discovery that the continuous version of the same model is not chaotic at all but has a perfectly stable, nontrivial fixed point.

All the same, nonlinear science cannot be reduced to entertainment, as the reasonable number of equations and formulae in this book testify. Mathematical formulae are certainly necessary for the precise description of physical phenomena. They can also be a didactic tool for clarifying an argument. For example, in chapter 4 of the book, catastrophes are classified for cases of up to 4 control parameters. As it turns out, there are 7 catastrophes (as Snow White is accompanied by 7 dwarfs, as the golden Hebrew lampstand has 7 branches!). As noted in the book, this classification can be done without writing the equations governing the systems, but the author's argument does rely on mathematical formulae. Although not strictly necessary, they certainly simplify the proof. A merit of this book lies in its clever use of mathematical formulae to help one understand the mechanism but avoiding complications when not absolutely necessary. Sometimes they are necessary: chapters 10 and 12 are really tough. However, as a whole, the reader has before them an enjoyable source of discoveries which mixes rigor and entertainment harmoniously.

Jacques Villain
Emeritus CEA Senior Researcher
Member of the French Academy of Sciences

Acknowledgments

I am indebted to Christiane Caroli and Philippe Nozières for their supportive attitudes which encouraged me to undertake the project of writing this book. I am also grateful to Michel Saint Jean who encouraged me from the outset of this project and Jacques Villain for many enlightening discussions, and I warmly thank the members of my research team in Grenoble, Dynamique des Fluides Complexes et Morphogenèse (DYFCOM), for the critical reading of the manuscript. I thank especially Pierre-Yves Gires who has gone into many details of the derivations of the model equations and their solutions, Christian Wojcik for preparing many plots, Richard Michel who produced from scratch many numerical results presented in this book and Alexander Farutin for checking the solutions to the exercises. I express my gratitude to Marc Rabaud for having redone the Saffman–Taylor experiment and providing me with the photos presented in chapter 15. During the elaboration of this project I have greatly benefitted from references [6, 9, 16, 23, 29, 32, 42, 58, 75, 77, 80, 87, 91, 99, 101].

Chaouqi Misbah

Contents

Chapter 1

Presentation of Main Ideas

Abstract *This chapter provides an overview of the main topics covered in this book, starting from systems with no spatial dimension (such as a point particle) and continuing to extended systems with patterns such as ripples in the sand, and stripes and spots on the furs of animals such as zebras or leopards.*

The twenty-first century saw with its emergence a new field of research: that of nonlinear phenomena. Far from being a fad or fashion, nonlinearity is, on the contrary, a problem underlying the scientific community in its entirety. This is reflected by its central position across all of the disciplines: physics, chemistry, mathematics, biology, geology, economy, social sciences, etc.

A fair portion of natural phenomena, such as the movement of the planets or the propagation of electromagnetic waves, can be described by linear theories via application of the superposition[1] principle. However, this approach is inadequate for most phenomena, in nature or everyday life, because of the prevalence of nonlinear effects. The stock market, for example, is a typical, daily incarnation of nonlinear behavior in the financial and economic worlds. At any given moment, the state of the stock market depends on an ensemble of decisions made by stockbrokers; that is, each stockbroker makes his decision to buy or sell stocks based on several factors, such as the global economic situation at that moment, current rumors on the financial health of one or another group of actionaries, or just on personal guesses about the decisions of other brokers. All in all, the stockbroker does not act as an isolated system, and this places his behavior beyond the scope of description by superposition.

---

1. The principle of superposition assumes that two coexisting effects are independent of each other and do not interfere with each other; thus, the result of the effects is the same as the sum of both.

© Springer Science+Business Media B.V. 2017
C. Misbah, *Complex Dynamics and Morphogenesis*,
DOI 10.1007/978-94-024-1020-4_1

Instead, we have the formidable problem of interaction and interdependence between phenomena.

The manifestations of nonlinearity, partially overlapping with complexity, is accompanied with a great diversity of behaviors. For example, the presence of temporal oscillations in phenomena (oscillating solutions); the appearance of chaotic states, characterized by an unpredictable evolution of the system, even if the system is completely deterministic and has, in principle, no outside interferences; and the birth of disordered or ordered spatial structures (morphogenesis) such as the stripes or spots on the furs of certain animals such as zebras or leopards.

Remarkably, this complexity will occur even in relatively simple systems (a fascinating fact for the scientists who study these phenomena). Both physics and chemistry – to name two scientific domains – are chock-full of these kinds of simple systems which, though modeled by relatively elementary equations, reveal an unsuspected complexity because of intrinsic nonlinear effects. Studying the complexity in these simple systems will help us establish a rough roadmap of useful concepts and methods for the study of more complex systems.

Now, the evolution of the stock market mentioned above is already a complex problem and difficult to reduce to a system of simple "enough" equations. Finding complexity in an already complex system is not especially enlightening. In addition, many complex systems cannot be easily modeled in terms of equations in a non-ambiguous way. Fortunately, we will see that we can reach several general conclusions without specifying the equations that govern systems. With catastrophe theory, for example, no explicit evocation of any set of evolution equations is necessary; one can reach general conclusions about the behavior and the state of these systems using relatively general considerations instead (see the three last sections of chapters 3 and 4).

One type of nonlinear phenomenon familiar to all of us, though not necessarily with the classification as such, is that of a chemical reaction. Here, the term "reaction" hints at the notions of "interaction", "non-additive", and thus of "nonlinear phenomena". Consider, as an example, the case of two substances S_A and S_B that react to create a substance S_C, as symbolized by the following equation:

$$S_A + S_B \longrightarrow S_C. \tag{1.1}$$

The speed of the reaction and creation of substance S_C depends, in the simplest case, on the product of the two concentrations, $A \times B$, of substances S_A and S_B. If the concentrations A and B are modified by the quantities δA and δB, the speed of the reaction changes in a way proportional to $(A + \delta A)(B + \delta B)$. The variation in speed is given by $(A + \delta A)(B + \delta B) - AB = A\delta B + B\delta A + \delta A\delta B$, and not by $\delta A + \delta B$, as would be the case if we were using the principle of superposition.

Both in the inert and in the living world, chemical reactions offer a rich panoply of fascinating behaviors. They preside over various biological regulations and evolutions. We will examine the various consequences of chemical reactions and their complex dynamics (see chapters 5 and 9).

It should be kept in mind, however, that the notion of "reaction" models (which are classically met in chemistry) is not limited to the domain of chemical species. The behavior of some populations, such as the prey-predator interaction (chapter 5) in biology, or in social science or economics, can be described by this type of model. It is thus likely that the behavior of these systems are governed by similar rules, even if, at first sight, there is no clear link between the two.

In general, a nonlinear system is described by a set of coupled equations, such as differential equations or partial differential equations. The presence of nonlinear terms creates, for the same system and the same value of parameters, a large, not to say infinite, number of possible solutions. This stands in stark contrast with linear systems, for which one unique[2] solution exists. A nonlinear system "chooses" one solution from the many possible ones; the question can be asked: is there a principle, a criterion, a law, that allows a system to choose a solution? Systems in equilibrium are governed by the second law of thermodynamics[3], but the absence of such a principle in non-equilibrium systems raises a great challenge regarding the choice of a solution among a large manifold of possibilities.

Even if a nonlinear system "chooses" an initial solution from the ensemble of all possible solutions, it is not given that this specific solution would be possible for any value of the system's control parameters[4]. As the parameters vary, the system may suddenly jump from one solution to another, say, from a static solution to one that oscillates. The new solution is qualitatively different from the first. The change, once the control parameter reaches its critical value, can happen suddenly and without any precursor. For example, a relatively stable climate zone can, following a fall in atmospheric pressure, give way to a large storm. This qualitative change is called a *bifurcation*.

---

2. We do not speak here of degenerate situations (which are, in principle, unphysical) such as linear systems of the form $x + y = 1$ and $2x + 2y = 2$, which allow for infinite solutions, since the two equations are not independent (degenerate system). In more formal language, a solution of a linear system is unique if the system satisfies the Fredholm alternative theorem (see chapter 6).

3. The second principle of thermodynamics, also called the second law, describes the irreversibility of an isolated system's evolution, especially after an exchange of heat. The principle stipulates that all spontaneous evolution of an isolated system creates entropy and the equilibrium is attained when entropy is maximal.

4. A parameter is called a control parameter if it is a quantity that, in principle, the experimentalist may act upon. More generally, a control parameter is a parameter that acts on the behavior of the solution of a nonlinear system.

In general, a bifurcation will appear either because the old solution (in the example above, the calm weather) becomes unstable, or simply because the stable solution stops existing (an example from classical mechanics: a beam reaches its limit – the equilibrium solution is no longer possible – this entails the collapse of the structure). In this book, we will explore the rich variety of bifurcations encountered in nonlinear systems (see chapters 2, 3, 5 and 6).

After observing the behavior of many natural systems, particularly those in biology, mathematician René Thom suggested the name *catastrophe* for describing the qualitative behavioral changes of a system. In the 1970s he constructed a new analytic approach to these systems in which the qualitatively different solutions of a system are represented in the space of its control parameters [101]. For example, imagine that we have a system with two control parameters that an experimentalist can change in order to induce one behavior or another. Thus, for some region of the control parameter space we may have a given behavior, while we have another behavior in a different region of that space. We can thus subdivide the parameter space into regions, each corresponding to a different behavior of the system (like the geographic lines separating countries on a map). It turns out that the ensemble of the frontier lines form a specific geometric figure, for example, a butterfly (see figure 1.1, also known as a catastrophe portrait). In this space, crossing a frontier line, or as they are called, a catastrophe line, corresponds to a qualitative change in the solution.

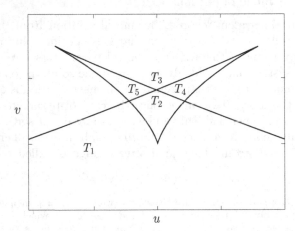

Figure 1.1 – A geometric figure of the "butterfly" type delimiting the regions corresponding to different types (T_1–T_5) of solutions. u and v are the dimensionless control parameters.

A fascinating feature is that for a given set of parameters we can only have a limited number of portraits. Each type of catastrophe has a unique portrait or "fingerprint", a specific geometric figure in parameter space. If we know the number of independent control parameters relevant to a system, it is possible to predict the number and nature of possible catastrophes without needing to know the evolution equations that govern the system. Knowing only the number of control parameters, we can identify seven basic (or elementary) catastrophes (and independently of the number of degrees of freedom), as we will see. In this book we will describe them from a relatively elementary and intuitive angle (see chapter 4).

An important component within the study of nonlinear phenomena is the study of stable solutions (i.e. solutions with a sufficiently long duration, not to say infinite). For example, the sun rises and sets regularly, and although the length of the day and night change with the seasons, the movement has a permanent character as it is repeated every year. Even a chaotic signal, though not repeated if we only look at the value of the signal, is repeated in an abstract (or for the moment, we can say, a statistical) sense that we will touch upon briefly (see chapter 8). In short, all that is not a transient phenomenon (i.e. lasts a small amount of time relative to the characteristic time of the system) will be called permanent, whatever the complexity of the movement.

From all the possible solutions, only those of a stable nature will endure, but we can sometimes observe unstable solutions[5] for short durations. For example, take a ball balanced on top of a hill (figure 1.2); this position will remain as a solution until, due to any small perturbation, the ball tumbles down. Since the top-of-the-hill solution is susceptible to weak perturbations, the solution is called linearly unstable[6]; the final state can be fundamentally changed after a perturbation. In contrast, stable solutions will remain the same in the face of perturbations of either intrinsic (say, uncontrolled fluctuations) or extrinsic (caused by an external agent) nature, returning to the initial state after a certain time following the perturbation (as would be the case for a ball located at the bottom of a valley).

When a system resists faint perturbations but gives way to stronger ones, the system is said to be linearly stable (i.e. stable against a small perturbation; figure 1.2) but nonlinearly unstable (which is to say there exists a threshold perturbation amplitude after which the perturbations throw the system out of its original equilibrium; i.e. one has to give a very strong kick to dislodge a ball from the depth of one valley to the depth of a neighboring valley).

---

5. The study of unstable solutions can sometimes be very interesting or essential in certain contexts, such as in chaos where it allows for a subtle comprehension of phenomena.

6. The term *linear* describes the method of analysis when studying the stability of a solution: the perturbation is weak enough that one may, in the vicinity of the studied solution, linearize the nonlinear equations associated with the system.

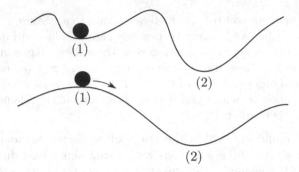

Figure 1.2 – In the top figure, a ball at point (1) has no chance of finding itself at point (2) due to any natural fluctuation. We call this position linearly stable. To change from point (1) to (2) a strong outside perturbation is necessary. Thus, position (1) is linearly stable but nonlinearly unstable. In the bottom figure, the ball is in an unstable position. In this case a small perturbation is enough to cause the transition from (1) to (2), and position (1) is linearly unstable.

The general and universal features of nonlinear phenomena can be represented in two different languages: catastrophe and bifurcation theory. Catastrophe theory deals with the determination of the geometric figure tracing out the boundaries of qualitatively differing solutions in parameter space (figure 1.1). Bifurcation theory deals instead with the specification of the evolution of one dynamic variable of the system (such as the equilibrium position of a mass in a mechanics problem) as a function of the control parameters (see figure 1.3, a bifurcation diagram).

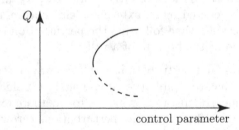

Figure 1.3 – A bifurcation diagram, representing a quantity Q characteristic of the nonlinear system studied as a function of a control parameter. The broken line represents an unstable branch, and the continuous line the stable branch.

Though two different representations, catastrophe and bifurcation describe the same reality; and especially, both search for universal behaviors independent of any specific model studied (whether it be a chemical, biological, or economic system, etc.). The goal of catastrophe theory is to determine the number of different types of catastrophes based purely on the number of independent control parameters. Similarly, bifurcation theory seeks the number of qualitatively different diagrams associated with the system; only symmetries and the number of control parameters matter, thus fixing the number of different scenarios. The two descriptions – catastrophe and bifurcation theory – contain the same information (elements) expressed differently, but catastrophe theory is particularly adapted to the study of structural *stability* in models.

In fact, catastrophe theory, unlike bifurcation theory, deals only with models that lead to solutions unaffected by small modifications (or perturbations) in the model. For example, suppose we adopt a mathematical model based on some a priori hypothesis. If, when studying the robustness of the model by introducing small perturbations, the bifurcation diagram changes qualitatively, we have a model that is *structurally unstable*: the model does not adequately describe the system, and must be discarded in favor of another that better fits the system's description.

It is useful to note that catastrophe theory assumes that the dynamics of a system are based on a potential. To be precise, by calling $x(t)$ a dynamic variable (this might be, for example, the position as a function of t in mechanics) the dynamics have a potential character, for which we can write down the following evolution equation:

$$\frac{dx}{dt} = -\frac{dV(x)}{dx}, \tag{1.2}$$

where $V(x)$ is the potential. Throughout the book we will see that in many cases a system that is out of equilibrium can eventually still be described this way. On the other hand, bifurcation theory has an even further reach: with it we can describe the form of the equations based on the system's symmetries without the necessity of an evolution equation written in a potential form.

We have mentioned the importance of the universality of approaches characterized by the ability to classify and enumerate different types of forms (bifurcations and catastrophes) without referencing a specific model. Nonetheless, when we speak of this universality, we do not mean to imply that the study of specific phenomena is unimportant. On the contrary, it is imperative to study specific cases in order to gain quantitative information – such as the precise value of a control parameter at which, in an experiment, one would see a bifurcation – in order to either avoid the bifurcation, take advantage of it, or simply study and add it to the repertory as such. Besides this, the study of a particular system could reveal the inadequacy of an adopted model if a contradiction between theory and experiment arises. Thus the study of specific systems plays an essential role for the development of evolution equations, establishing the validity of a particular hypothesis with respect to the observed phenomena, etc. The universal nature of the analysis (offered by bifurcation and catastrophe theory), besides being intellectually satisfying, gives certainty to our results, since all bifurcations belong to well-defined classes. To summarize, we need both approaches (knowledge of the specific system as well as the classification of possible catastrophes and bifurcations) to guarantee a robust scientific foundation. The universal classification allows us to unravel the general behavior of a system even when no adequate set of equations exists to describe the system, but without giving the specific knowledge of *when* one or another type of catastrophe could occur in a given experiment.

We will begin with some simple examples before going over the more technical aspects, where nevertheless the mathematics will remain elementary. The questions will increase in difficulty in order to complete, illustrate, and add to our general propositions. The explicit examples (from mechanics, for example) will also serve to progressively introduce the ideas and vocabulary of nonlinear dynamics. We will gradually generalize the main notions.

First we discuss steady-state solutions, then continue with oscillatory solutions. Stable solutions can sometimes lose their stability in favor of more complex solutions such as aperiodic solutions, which tend to be of a more erratic or chaotic appearance; we will not explore this class of chaotic solution in depth because many books already treat the subject (see especially [9, 75, 80, 93]). We will elaborate on the subject only briefly, presenting the essential concepts.

Now, the time evolution of a solution is not the only important aspect of nonlinear phenomena. Natural systems often have a spatial dimension, introducing an extra layer of richness. We are speaking about the fascinating creation of patterns in nature, or *morphogenesis*. Nature offers innumerable examples of spatial organization, in diverse domains, ranging from inert to living material, passing through social and economic sciences. Spatial order is easily visible to the naked eye. The examples around us are abundant (see [5] for a highly varied series of examples and images). Who amongst us has not admired the attractive and more or less regular pattern of cumulonimbus clouds (figure 1.4), or that of sand ripples in the desert or at the edge of an ocean in the declining tide – structures that organize themselves over long distances (figure 1.5). If one takes a moment to think about it, the existence of so much order is troubling when we identify that the cause of this pattern (in the desert, the wind) is often in a disordered state (turbulent wind). On the other hand, one should not forget that a naturally ordered system can revert to a disordered state, into a spatio-temporal chaos[7], upon variation of the control parameters.

---

7. Spatio-temporal chaos is associated to disorder in space and a loss of time-memory.

Figure 1.4 – Organized structure in a cumulonimbus. [© NOAA Photo Library, NOAA Central Library; OAR/ERL/National Severe Storms Laboratory (NSSL)]

Figure 1.5 – Wind-formed sand ripples. [© Greg Pickens #631279 Fotolia]

Hydrodynamics (also known as fluid mechanics) is another field that reveals an abundance of ordered states (see the very accessible books by Guyon et al. [41] and by Varlamov et al. [107]). For example the free surface of a liquid heated at its base develops what is known as Bénard–Marangoni cells (figure 1.6). Similarly, in geophysics, basalt columns spontaneously form as lava crystalizes (figure 1.7).

The most famous example of spatial organization is probably that of chemical reaction-diffusion systems initially studied by the English mathematician Alan Turing (1912–1954) [104]. He showed that in a system with two chemical substances, one an activator, S_A (which assures its growth), and the other an inhibitor, S_B (moderating the growth of its antagonist A), a homogeneous solution in which the two species are perfectly mixed can become unstable in favor of the creation of a spatial order. Such an instability occurs if S_B diffuses quickly enough with respect to substance S_A. The spatial order can take the shape of bands, hexagons (honeycomb), etc. The system shows an alternation between regions strongly concentrated in S_A and those rich in S_B (see chapter 9 for experimental results). Alan Turing furthermore suggested that such organization, and all spatial organized structures in general, could lead to a certain understanding of the morphogenesis of living beings. The computer generated solutions of simple reaction-diffusion systems of two variables can create patterns that are surprisingly like those found in nature (such as seashells [11]). In modern biological concepts of development, it is a commonly held point of view that the amount of information found in a living being is too much to be entirely encoded within the genome. Thus, purely phenotypical processes, without any genetic influence, could make a significant contribution to the emergence of patterns and forms in nature. Even though a direct proof of a Turing-type mechanism remains to be demonstrated in living systems, many recent advances seem to confirm Turing's ideas. We will discuss these aspects in chapter 9.

One common property of the nonlinear systems discussed so far (and of many others!) is their ability to generate an ordered spatial structure at (and beyond) a critical value of a control parameter. The pattern-forming systems undergo a bifurcation from a homogeneous to an ordered state. This bifurcation happens whenever the homogeneous solution loses stability or completely ceases to exist. Most of the systems we are referring to here are in a state of non-equilibrium; a constant source of energy (or its analog) is being added to the system (the energy of wind on the sand, the addition of a reactant in a chemical reaction, nutrients in biology, etc.). The change from a homogeneous state to the ordered state is called a *primary bifurcation* because, in fact, new (secondary) bifurcations can occur to modify this ordered state as the system continues to receive more energy. Eventually, with more and more energy, the system will enter into a regime of spatio-temporal chaos. We will briefly cover this in chapter 12.

Figure 1.6 – A structured pattern of the Bénard–Marangoni type on a free surface is obtained by heating the liquid from below. This structure is induced by the surface tension, which is sensitive to the temperature. This experiment can be reproduced at home. One must simply heat some cooking oil in a pan. Then add a small amount of aluminum powder to the oil. The powder is for the visualization of the convection cells. The liquid rises to the center of the cells (like a fountain) and sinks again at the borders, but this movement keeps intact the hexagonal structure of the convection cells. [© M.G. Velarde]

Figure 1.7 – An organized pattern of the Bénard–Marangoni type on the surface of Earth: basalt pillars. [© Michael Stüning #876669 Fotolia]

Bifurcations are often accompanied by a loss of symmetry. A bed of sand is flat before it becomes sculpted by ripples, so the topography from any two points is indistinguishable. As ripples form, the topography between crest and trough becomes starkly distinguished. Now one must travel a distance equal to the repetition length of the motif (the wavelength of the ripples) to find the same landscape (such as a crest). This is an example of translation symmetry breaking; after the formation of the folds, the symmetry has become discrete, and so one must travel an integer number of wavelengths to find the same topography.

In the neighborhood of a bifurcation point (i.e. when the value of the control parameter is close enough to the critical value needed to give rise to order), all systems are described by a universal equation called the *amplitude equation*. Once a system is far from thermal equilibrium (for example, if we add more and more energy), the ordered state can, in turn, become unstable and undergo a secondary bifurcation. Each new bifurcation leads to a new loss of symmetry. In general, the final state of each system is temporal and spatial disorder, namely spatio-temporal chaos or turbulence: the system then loses all temporal and spatial correlations.

A physicist may be tempted to draw parallels between primary bifurcations (the emergence of order) and the theory of phase transitions (for example, that of the liquid-solid transition occurring at a low enough temperature). We will see that this is a valid analogy so long as the system is studied in the neighborhood of the instability threshold, and if the instability is stationary (i.e. has no time oscillation). This analogy quickly reaches its limits, and non-equilibrium processes come to the fore, taking the studied systems to new depths. The dynamics become complex, and have no resemblance to phase transitions. Non-equilibrium systems may have a wide variety of dynamic behaviors – something not allowed for by the thermodynamics of a system globally in equilibrium (see chapter 12).

Systems in equilibrium[8] are governed by principles that are generally well-established: for example, they are characterized by a minimum value of energy, or a maximum of entropy, according to the exact configuration considered (e.g. in thermodynamics, processes having to do with a constant temperature or a constant pressure are evoked). These principles lead to what we call variational models[9]. Systems that are not in equilibrium are non-variational, by virtue of the fact that their state, or their dynamics, cannot

---

8. In physics, we may speak of thermodynamic equilibrium, but the notion of non-equilibrium is more general, as we have mentioned before.

9. This name references the fact that the state that is selected is either a minimum or a maximum of a certain quantity, such as energy. The extremum is obtained by taking the zero of the derivative of a certain quantity, such as energy (or of a functional, that is, a function of a function).

be found through attainment of an extremum, such as the minimum energy. At present, we do not have any criterion such as the one for thermodynamic systems (or any system globally in equilibrium) that allows us to predict how, and by which principle, a system differentiates between two (or, sometimes, an infinite number of) possible solutions. Why does nature "prefer" one solution over any other? Is there a selection criterion? The non-variational character of these systems allows for a large variety of behaviors and dynamics, and we are often incapable of predicting what the final solution adapted by the system will be. Another characteristic of the systems studied in this book has to do with dissipation; for example, chemical reactors combine and consume themselves, energy is consumed by systems, or by organisms. These are dissipative non-equilibrium systems which are equally irreversible. For example, as a bike goes down a path, the wheels lose energy through friction against the ground. If we take the same path in reverse, the lost energy is not regained: the phenomenon is irreversible. Non-equilibrium or dissipative and irreversible processes are predominant in natural processes.

The first three chapters are essential to the understanding of the rest of the book. They introduce terminology and basic concepts. These three chapters are accessible to undergraduate students, and also to high schoolers with a scientific background. The mathematical level needed is elementary (single variable functions, derivatives, first and second order differential equations). The difficulty augments progressively throughout the book. We estimate most of the other chapters will be accessible for master students in physics, chemistry, mathematics, biomathematics, mechanics, geophysics, economics, and social sciences. The book is equally addressed to established researchers who are looking for an introduction to the fabulous field of nonlinear sciences.

Chapter 2

Basic Introduction to Bifurcations in 1-D

Abstract *This chapter introduces a simple example, that of a mass attached to a spring where the mass slides along a horizontal bar. We show that this system has several equilibrium solutions and exhibits one of the usual bifurcations, the pitchfork bifurcation. We will paint an intuitive and simple picture explaining the existence of this bifurcation and introduce the universal amplitude equation, as well as some general notions such as symmetry breaking, adiabatic elimination, and the analogy with systems in thermal equilibrium. We conclude with an exact solution to the amplitude equation, confirming classical stability studies.*

Bifurcation theory is the study of qualitative changes in solutions of nonlinear systems as a function of their control parameters. A change is called qualitative when, for some critical value of a control parameter, the solution for a system changes abruptly from one type to another, for example, from a stationary to an oscillatory solution.

In this and the following chapter, we undertake the study of bifurcation theory through elementary examples in point particle mechanics, so that we can introduce the conceptual framework and vocabulary appropriate for nonlinear science while also benefitting from the simple and intuitive visualizations native to point particle mechanics, allowing us to go on to study more complex systems.

© Springer Science+Business Media B.V. 2017
C. Misbah, *Complex Dynamics and Morphogenesis*,
DOI 10.1007/978-94-024-1020-4_2

15

2.1. A Simple Mechanical Example

2.1.1. Potential Energy and Equilibrium

Our first example is the simple mechanical system of a mass suspended on a spring of spring constant (elasticity) k, and length, when at rest, ℓ_0 (called the rest length); the spring is fixed at a point A, and the mass is in the shape of a ring which slides along a horizontal rod (figure 2.1).

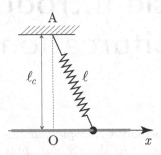

Figure 2.1 – Schematic view of the system. The rest length of the spring is ℓ_0 (not seen on the figure). ℓ_0 can be smaller than ℓ_c (spring is then stretched when at the vertical) or larger than ℓ_c (spring compressed when at the vertical).

We pull the mass along this horizontal axis a distance x from the point of origin, O ($x = 0$), located directly below point A, and let it go. If the movement has friction, the system will evolve naturally until it reaches a state of equilibrium; what is this state? Instinctively, we are tempted to say that the mass M returns to its original location, O. However, this is only one particular solution, while the general solution encompasses several coexisting equilibrium states. The state chosen by the system depends on the length ℓ_c (which will be our control parameter), or more specifically, the value of ℓ_c compared to ℓ_0, the rest length. Thus our example, seemingly as simple as a standard harmonic oscillator, reveals an underlying complexity of solutions and we discover one of the most common bifurcations: the pitchfork bifurcation.

To get a quantitative assessment of this system we evaluate the potential energy $E_p(x)$, dependent on the elasticity constant k and the spring extension (or compression) $\Delta\ell = \ell - \ell_0$, where ℓ is the instantaneous length of the spring, in the following way: $E_p = k\Delta\ell^2/2$ (see figure 2.1). According to the Pythagorean theorem, the displacement x along the horizontal axis – the degree of freedom of interest to us – is related to lengths ℓ_c and ℓ by

$\ell = \sqrt{x^2 + \ell_c^2}$. We can thus write the potential energy as:

$$E_p(x) = \frac{1}{2}k\left[\sqrt{x^2 + \ell_c^2} - \ell_0\right]^2. \tag{2.1}$$

For small x, we can use a second order Taylor expansion to approximate the potential energy:

$$E_p(x) = C + k\frac{\ell_c - \ell_0}{2\ell_c}x^2 \tag{2.2}$$

where $C \equiv k(\ell_c - \ell_0)^2/2$ is a constant; we will drop it from now on.

This expression of the potential energy E_p is similar to the potential energy expression for any regular spring. However, the physical configuration of the system studied here is fundamentally different (see figure 2.1): the position $x = 0$ is not always a stable solution.

If we fix ℓ_c such that $\ell_c > \ell_0$, the factor $(\ell_c - \ell_0)$ in front of x^2 is positive and $E_p(x)$ has a minimum at $x = 0$, corresponding to a stable[1] position. In the opposite situation ($\ell_c < \ell_0$), E_p has a maximum at $x = 0$, corresponding to an unstable position. The equality $\ell_c = \ell_0$ thus corresponds to the critical condition where stability is lost at $x = 0$ (see figure 2.2).

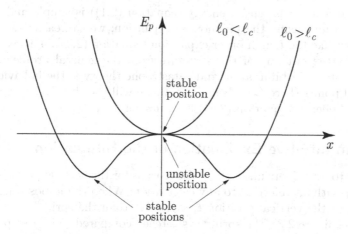

Figure 2.2 – Qualitative behavior of the potential energy.

When the solution is unstable at $x = 0$ (i.e. $E_p(x = 0)$ is a maximum), the mass will move to either side. As x is displaced farther and farther from the origin, the second order Taylor expansion of the potential energy

---

1. By way of analogy, a ball at the bottom of a valley remains "stuck" in this position due to the force of gravity (gravitational potential well), even if we apply a small perturbation; it is occupying, at this depth, a stable position.

(equation (2.2)) ceases to be valid. We must look to the next order expansion:

$$E_p = \frac{k}{2}\left[\frac{\ell_c - \ell_0}{\ell_c}x^2 + \ell_0\ell_c\frac{x^4}{4\ell_c^4}\right].$$

(2.3)

If we are in the range where $\ell_c \simeq \ell_0$, we can make the following reduction:

$$E_p = \frac{k}{2}\left[\frac{\ell_c - \ell_0}{\ell_c}x^2 + \frac{x^4}{4\ell_0^2}\right]$$

(2.4)

where we have replaced ℓ_c by ℓ_0 in the coefficient of x^4. The contribution to the potential energy given by the new x^4 term is always positive.

We can easily verify that if the length imposed on the spring when in the vertical position ℓ_c is smaller than the rest length of the spring ($\ell_c < \ell_0$, $x = 0$ is unstable), the potential energy $E_p(x)$ has a local maximum at $x = 0$ and possesses two minima, one at either side of $x = 0$ (see figure 2.2). It follows that the mass, leaving the initially unstable position of $x = 0$, will end up in the x position corresponding to one of the two minima of $E_p(x)$. In principle, the mass is as likely to go to one as to the other (positive or negative x); however, in real systems there is always a slight imperfection that gives precedence to one of the two directions.

The general form of potential energy (equation (2.1)) is simple and can be analyzed without using the polynomial expansion; we can easily see that it behaves just like the fourth order expansion (equation (2.4)). The expansion does not flatten out any of the structure from the general problem. Since what interests us in bifurcation and catastrophe theory is the behavior near the critical points (here $\ell_c \simeq \ell_0$ and $x \simeq 0$) we will use the Taylor expansion of potential energy (equation (2.4)) from now on.

2.1.2. An Intuitive Explanation of the Bifurcation

It is useful to intuitively understand the reasons for the vertical position's loss of stability without referring to any calculation. When the imposed length of the spring in the vertical position ℓ_c is greater than the spring's rest length ($\ell_c > \ell_0$, see figure 2.3), the spring is stretched compared to its rest position. Every displacement away from the original equilibrium position $x = 0$ is penalized, since it requires further stretching, and raises the total potential energy. Thus, in this case, $x = 0$ is a stable position.

By contrast, as soon as the imposed length of the spring in the vertical position ℓ_c is smaller than the rest length ($\ell_c < \ell_0$) the spring is compressed at position $x = 0$. A displacement with respect to the vertical contributes to the release of the spring, and thus a reduction of the potential energy. The new equilibrium position is reached when the original compression of the spring is optimal; that is, when the mass is in one of the two positions minimizing the potential energy $E_p(x)$ (see figure 2.2).

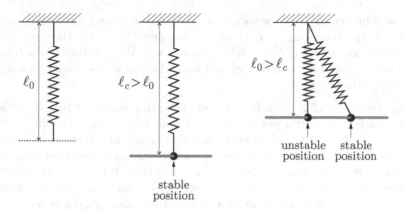

Figure 2.3 – A schematic view of different configurations. From left to right: rest length; stretched spring in vertical position (the spring resists moving away due to increased stretching); and compressed spring in vertical position (a deviation from that position is favorable since it reduces stretching).

2.1.3. Analysis of the Bifurcation

So, the $x = 0$ solution of the system studied in this chapter (see figure 2.1) becomes unstable as soon as the imposed length of the spring at the vertical position ℓ_c is smaller than its rest length ($\ell_c < \ell_0$). Let us now determine the new equilibrium position when in the vicinity of the instability threshold[2] i.e. $\ell_c \simeq \ell_0$. Once the threshold $\ell_c = \ell_0$ is reached, the mass displaces from its vertical position $x = 0$ and the nonlinear effects of the higher order terms (x^4) intervene to ensure saturation.

Using the condition $dE_p/dx = 0$ to define the extremas of energy and their stability relative to the potential energy, we determine the equilibrium solutions: (i) when $\ell_c > \ell_0$, solution $x = 0$ is stable; (ii) whereas when $\ell_c < \ell_0$, solution $x = 0$ is unstable (see figure 2.2). From the fourth order Taylor expansion of the potential energy (equation (2.4)) we find the following solutions (subscript 0 in x_0 indicates an equilibrium solution):

$$x_0 = 0 \quad \text{and} \quad x_0 = \pm\ell_0\sqrt{2\frac{(\ell_0 - \ell_c)}{\ell_c}}. \tag{2.5}$$

---

2. Though the system studied in this chapter can be determined entirely with an analytic calculation, most nonlinear systems cannot. It is thus necessary to take up linear expansions to analyze the system. Focusing on the vicinity of the instability threshold, besides justifying the Taylor expansions of functions related to the system, gives the problem a universal character, independent of the nature of the system and thus of the specific mechanisms that govern it.

The first solution, $x_0 = 0$, always exists, whereas the second solution requires that $\ell_c < \ell_0$. When $x_0 = 0$ becomes unstable ($\ell_c < \ell_0$) the new double solution $x_0 = \pm\ell_0\sqrt{2(\ell_0 - \ell_c)/\ell_c}$ takes over. We say there is a bifurcation of the stationary solution $x_0 = 0$ toward a pair of stationary solutions $x_0 = \pm\ell_0\sqrt{2(\ell_0 - \ell_c)/\ell_c}$.

Plotting the value of x_0 as a function of the spring length at rest ℓ_0, which in this case is our control parameter, we obtain what is called a bifurcation diagram (see figure 2.4), which is simply the portrait of the stationary solutions (equation (2.5)). The broken curve on figure 2.4 represents the unstable solution. This is the usual convention in bifurcation theory. Let us mention that the bifurcation in question, called pitchfork bifurcation because of the shape of the curve shown in figure 2.4, is sometimes also called a *supercritical bifurcation* (see section 3.3). Furthermore, it belongs to the family of *cusp* catastrophes, which we will discuss further in chapter 4.

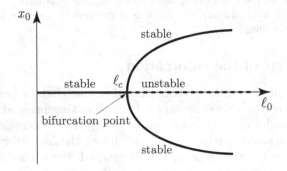

Figure 2.4 – A pitchfork bifurcation diagram. Continuous lines correspond to stable branches, and the discontinuous line to unstable branches.

2.1.4. Universality in the Vicinity of a Bifurcation Point

In the study of any nonlinear system, the first major steps are to search for bifurcation points and to look at polynomial expansions in the vicinity of these points.

The full potential energy $E_p(x)$ (equation (2.1)) is simple enough that we could solve it without using a polynomial approximation. It can easily be checked that the qualitative behavior found with the fourth order approximation (see figure 2.2) is equivalent to the behavior found by analysis of the exact expression (equation (2.1)).

The resemblance in the potential energy's qualitative behavior when using both the exact and approximate forms is indicative of the fact that the qualitative change in solutions only occurs near $x = 0$; that is, once the length of the spring at the vertical is of the same order as the rest length ($\ell_c \simeq \ell_0$). Looking at the potential energy around the point $x = 0$ is thus especially interesting. More generally, the approach consisting of analyzing equations in the vicinity of bifurcation points allows for the extraction of generic qualitative behaviors, thus allowing one to obtain a universal form of equations around these points. This is achieved by introducing a small parameter that measures the distance from the threshold value needed for bifurcation.

We can illustrate the importance of these bifurcation points and their neighborhoods by taking up the previous simple mechanical spring system again, and introducing a small parameter ϵ to define the distance from the threshold value needed for bifurcation as follows:

$$\epsilon = \frac{\ell_0 - \ell_c}{\ell_c}. \tag{2.6}$$

The fourth order potential energy equation (2.4) can be rewritten as follows:

$$E_p = \frac{k}{2}\left[-\epsilon x^2 + \frac{x^4}{4\ell_0^2}\right]. \tag{2.7}$$

Since x^2 is multiplied by ϵ, the fourth order term, though a priori of less importance if x is small, becomes of the same order of magnitude as the quadratic term. Looking at the stationary solutions in the regime where the spring's length in the vertical position is smaller than when at rest (equation (2.5)) we find that $x_0 \sim \sqrt{\epsilon}$. Thus, both of the terms that remain in the potential energy equation above (equation (2.7)) are of the same order ϵ^2.

The expression of potential energy as a function of the parameter measuring the proximity to the bifurcation threshold (equation (2.7)) is a general form for all systems whose equilibrium solution loses stability in favor of another equilibrium solution. We will introduce other general forms (corresponding to other types of bifurcation) as we proceed. Instead of equilibrium, we will speak more generally of stationary solutions, to emphasize that not all systems of this type are described by a mechanical energy. Thus, variable x, which here represents a position as defined in classical mechanics, might represent other physical quantities in other systems. In chemical reaction-diffusion systems, x refers to the amplitude of a chemical concentration; in other systems, x can be associated with the velocity of a fluid in convection, etc. The system changes, the physical mechanisms too, but the form of the evolution equation near a bifurcation point, also referred to as the amplitude equation (see subsection 2.3.4), remains identical. This general quality in amplitude equations defines the universality of behavior near bifurcation points.

2.2. An Analogy between Pitchfork Bifurcations and Second Order Phase Transitions

The expression for potential energy (equation (2.7)) as a function of parameter ϵ (distance to the bifurcation point) is in fact similar to the Landau free energy associated with second order phase transitions [98]. A classic example of a second order transition is found in ferromagnetism. Many materials are potentially magnetic; generally, at high temperature, they display no magnetization, but some finite value of magnetization is found when the material is cooled below a certain temperature. This type of material has within it small *dipoles* (or magnetic moments), carried by the atoms. At high temperatures, thermal vibrations induce disorder between the dipoles, hence the net lack of magnetization. However, at low temperatures, the interaction between the dipoles becomes comparable to and surpasses that of the thermal vibrations. The dipoles then align themselves in the same direction, and their individual fields add together to create a macroscopically visible magnetization. This is called the ferromagnetic phase transition. The order parameter of the transition is magnetization M, which is null at high temperatures ($M = 0$) and non-zero at low temperatures ($M \neq 0$). The term "order" in order parameter refers to the alignment of the magnetic moments. The transition between high (disordered) and low (ordered) temperature takes place at critical temperature $T = T_c$. For $T < T_c$ (and T close enough to T_c), $M \sim \pm\sqrt{T_c - T}$. Thus, the variable x in the mechanics problem plays the same role as the order parameter M does here. The parameter ϵ is analogous to $T - T_c$. The Landau energy associated with the transition is given by $E \sim \epsilon M^2 + M^4$, the same as the potential energy for the mechanical example (equation (2.7)).

2.3. Dynamical Considerations

Up until now, we have considered the stationary states of dynamical systems. However, systems evolve in time and, even if a final stationary state is attained, the evolution toward this state has some duration dependent on different mechanisms at play (for example friction in mechanics, or viscosity in hydrodynamics). In addition, besides the stationary behavior, dynamical systems often show permanent oscillations of a more or less complex nature, sometimes leading to chaotic behavior. We will now illustrate the notion of temporal evolution for our mechanical system.

Using Newton's second law we can write:

$$m\ddot{x} + \mu\dot{x} + \frac{\partial E_p(x)}{\partial x} = 0, \qquad (2.8)$$

where m and μ represent, respectively, the suspended mass from the spring and the friction coefficient associated with the mass's movement along the horizontal rod. The last term, $\partial E_p(x)/\partial x$, of the above equation (2.8) is none other than the force, F, which comes from the anharmonic[3] potential $E_p(x)$, $F = -\partial E_p(x)/\partial x$. The stationary states that we have studied are more generally called fixed points. We can determine these points using the equation of motion (2.8) and applying conditions $\dot{x} = 0$, $\ddot{x} = 0$, implying $\partial E_p/\partial x = 0$.

From the fourth order potential energy equation which incorporates parameter ϵ (equation (2.7)), we obtain the following fixed points:

$$x_0 = 0 \quad \text{and} \quad x_0 = \pm\ell_0\sqrt{2\epsilon} \qquad (2.9)$$

which are the same two that we previously found (equation (2.5)).

2.3.1. Analysis of Linear Stability

Once we have determined the fixed points, also referred to as the stationary states of the system, we continue with our analysis by studying their stability. Will the system return to its original state after small perturbations $x_1(t)$, or will it stray to a different solution? To answer this question, let us superimpose stationary state x_0 with perturbation $x_1(t)$:

$$x(t) = x_0 + x_1(t), \qquad (2.10)$$

and now determine the evolution of perturbation $x_1(t)$ over time. If $x_1(t)$ decreases with time, stationary state x_0 is stable, and if, on the contrary, $x_1(t)$ increases over time, stationary state x_0 is unstable.

We will make the assumption that perturbation $x_1(t)$ is small in comparison to x_0 ($x_1(t) \ll x_0$) so that we may linearize the spring system's equation of motion (2.8) and a priori limit our Taylor expansion to first order in x_1. Accordingly, we obtain the following approximation for the linearized equation of motion:

$$m\ddot{x}_1 + \mu\dot{x}_1 + x_1 E_p''(x_0) = 0, \qquad (2.11)$$

where $E_p''(x_0) \equiv (\partial^2 E_p/\partial x^2)_{x=x_0}$ (from now on we omit the argument of E_p, so long as the abbreviation is clear). In reference to the linearization of the equation, this is called a *linear stability analysis*.

---

3. The quadratic form (second degree homogeneous polynomial) of the energy is often called a harmonic potential. In the present case, $E_p(x)$ is not quadratic in x and thus we call the potential anharmonic.

Now, to solve the spring's linear equation of motion (2.11), let us look for solutions of the type $x_1(t) \sim e^{\omega t}$, which will give us a second order equation in ω:

$$mw^2 + \mu\omega + E_p'' = 0. \tag{2.12}$$

We posit discriminant $\mu^2 - 4mE_p''$ as positive, a condition that is satisfied if the friction coefficient μ is high enough[4]. Thus, the solutions to equation (2.12) are:

$$\omega_{\frac{1}{2}} = \frac{1}{2m}\left[-\mu \pm \sqrt{\mu^2 - 4mE_p''}\right] \tag{2.13}$$

and the solution to the linearized equation of motion (2.11) can be written in its most general form:

$$x_1(t) = ae^{\omega_1 t} + be^{\omega_2 t}, \tag{2.14}$$

where a and b are integration constants. Note that the subscripts 1 and 2 in $\omega_{\frac{1}{2}}$ refer to the solution with $+$ and $-$ signs, respectively. The temporal evolution of perturbation $x_1(t)$ thus depends on the sign of $\omega_{\frac{1}{2}}$ in the exponentials. ω_2 is always negative (see equation (2.13)), indicating an exponential decrease; the linear stability of the system at x_0 thus depends exclusively on ω_1 and as such, on the sign of the potential energy's second derivative, E_p'', at point x_0 (in other words, the concavity of the potential energy at x_0). Two possible scenarios exist:

1. $E_p'' > 0$ and thus $\omega_1 < 0$: $x_1(t)$ decreases exponentially with time and x_0 is a stable solution;

2. $E_p'' < 0$ and thus $\omega_1 > 0$: $x_1(t)$ grows exponentially as $e^{\omega_1 t}$ with time and solution x_0 is unstable.

These results, obtained by a dynamical analysis of the system, correspond exactly to the results we found in the preceding analysis of the concavity of E_p (see subsection 2.1.1). In general, when we cannot formulate a problem in terms of its potential energy, we may instead use a linear stability analysis of the dynamical system, as discussed above.

2.3.2. Critical Slowing Down

The time evolution of the perturbation applied to the dynamical spring system studied earlier in the chapter depends on the potential energy's second derivative (equation (2.13)) at x_0; differentiating the potential energy (equation (2.7)) twice with respect to x we obtain:

$$E_p'' = \frac{k}{2}\left[-2\epsilon + 3\frac{x_0^2}{\ell_0^2}\right]. \tag{2.15}$$

---

4. This hypothesis lets us eliminate solutions that have temporal oscillations of no interest to us in this discussion.

For stationary solution $x_0 = 0$, this equation takes the form:

$$E''_p = -k\epsilon. \tag{2.16}$$

Using this expression at $x_0 = 0$, we can rewrite the expressions for $\omega_{\frac{1}{2}}$ (equation (2.13)), the exponential terms of the perturbation $x_1(t)$. Remaining in the <u>high friction</u> regime ($\mu^2 + 4mk\epsilon > 0$), we obtain $\omega_{\frac{1}{2}} = \mu/2m[-1 \pm \sqrt{1 + 4mk\epsilon/\mu^2}]$. In the neighborhood of the pitchfork bifurcation's critical point ($\epsilon \to 0$), we can make a Taylor expansion to the first order in ϵ in order to obtain the following expressions of $\omega_{\frac{1}{2}}$:

$$\omega_1 \simeq \frac{k}{\mu}\epsilon \quad \text{and} \quad \omega_2 \simeq -\frac{\mu}{m}. \tag{2.17}$$

ω_2 is a negative constant (in accordance with the analysis of the last section), so $e^{\omega_2 t}$ decreases exponentially with time. The characteristic decay time associated with this term is of order $|1/\omega_2| = m/\mu$.

ω_1 is proportional to the algebraic distance ϵ from the bifurcation threshold; it is thus also negative (meaning an exponential decay of the term $e^{\omega_1 t}$) when ϵ is negative ($\epsilon = (\ell_0 - \ell_c)/\ell_c$, ℓ_c is greater than rest length ℓ_0).

However, if the length of the spring at the vertical is smaller than its rest length $\ell_c < \ell_0$, ω_1 is positive, and the perturbation grows exponentially. The growth rate of perturbation $e^{\omega_1 t}$ is of order $1/\omega_1 \sim 1/\epsilon$; this characteristic time becomes longer and longer as the critical condition of the bifurcation is approached ($\epsilon \to 0$). This phenomenon is known as critical slowing down: the dynamics of the system become slower and slower as the system approaches the critical threshold.

2.3.3. Adiabatic Elimination of Fast Modes: Reducing the Number of Degrees of Freedom

The study of the dynamics near fixed points has permitted us to, on the one hand, determine the stability of these stationary states and, on the other, to identify the different response modes of the system for small perturbations. The mode associated with ω_2 corresponds to a stable and rapid mode (exponential decrease of the perturbation) while ω_1 (with $\epsilon > 0$) corresponds to an unstable mode whose exponential growth slows down in the neighborhood of the critical point.

As time passes, the stable mode, following a quick evolution, decays exponentially in time, while the slow mode persists; the dynamics are thereby governed by slow modes. This is often the case in the study of dynamical systems, which simplifies the analysis. For systems with a large number of modes (or degrees of freedom), only slow modes persist in the vicinity of the bifurcation; the fast modes are thus *adiabatically*[5] slaved to the slow modes. The elimination of rapid modes to the benefit of slow ones is called *adiabatic elimination*; the rapid modes adapt *adiabatically* to the dynamics of the slow modes.

So to make an adiabatic elimination in the dynamics of the system studied here, we take the general form describing the evolution of perturbation $x_1(t)$ (equation (2.14)) and keep only the slowly evolving mode (since the other mode extinguishes quickly):

$$x_1(t) \simeq ae^{\omega_1 t} = ae^{(k/\mu)\epsilon t}. \tag{2.18}$$

Notice that t appears as part of a product with ϵ, the distance from the bifurcation threshold. It is useful to introduce a new variable T, called the slow variable, defined as:

$$T = \epsilon t, \tag{2.19}$$

the time derivative of which reads:

$$\frac{d}{dt} = \epsilon \frac{d}{dT}. \tag{2.20}$$

This way, we can frame the evolution of perturbation $x_1(t)$, which takes the form $1/\epsilon$, through the use of this new variable T, making it of order one. This is not just a convenient change of variables; the slow variables are crucial for simplifying the analysis of dynamical systems with various evolution mechanisms working at very different timescales. By defining this kind of variable we are able to establish a systematic classification of slow or fast evolutions (dominant and subdominant); called multiscale analysis, we will encounter and use it throughout the book.

Having introduced the new variable T in equation (2.18), we can write the time evolution of the perturbation as:

$$x_1(t) = ae^{(k/\mu)T}, \tag{2.21}$$

to illustrate that the perturbation x_1 does not depend separately on the time t and distance ϵ, but rather on their product, T.

---

5. The term adiabatic, used in thermodynamics, comes from the Greek term *adiabatos* which signifies insurmountable. A system undergoes an adiabatic transformation when, during the transformation, it is thermally isolated from the exterior environment. The system may exchange mechanical energy with the exterior but heat exchange is prohibited (through insulating boundaries called adiabatic). If the boundaries of the system conduct heat, the transformation can still be considered adiabatic if it occurs very quickly, that is, if there is not enough time for an exchange of heat to occur during the transformation. It is in this spirit that we use the term adiabatic: the mode associated with ω_2 very quickly (adiabatically) adapts to the slow mode.

Keep in mind that the expression for perturbation $x_1(t)$ (equations (2.14) and (2.21)) is a solution of the linear approximation of the dynamic equation (2.11). In this linear regime, constant a cannot be determined: a multiplication of the constant by another arbitrary constant would also be a solution to equation (2.21). Also, for $\epsilon > 0$, the solution increases exponentially over time, making the linear approximation invalid after some time: x_1 becomes so large that we can no longer ignore the nonlinear terms of the original dynamics equation (2.8). Nevertheless, the linear solution is a good approximation for short timescales, and it will often be used in order to guide our analysis in the nonlinear regime.

2.3.4. Amplitude Equation

We now propose to investigate the nonlinear evolution, restricting ourselves to the situation in the neighborhood of critical condition $\epsilon = 0$. The general strategy consists of expanding the original equation in a power series around a small parameter (here ϵ) and then deducing the contributions order by order; thus one of the first steps is to determine the first relevant power in the series expansion. A good approximation will often assume the form of a positive power series with respect to small parameter ϵ. Given that the amplitude of the stationary solution (equation (2.9)) takes the form $\epsilon^{1/2}$, it is logical to look for a solution to the full nonlinear equation (2.8) as a series of $\epsilon^{1/2}$:

$$x(t) = \epsilon^{1/2} A_0(T) + \epsilon A_1(T) + \cdots \qquad (2.22)$$

where functions $A_0(T)$, $A_1(T)$, ... are the solutions of the general nonlinear equation (2.8) to be determined later.

Inserting the $\epsilon^{1/2}$ power expansion (equation (2.22)) into the general dynamics equation (nonlinear equation (2.8)) we obtain, to power $\epsilon^{3/2}$, the following *amplitude equation*:

$$\frac{dA_0}{dT} = \frac{k}{\mu} A_0 - \frac{k}{2\mu \ell_c^2} A_0^3. \qquad (2.23)$$

The form of this amplitude equation, also referred to as the *normal form* of a pitchfork bifurcation, does not depend on the precise form of E_p, provided that E_p be even with respect to x.

Note that to leading order the second temporal derivative does not enter the amplitude equation. In other words, there is a reduction in the number of degrees of freedom (since we go from a second order to a first order equation). The second derivative, to this leading order, is practically zero, $\ddot{x} \simeq 0$. This is equivalent to adiabatic elimination discussed earlier.

If we ignore the nonlinear term $-(k/2\mu\ell_c^2)A_0^3$ in the amplitude equation (2.23), our solution is given by $A_0(T) = ae^{(k/\mu)T}$ (where a is an integration factor) which is also exactly the solution for the linear equation (2.18) with only the slow mode included. The inclusion of the nonlinear term in the amplitude equation (2.23) contributes an increasing opposition to the growth of amplitude $A_0(t)$ (due to its negative sign) until completely saturating the solution[6], i.e. until the right-hand side of equation (2.23) vanishes, leading to a given value of the amplitude (besides the trivial solution $A = 0$). This is the saturation amplitude which is the result of an interplay between the linear term, which acts in favor of a temporal increase, and the nonlinear term which has an antagonist effect.

2.3.5. Canonical Form of the Amplitude Equation

The universality of the amplitude equation (2.23) becomes apparent with a new change of variables. Using $A_0 = A\ell_0\sqrt{2}$ and $\tau = kT/\mu$, we can write the amplitude in its *canonical* form:

$$\frac{dA}{d\tau} = A - A^3, \qquad (2.24)$$

thus named for its independence from any physical parameters of the studied system.

The amplitude equation (2.24) can also be described by the potential $V(A) = -A^2/2 + A^4/4$ (lending the dynamics a potential character):

$$\frac{dA}{d\tau} = -\frac{dV(A)}{dA}. \qquad (2.25)$$

To continue the game of rewriting equations, let us substitute A by $A/\sqrt{\epsilon}$ and τ by $\tau\epsilon$. This will allow us to explicitly write down (the small and important) variable ϵ (distance to the instability threshold), giving us a new expression for potential $V(A) = -\epsilon A^2/2 + A^4/4$ and, accordingly, an alternative form of the amplitude equation:

$$\frac{dA}{d\tau} = \epsilon A - A^3. \qquad (2.26)$$

Here the amplitude equation contains control parameter ϵ in an adimensional form, enabling us to refer directly to the value of the distance from the instability threshold, if need be.

From equation (2.24), the fixed points are given by $A = 0$ and $A = \pm\sqrt{\epsilon}$, while the time-dependent solution in the linear regime (equation (2.26)) takes the form $A = ae^{\epsilon\tau}$, with a an integration constant.

---

6. The way this first nonlinear term acts against the linear amplitude is not a general rule. It may instead be the following term (or another, higher order term) in the amplitude equation – such as A^5 – that leads to the nonlinear saturation. In certain models, all nonlinear terms act in the same way as the linear term.

To summarize, at $A = 0$ the system bifurcates to a new stationary solution that varies with $\sqrt{\epsilon}$, while the characteristic time of the solution varies with ϵ^{-1}. $A = 0$ is stable for $\epsilon < 0$ and unstable for $\epsilon > 0$: the loss of stability is accompanied by the emergence of two solutions, $A = \pm\sqrt{\epsilon}$, in the form of a pitchfork bifurcation.

2.3.6. Attractors

Though nonlinear, amplitude equation (2.24) has an exact solution. To find it we first rewrite the canonical amplitude equation (2.24) as:

$$\frac{dA}{A(1 - A^2)} = \frac{dA}{A} + \frac{1}{2}\frac{dA}{(1 - A)} - \frac{1}{2}\frac{dA}{(1 + A)} = d\tau, \qquad (2.27)$$

which we can immediately integrate to give:

$$\frac{A}{\sqrt{|1 - A^2|}} = \frac{A_i}{\sqrt{|1 - A_i^2|}}e^{\tau}, \qquad (2.28)$$

where A_i is an integration constant corresponding to the initial amplitude (at $\tau = 0$).

This solution is represented in figure 2.5, illustrating the transient behavior before the reaching of a stationary state, $A = +1$ or $A = -1$. The final state depends on the initial condition: if it is positive, the solution is $A = 1$, and if negative, $A = -1$.

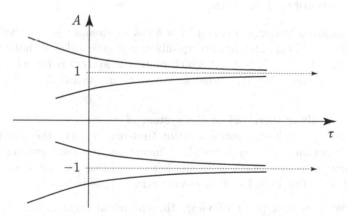

Figure 2.5 – Amplitude as a function of time for different initial conditions. The horizontal finely dotted lines correspond to fixed points ± 1.

The two solutions $A = 1$ and $A = -1$ constitute two possible *attractors* of the dynamical system. The first attractor claims the range $[0, \infty]$ as its basin of attraction, defined as the set of initial conditions which tends (at infinite time) toward the solution $A = 1$, while the second attractor's basin is given by the interval $[-\infty, 0]$.

To conclude, note that solutions $A = \pm 1$ are equivalent, a fact that simply reflects the symmetry $A \to -A$ of the canonical evolution equation (2.24).

2.3.7. Lyapunov Function

The amplitude equation (2.25) derived from the potential $V(A)$ has a particular property. One can write:

$$\frac{dV(A)}{d\tau} = \frac{dV}{dA}\frac{dA}{d\tau} = -\left[\frac{dV}{dA}\right]^2 \leqslant 0. \tag{2.29}$$

In other words, the potential $V(A)$ is a monotonic function that decreases over time, until reaching a stationary state (either $dV(A)/d\tau = 0$ or $dA/dt = 0$). The potential $V(A)$ is called a *Lyapunov function*.

This is reminiscent of the monotonic evolution properties of systems that are globally in thermodynamic equilibrium – properties that in general are not seen in non-equilibrium systems. However, we shall see throughout this book that some non-equilibrium examples can be recast, in the vicinity of a bifurcation, into a variational (or potential) form (as in equation (2.25)) while some other examples cannot.

2.3.8. Symmetry Breaking

Bifurcations are often accompanied by a *break of symmetry*. The system we have studied in this chapter has an equilibrium position $A = 0$ that becomes unstable when $\ell_c < \ell_0$ ($\epsilon > 0$), at which point the system is forced to evolve toward one of the other two possible equilibrium states: $A \sim \pm\sqrt{\ell_0 - \ell_c} \sim \pm\sqrt{\epsilon}$.

Intuitively speaking then, before the system bifurcates, the system, in stationary state $A = 0$, has a central solution that privileges neither the left nor the right direction. However, after the bifurcation, the mass evolves toward a stationary state that is either to the left or right of the initial stationary solution $A = 0$. The initial system's symmetry is thus reduced.

From a mathematical point of view, the canonical form of the dynamics equations in the neighborhood of the bifurcation (equation (2.24)) possesses the symmetry $A \to -A$, as does the solution $A = 0$, but neither of the two solutions that exist after bifurcation, $A = \sqrt{\epsilon}$ and $A = -\sqrt{\epsilon}$, conserve this symmetry. Each solution taken separately is not invariant under the operation $A \to -A$. From this perspective, it is indeed a break of symmetry.

Nonetheless, the *set* of solutions $\{A = \sqrt{\epsilon}, A = -\sqrt{\epsilon}\}$ is invariant under the symmetry transformation $A \to -A$ (the two solutions are permuted, and the pair of solutions remain altogether the same). In other words, in considering the set of solutions, invariance is conserved, which is a consequence of the symmetry invariance of the canonical amplitude equation (2.24).

2.4. Dynamical Systems and the Importance of the Pitchfork Bifurcation

In this chapter, we have studied the pitchfork bifurcation through a mechanical example that is simple but essential to the understanding of several other dynamical systems such as those observed in chemistry (reaction and diffusion of chemical elements), in physics (velocity fields in hydrodynamics such as convective Bernard rolls), and so on.

Beside these concrete examples, for all systems that have a one dimensional stationary instability – in which the perturbed system departs from the equilibrium solution (time-independent) to evolve toward another stationary solution – we can show that the amplitude of the field (or fields) representing the dynamics obeys an equation of the same type as the canonical equation (2.24) characteristic of the pitchfork bifurcation, belonging to the family of *cusp catastrophes* (see chapter 4). We shall encounter several examples throughout this book.

2.5. Exercises

2.1

Let us consider a device that is slightly different from that of figure 2.1, where now the rod is tilted by an angle θ with respect to the horizontal axis. Due to this inclination, the device is affected by gravity (figure 2.6).

1. Write down the potential energy where the mass is supposed to be at an arbitrary position x. Determine the fixed points and their stability.

2. Does the system exhibit a bifurcation? If yes, which type of bifurcation? Justify your answer.

3. Determine the expression of the potential energy in the vicinity of the bifurcation point. Which bifurcation do we have?

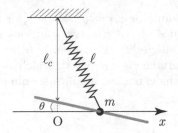

Figure 2.6 – Schematic view of the mechanical device.

2.2

Consider the following nonlinear equation:

$$\dot{x} = \epsilon x + x^3, \tag{2.30}$$

where ϵ is a real number.

1. Determine the fixed points and their existence condition.

2. Express equation (2.30) in a potential (variational) form. Denote by $V(x)$ the potential. Plot $V(x)$ for different values of ϵ. With the help of these figures determine the stability of fixed points.

3. Plot the bifurcation diagram by using the convention whereby unstable branches are represented by dashed lines, and stable branches by solid ones.

4. Let x_0 denote any fixed point of equation (2.30). Let us set $x(t) = x_0 + x_1(t)$. By keeping only linear terms in $x_1(t)$ in equation (2.30), determine the equation obeyed by $x_1(t)$. For each fixed point x_0 determined above in question 1, write the solution for $x_1(t)$; we denote by $x_1(0)$ the initial value. How does $x_1(t)$ behave in the course of time? Comment on the results. How do these results compare to those found in question 3?

5. If $\epsilon > 0$, are there stable fixed points? If yes, specify them; if no, how should one modify equation (2.30), still keeping the $x \to -x$ invariance, in order to obtain stable fixed points? Justify your answer (several answers are possible).

6. Equation (2.30) admits, in fact, an exact solution. Find it in a similar way to which an exact solution is found in the chapter, or use an alternative method. Plot $x(t)$ as a function of t. Which conclusion could you draw? How do these results compare to those found above from analysis of the potential?

2.3

Let us consider a device which is slightly more complex than that of figure 2.1, where the mass slides along a rod having a circular shape (figure 2.7).

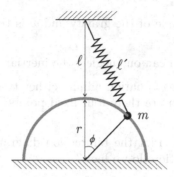

Figure 2.7 – Schematic view of the mechanical device.

1. Write down the potential energy E of the mass as a function of r, ϕ, ℓ and ℓ_0.

2. Analyze the fixed points and their stability by taking ϕ as a degree of freedom.

3. Is there a bifurcation? If yes, please specify the type of bifurcation.

4. By focusing on the vicinity of the bifurcation point, write E to order four in ϕ. What conclusion do you draw?

2.4

Let us consider the motion of a mass m attached to a hoop undergoing rotation (figure 2.8). The mass is subject to gravitational as well as centrifugical forces. In addition, the mass is subject to a friction force denoted by $\mu\dot{\phi}$ where μ is the friction coefficient.

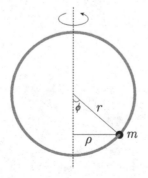

Figure 2.8 – Mechanical device.

1. By using Newton's law, show that ϕ obeys the following nonlinear equation:

$$mr\ddot{\phi} = -mg\sin(\phi) + mr\omega^2 \sin(\phi)\cos(\phi) - \mu\dot{\phi}, \qquad (2.31)$$

   where g is the amplitude of the gravity and ω is the rotation speed (the pulsation) of the hoop.

2. Under which condition can one neglect the inertial term?

3. Show that the resulting equation admits either two or four fixed points. Under which condition are there four fixed points? Write down their explicit forms.

4. Let us set $\nu = r\omega^2/g$. Plot the bifurcation diagram in the $(\phi\text{-}\nu)$-plane. Which type of bifurcation do we have?

5. Provide an intuitive explanation to the existence of a bifurcation.

6. By neglecting inertia write equation (2.31) in a potential (or variational) form.

7. Expand $V(\phi)$ in the vicinity of a bifurcation point. What conclusion could you draw?

8. By using results obtained above, write the differential equation obeyed by ϕ in the vicinity of the bifurcation point (neglect inertia).

Chapter 3

The Other Generic Bifurcations

Abstract *In this chapter we present a detailed study of the generic bifurcations. In addition to the pitchfork bifurcation seen in the last chapter, we introduce the saddle-node bifurcation (through the example of a simple pendulum), the imperfect pitchfork bifurcation, the subcritical bifurcation (characterized by hysteresis), and the transcritical bifurcation. We will then introduce and illustrate catastrophe theory by way of a simple example.*

Having dealt exclusively with the pitchfork bifurcation in the last chapter, we now turn to other bifurcations commonly observed in nonlinear systems.

3.1. Saddle-Node Bifurcation

We would like to introduce the saddle-node bifurcation directly through a classical mechanics example, the pendulum.

3.1.1. Review: Simple Pendulum

First, let us review the classic problem of a simple pendulum: a mass m is suspended from a string of length ℓ attached at a point O (see figure 3.1). The evolution equation for the mass is derived from the law of inertia, the fundamental law of dynamics. Taking I to be the moment of inertia and θ the angle between the pendulum and the vertical axis, the law of inertia states the following:

$$I\ddot{\theta} = J, \qquad (3.1)$$

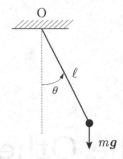

Figure 3.1 – Schema of the pendulum system.

where J is the torque. In vector notation, the torque is expressed as $\boldsymbol{J} = \boldsymbol{r} \times \boldsymbol{F}$, where $\boldsymbol{r}$ is the vector position, with the point of suspension O as its origin, and $\boldsymbol{F}$ is the applied force. In the case of the simple pendulum, the only applied force is that of its weight, $\boldsymbol{F} = \boldsymbol{p} = m\boldsymbol{g}$, where $\boldsymbol{g}$ is the gravitational vector. The moment of inertia is given by the product of mass m and the pendulum's length ℓ squared ($I = m\ell^2$). Taking into account that $\boldsymbol{g}$ is in the plane of movement, torque J is given by $J = -mg\ell \sin \theta$. The equation of motion takes on the form:

$$\ddot{\theta} + \frac{g}{\ell} \sin \theta = 0. \qquad (3.2)$$

This equation can also be obtained using the original form of Newton's law, the fundamental law of dynamics, which states that the product of mass and acceleration is equal to the sum of the forces. If we apply the law along the tangent of the trajectory (the mass traces a small section of a circle as it oscillates around the suspension point), giving us $mg \sin \theta$, and equate it to acceleration $m\ell\ddot{\theta}$, we obtain equation (3.2) again. The fixed points satisfy $\sin \theta_0 = 0$ (index 0 refers to a fixed point). Two distinct solutions are possible: $\theta_0 = 0$ and $\theta_0 = \pi$. The first fixed point corresponds to the mass being located at the bottom, while the second corresponds to the mass assuming an upright position. Intuitively, the first point is stable and the second unstable. If friction with air is included, one has to add a term proportional to $\dot{\theta}$ to equation (3.2). The dynamics of such a system is trivial: whatever the initial point, its final state is the fixed point $\theta_0 = 0$ after energy has been dissipated.

3.1.2. Driven Simple Pendulum with Dissipation

Let us now add a force $\boldsymbol{f}$ to the pendulum at the location of the mass; the total force becomes $\boldsymbol{F} = \boldsymbol{p} + \boldsymbol{f}$. Given that the force has a constant amplitude and is directed along the trajectory of the mass (that is, the force is tangent to the circle drawn by the pendulum's trajectory, see figure 3.2), the magnitude of the torque provided by the force is given by $f\ell$. Let us also include a frictional force of coefficient 2λ. This system could be physically built by

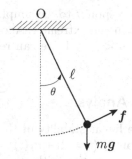

Figure 3.2 – The pendulum in the presence of a force f.

adding a small motor, which provides a torque equal to $f\ell$, to a pendulum at suspension point O. With the new terms, the pendulum equation reads:

$$\ddot{\theta} + 2\lambda\dot{\theta} + \frac{g}{\ell}\sin\theta - \frac{f}{m\ell} = 0. \tag{3.3}$$

This equation could also be derived directly from Newton's law.

We can rewrite the evolution equation using the following change of variables[1]: $t \to t\sqrt{g/\ell}$, and defining $\mu = \lambda(\ell/g)^{1/2}$ as well as $\nu = f/(mg)$:

$$\ddot{\theta} + 2\mu\dot{\theta} + \sin\theta - \nu = 0. \tag{3.4}$$

We can find two distinct values of θ_0:

$$\theta_0 = \arcsin(\nu) \equiv \Omega, \quad \theta_0 = \pi - \Omega. \tag{3.5}$$

The solutions of this equation must satisfy $\sin\theta_0 = \nu < 1$. A schematic view of the solutions is given in figure 3.3.

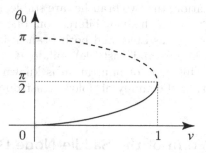

Figure 3.3 – Curve of fixed points (bifurcation diagram) of the pendulum in the presence of a force f.

---

1. Note that the operation $\theta \to -\theta$ is equivalent to a sign change in front of ν. This is why it will suffice to consider the case $\nu > 0$, since the case of $\nu < 0$ can be deduced by changing θ to $-\theta$ in the final results.

The particular case $\nu = 0$ corresponds to the simple pendulum, with the two solutions $\theta_0 = 0$ and $\theta_0 = \pi$. In the extreme case of $\nu = 1$, the two solutions converge to $\theta_0 = \pi/2$. Since $0 < \nu < 1$, we can restrict our analysis to the domain $0 < \Omega < \pi/2$.

3.1.3. Linear Stability Analysis

Next we must determine the linear stability of the fixed points. To do this, we apply a small perturbation θ_1 ($\theta = \theta_0 + \theta_1(t)$) to the stationary state θ_0, replace any θ in the evolution equation with this expression, and keep only terms that are linear in θ_1. Assuming an exponential form $\theta_1 \sim e^{\omega t}$ for the solution, we obtain the following:

$$\omega^2 + 2\mu\omega \pm \cos(\Omega) = 0. \tag{3.6}$$

Here, the "$+$" form corresponds to the case where $\theta_0 = \Omega$, and the "$-$" to the case $\theta_0 = \pi - \Omega$. The two eigenvalues associated with the first fixed point ($\theta_0 = \Omega$) are given by:

$$\omega_\pm = -\mu \pm \sqrt{\mu^2 - \cos(\Omega)}. \tag{3.7}$$

If the friction[2] is strong enough such that $\mu^2 \gg \cos(\Omega)$, we find that $\omega_+ \simeq -\cos(\Omega)/(2\mu^2) < 0$ and $\omega_- \simeq -2\mu < 0$. In other words, the two values are negative, imposing a steady exponential decrease of θ_1 over time, which means the fixed point is stable.

A similar calculation for the other fixed point $\theta_0 = \pi - \Omega$ leads to $\omega_+ = \cos(\Omega)/(2\mu^2) > 0$, and $\omega_- = -2\mu < 0$. Since for $\omega_+ > 0$, θ_1 grows exponentially over time t, it is an unstable fixed point. Comparing the bifurcation diagrams for this system (see figure 3.3) with that of the pitchfork bifurcation (figure 2.4), we see a clear difference.

In the pitchfork bifurcation, the two branches are stable, bifurcating from the state $x_0 = 0$ once $\ell_0 > \ell_c$. In this new bifurcation, one of the two branches is stable and the other is unstable, and both branches cease to exist once $\nu > 1$: both branches (stable and unstable) collide at $\nu = 1$. This is a new qualitative behavior. This type of bifurcation is known as the *saddle-node* bifurcation, and is part of the family of "fold" catastrophes, as we will see in chapter 4.

3.1.4. Universal Form of the Saddle-Node Bifurcation

Near the critical point, a bifurcation will always acquire a universal character. To see how this happens in the saddle-node bifurcation, we will now analyze the equations we just derived in this neighborhood.

---

2. This hypothesis does not affect the general results, but simplifies the algebra.

By convention, we make a preliminary change of variables in order to calibrate the value of the critical point to $\theta_0 = 0$ (and not $\pi/2$), in line with the current prevalent notation in the nonlinear sciences. Accordingly, we define $\theta = \pi/2 + \psi$, and the evolution equation of the pendulum (3.4) becomes:

$$2\mu\dot{\psi} + \cos\psi - \nu = 0, \tag{3.8}$$

where we have used the approximation of small $\ddot{\psi}$ when close to the critical point (recall that close to bifurcation ω is small, and $\ddot{\psi}$ is small compared to $\dot{\psi}$). Angle ψ is also small in the vicinity of the bifurcation point, so we can make a polynomial expansion of the pendulum equation (3.8) for small ψ. To leading order, we obtain the evolution equation:

$$\dot{\psi} = \psi^2 - \epsilon, \tag{3.9}$$

where $\epsilon = 2(1 - \nu) > 0$ (remember that the bifurcation corresponds to $\nu = 1$ and thus ϵ is small in that neighborhood). Evolution equation (3.9) is the associated amplitude equation for the generic form of a saddle-node bifurcation.

The two fixed points, which exist for $\epsilon > 0$, are $\psi_0 = \pm\sqrt{\epsilon}$. We can easily verify that $\sqrt{\epsilon}$ is unstable, and $-\sqrt{\epsilon}$ stable, by rewriting the equation in the form of a potential:

$$\dot{\psi} = -\frac{dV}{d\psi}, \quad V = \epsilon\psi - \frac{\psi^3}{3}. \tag{3.10}$$

For $\epsilon > 0$, potential $V(\psi)$ has a minimum (stable solution) and a maximum (unstable solution) (see figure 3.4); for $\epsilon = 0$ the minimum and maximum merge and disappear, and we have instead a horizontal inflection point. For $\epsilon < 0$, V has no extremum, indicating the disappearance of a fixed point; all of this restates the results of the previous section in a different language.

3.1.5. Origin of the Name "Saddle-Node"

For $\epsilon > 0$, the potential $V(\psi)$ of the saddle-node bifurcation has a minimum (see figure 3.4); in this domain, the system tends toward a stationary stable solution over time t $(t \to \infty)$, whatever the initial conditions of the system. In visual terms we can describe it as a stationary state which "attracts" all trajectories, which is why, in this context, the term "node" is used for the minimum of the potential $V(\psi)$.

For the same range of parameters $(\epsilon > 0)$, the configuration corresponding to the maximum of the potential (see figure 3.4) is a stationary solution but which is unstable, and all initial conditions close to this point will diverge toward the neighboring minimum of the potential in order to converge on the stable state. The usage of the term "saddle" here is due to the implicit

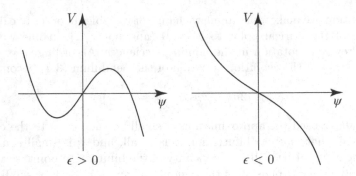

$\epsilon > 0$ $\epsilon < 0$

Figure 3.4 – Evolution of the potential for the saddle-node bifurcation.

presence of a saddle point in the potential, a saddle point characterized by the presence, in the phase space diagram, of a maximum in the ψ direction, and a minimum in the other direction, which we will now describe.

If we conserve the second order time derivative in the original formulation of the problem, the stable mode can be seen explicitly, and it introduces a second dynamic variable of the system (giving us a two dimensional phase space). Actually, we are speaking of the two eigenmodes associated with the quadratic dispersion relation, equation (3.6). Since, as seen above, the dispersion relation has a second eigenvalue which is always negative, the associated eigenmode decays exponentially with time. Let that mode be denoted by ψ_s (the subscript s refers to "stable"), we have then $\dot{\psi}_s = -\psi_s$ (we have set the prefactor to unity which is always possible with an appropriate rescaling of time). In summary we have

$$\dot{\psi}_s = -\psi_s$$
$$\dot{\psi} = \psi^2 - \epsilon. \qquad (3.11)$$

By introducing the amplitude vector $\boldsymbol{A} = (\psi_s, \psi)$, we can describe the evolution equation of the system in a compact form (since the amplitude vector is derived from a scalar potential):

$$\dot{\boldsymbol{A}} = -\nabla V(\psi_s, \psi) \qquad (3.12)$$

with the potential defined as:

$$V = \frac{\psi_s^2}{2} + \epsilon\psi - \frac{\psi^3}{3}. \qquad (3.13)$$

The components of the gradient operator ∇ are ∂_{ψ_s} and ∂_ψ. Along the ψ_s direction we have a maximum of V at $\psi_s = 0$, while along the ψ direction we have a maximum and a minimum of V for $\epsilon < 0$. When $\epsilon \to 0$, we have a minimum in both directions (node) around the fixed stable point $\psi \sim 0$, and a maximum and minimum (saddle) at the unstable point.

3.1.6. An Intuitive Picture of the Saddle-Node Bifurcation

We can use a geometric approach to analyze the stability of the forced pendulum's dynamical fixed points without reference to equations.

In this system a fixed point exists anytime the two forces which are present are at equilibrium; in this case, this means that the projection of the weight $p = mg$ along the trajectory and the constant force f, by definition tangential to the movement, are of equal amplitude but opposite sign (i.e. $p_t = -f$, with p_t being the projection of the weight p). This equilibrium happens at two points, one stable and the other unstable (see figure 3.5); the first point is defined by the angle of the pendulum to the vertical $\theta = \theta_0$, and the second by the angle $\theta = \pi - \theta_0$.

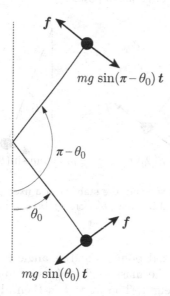

Figure 3.5 – Schema of two possible equilibrium positions; t is the unit vector tangent to the trajectory.

For the first point, by displacing the mass an angle θ_0' away from the value θ_0 (see figure 3.6, right-hand side), the projection of the weight p_t becomes greater than the force f ($|p_t| > |f|$); the mass moves back to equilibrium position θ_0. Displacing the mass any amount $\theta_0' < \theta_0$ also ends with a return of the mass to its original equilibrium position (see figure 3.7). In this case the force f is the one which is greater than the projected weight ($|f| > |p_t|$) and exerts the force which returns the system to its original stable equilibrium position, θ_0.

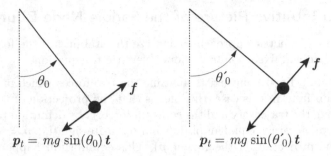

Figure 3.6 − Schema showing the stability of a fixed point. If the mass moves upwards, angle θ_0 becomes θ_0' (to the right), and the weight brings the mass back to its initial position.

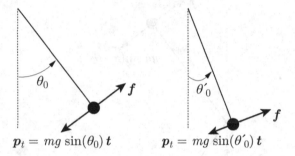

Figure 3.7 − Schema showing the stability of a fixed point. If the initial angle θ_0 decreases and becomes θ_0', force f brings the mass back to its initial position.

In contrast the second fixed point, found at angle $\theta = \pi - \theta_0$, is unstable: an upwards deviation of the mass at $\theta = \theta_0'$ (see figure 3.8) leads this time to a weaker projected weight. The force f is then always greater than the projected weight, and so the mass moves continuously farther away from the initial position. The mass follows a counterclockwise trajectory until it attains a stable position, its energy having been dissipated by friction[3]. The reverse happens when θ_0 is deviated downwards: the projected weight exceeds the force, and the mass is directed in a clockwise movement toward the stable equilibrium point (see figure 3.9).

---

3. In a frictionless system, the movement would continue indefinitely due to the principle of inertia.

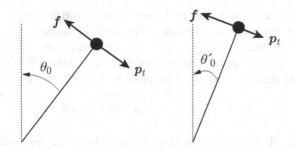

Figure 3.8 – Schema showing the instability of a fixed point. If the mass moves upwards (θ_0 becomes θ'_0), force f overrides the weight, and the mass gets farther away from its initial position. It will rotate in a counterclockwise direction.

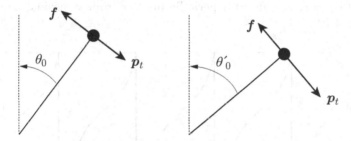

Figure 3.9 – Schema showing the instability of a fixed point. If the mass moves downwards (θ_0 becomes θ'_0), the weight overrides force f, and the mass continues its trajectory in a clockwise motion.

3.1.7. Topple Movement (or Tumbling)

When the weight component is equal to that of force f, the parameter $\nu = f/mg$ is equal to one ($\nu = 1$ meaning that $f = mg$). The two solutions, stable and unstable, coincide; the dynamics of the system converge at a fixed point where $\theta = \pi/2$. If the magnitude of the force f is further augmented, ν exceeds unity ($\nu > 1$); force f is now always stronger than the weight, and will permanently drive mass m in a counterclockwise direction. We call this type of movement *tumbling*.

We write the exact solution for the pendulum's evolution equation (3.9) in the neighborhood of the critical point:

$$\psi(t) = -\sqrt{\epsilon}\tanh\left[\sqrt{\epsilon}(t + C_1)\right] \tag{3.14}$$

where C_1 is an integration constant determined by the initial conditions. As time tends to infinity ($t \to \infty$), the solution tends toward the stable position $\psi = -\sqrt{\epsilon}$. This equation is valid in the neighborhood of the critical point, and for $\epsilon > 0$ (or $\nu < 1$), in which case there is a fixed point. However, when $\epsilon < 0$

(or $\nu > 1$) there is no fixed point, and there is no reason to focus on the regime of $\psi = 0$ (which defined a fixed point). We must look again to the complete equation (3.4). Assuming the friction is strong enough to overcome the inertia of the system, we can suppress the second time derivative of angular variable ψ ($\ddot{\psi} \to 0$), and make a change of variable in t to rid us of constant 2μ, to obtain

$$\dot{\theta} + \sin\theta - \nu = 0. \tag{3.15}$$

It can be shown that this nonlinear equation admits the following exact solution:

$$\theta(t) = -2\arctan\left(\frac{\tan\left[\frac{t}{2}\sqrt{\nu^2 - 1} + \frac{C_1}{2}\sqrt{\nu^2 - 1}\right]\sqrt{\nu^2 - 1} - 1}{\nu}\right). \tag{3.16}$$

This solution corresponds to a periodic but far from sinusoidal movement (see figure 3.10).

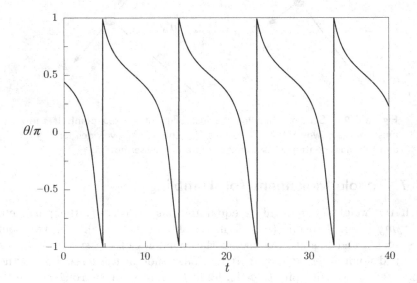

Figure 3.10 − Behavior of angle $\theta(t)$ in the *tumbling* regime.

3.1.8. Digressions in the Field of Biology: Tumbling Motion of a Red Blood Cell

The simple mechanical model of the forced pendulum is very useful for the description of certain biological dynamic behaviors, notably those of red blood cells, the major cellular components of our blood. When the blood is circulating in a blood vessel, the red blood cells are often subject to a shear flow. At the scale of the globule, the velocity of the flow varies quasi-linearly with distance from the blood vessel wall. Therefore, the two sides of a red blood cell are subject to different velocity amplitudes, creating a shear flow resulting in a hydrodynamic torque, causing the cell to rotate, or tumble[4], around its own axis. Now, these globules may avoid the tumbling if they transmit this energy to the cytoplasmic membrane, which has the property of being able to slide like tank's conveyor wheels (or the chain of a bicycle). Thus, if a globule transmits enough of the energy to the membrane, it avoids turning around its axis (tumbling). In this scenario, the globule orients itself in the flow at a fixed angle (a fixed point) as in the case of the forced pendulum at equilibrium (i.e. a steady and stable fixed point); the equilibrium solution comes from a balance between the flow which makes it turn and the conveyor belt movement (also called tank-treading motion) of the membrane opposing it, by absorbing part of the energy due to shear flow.

This conveyor movement of the membrane (or tank-treading) drags along the internal liquid of the red blood globule, the hemoglobin, whose function is to regulate the oxygen. If the internal viscosity of the globule augments, as is the case in vivo for people with a pathology such as sickle-cell anemia, the tank-treading movement of the membrane is reduced, because the membrane has to drag an internal liquid that has become too viscous. There exists a critical internal viscosity above which the tank-treading movement of the membrane is too energetically expensive, and the globule switches to a stiff-like tumbling movement.

We have visuals of the dynamic behavior of such objects from laboratory experiments (see figure 3.11), which show that the orientation angle adopted by the globules is indeed similar to that adopted by the forced pendulum (figure 3.10); see [89].

Figure 3.11 – Photographs showing the behavior, in successive moments, of two model globules (two giant liposomes fabricated in the laboratory) undergoing a *tumbling* movement. A red cell presents the same type of movement. [From [66] M.-A. Mader, V. Vitkova, M. Abkarian, A. Viallat & T. Podgorski. Dynamics of viscous vesicles in shear flow, *Eur. Phys. J. E*, **19**: 389397, 2006, with kind permission from Springer Science and Business Media]

---

4. A globule is more likely to tumble if it is less inflated than a regular globule. A flatter globule will stretch further with the shear flow, creating a greater torque and thus a propensity for tumbling.

Thus, in standard biological conditions, and despite the greater complexity of the biophysical mechanisms that contribute to the globule's behavior, we can say that the dynamics of the red blood cell follow the same evolution equation of a forced pendulum (3.4); for a more detailed treatment, see [89].

3.2. The Imperfect Bifurcation; Extrinsic Symmetry Breaking

To lead us into our discussion of imperfect bifurcations, let us briefly recall the system studied in the last chapter. A mass in the form of a ring is suspended vertically from a spring of constant k and initial length ℓ_c, which can be larger or smaller than its rest length, ℓ_0; this mass is moved a distance x along a horizontal rod (see figure 2.1). The amplitude equation (2.24) associated with the dynamic evolution of the mechanical system is invariant under the change of variable $A \to -A$; thus, for each solution A of the amplitude equation, $-A$ is also a solution.

This symmetry in the solutions is the result of an assumed symmetry between right and left in the mechanical system; we have effectively considered the right side position of the spring to be exactly equivalent to its left side position (see figure 2.1). However, real systems always have a degree of imperfection resulting in a light asymmetry, the effect of which can be seen explicitly in the following way. Let us perform the experiment in which the mass is initially put at position x which is assumed to be an unstable solution, and apply a very small unbiased perturbation to the system. The mass will depart from its initial position $x = 0$ and move along the horizontal axis until it reaches the new equilibrium stable position $x \neq 0$. If we repeat the same experiment a large number of times, and count the number of times the mass ends either left or right of its original position, we would see that for the ensemble of tries the number of final "right" and "left" solutions is almost never identical. There always exists at least one weak bias (real systems are never absolutely perfect) and the system will, in the end, tend to favor one of the two equilibrium solutions. This bias can be represented, in the mechanical spring system we detailed up until now, by the addition of a force f, applied to the mass of the system, where $f > 0$ when oriented in the direction of positive x (figure 3.12); this introduces a bias toward the "right" side solutions of the system.

To describe the "imperfect" dynamics of this system, the evolution equation must incorporate force f, and becomes $m\ddot{x} + \mu\dot{x} + \partial E_p/\partial x - f = 0$, where we recall that m is the sliding mass, μ the friction coefficient and E_p the potential energy of the system. If we designate the "imperfect" potential energy $\tilde{E}_p$ of

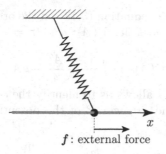

f : external force

Figure 3.12 – Schema of a spring system in the presence of a bias as represented by force f.

the system as the sum of the "perfect" potential energy of the system E_p and the work $-fx$ of the force f ($\tilde{E}_p = E_p - fx$), the evolution equation again takes on a generalized form similar to that obtained for the perfect system: $m\ddot{x} + \mu\dot{x} + \partial\tilde{E}_p/\partial x = 0$ (see equation (2.8)). Expanding the potential energy equation to fourth order, we obtain:

$$\tilde{E}_p = \frac{k}{2}\left[-\epsilon x^2 + \frac{x^4}{4\ell_0^2}\right] - fx, \tag{3.17}$$

which may be compared with the potential energy equation for the "perfect" system (equation (2.7)). Distance ϵ to the bifurcation threshold is defined, as in the previous chapter, by $\epsilon = (\ell_0 - \ell_c)/\ell_c$.

Following the same procedure as in the previous chapter (looking for fixed points, analyzing the linear stability around the fixed points, and finding solutions (expanding the series in ϵ) to the complete nonlinear equation in order to find the amplitude equation of the perfect system, equation (2.26)), we arrive at the following amplitude equation for the "imperfect" system:

$$\frac{dA}{d\tau} = \epsilon A - A^3 + \nu \tag{3.18}$$

where $\tau = k(\epsilon t)/\nu$ and $\nu = \bar{f}/(\ell_c k)$, with $\bar{f} = f/\epsilon^{3/2}$.

Note that in order to get this result, we have made the implicit assumption that the force f is large enough to have an effect of the same order as the dominant term in the dynamical evolution equation; that is, that f is on the order[5] of $\epsilon^{3/2}$. This hypothesis of $f \sim \epsilon^{3/2}$ justifies that we only expanded the potential energy equation of the "imperfect" system to fourth order (equation (3.17)); then all the considered terms are of the same order. It also ensures that the new term ν, describing the imperfection in the system, also

---

5. To obtain this result, we remember that solution $x(t)$ in the neighborhood of the bifurcation threshold ($\epsilon = 0$) is found in the form of a series of order $\epsilon^{1/2}$, see equation (2.22).

intervenes in the amplitude equation (3.18) on the order[6] of ϵ. Let us rewrite this equation using the potential $V(A)$: $dA/d\tau = -\partial V/\partial A$. The potential has the form:

$$V(A) = -\frac{\epsilon A^2}{2} + \frac{A^4}{4} - \nu A. \tag{3.19}$$

This form of the potential allows us to identify the equilibrium states of the system by looking for the *extrema* P of the potential V which satisfy the following cubic equation:

$$P(A) \equiv \epsilon A - A^3 + \nu = -\frac{\partial V}{\partial A} = 0. \tag{3.20}$$

A maximum in the potential V corresponds to an unstable equilibrium state, and a minimum corresponds to a stable equilibrium state. A cubic equation has two generic types of solution: its three roots are either all real numbers, or one may be a real number and the other two complex (i.e. with a non-zero imaginary component).

Instead of explicitly solving the cubic equation let us adopt a simple line of reasoning, founded on physical arguments and on an analysis of the properties and asymptotic behavior of the curve $P(A)$. For zero amplitude ($A = 0$), the equation for the extrema P (3.20) gives us a value equal to ν ($P(0) = \nu$); this value $P(0)$ is positive because ν is directly proportional to force f, which is oriented in the positive x direction.

Now consider the behavior of the extrema $P(A)$ a little further; this time, in the neighborhood of the point $A = 0$. Still using equation (3.20) above, we obtain, for small amplitude A, an approximate expression for the extrema, dominated by the weakest power term in A; that is, $P(A) \sim \epsilon A$. The derivative of $P(A)$ at this point is thus of order ϵ, which implies that in this neighborhood, function $P(A)$ grows for values of $\epsilon > 0$ and decreases for $\epsilon < 0$. However, the asymptotic behavior of $P(A)$ at large amplitudes is given by the approximate expression at the extrema as dominated by the highest power term in A, in this case $P(A) \sim -A^3$, which tends toward $-\infty$ for $A \to +\infty$ and $+\infty$ for $A \to -\infty$.

With the goal of clearly enumerating and identifying the extrema of the system's potential $V(A)$, let us distinguish two cases, as determined by the sign of parameter ϵ.

1. $\epsilon < 0$. The length of the spring, when vertical, is greater than the resting length of the spring ($\ell_c > \ell_0$). Looking at equation (3.20), we find that $P(A)$ is a constantly decreasing function, since the derivative of P with

---

6. In the hypothesis where f is of smaller order than $\epsilon^{3/2}$, the imperfection term ν appears naturally in a higher order in the Taylor expansion of potential energy; but we will see that in the end the qualitative behavior of the imperfect system does not depend on the amplitude of the imperfection term ν.

respect to A is always negative $(dP/dA = \epsilon - 3A^2 < 0)$. Considering the previously discussed properties of curve $P(A)$, it follows that equation $P = 0$ has one real solution for $A > 0$ (see figure 3.13).

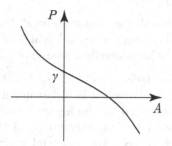

Figure 3.13 – Generic behavior for function P when $\epsilon < 0$.

2. $\epsilon > 0$. The length of the spring, when vertical, is smaller than its resting length ($\ell_c < \ell_0$). In this case, the derivative of the extrema, given by $dP/dA = \epsilon - 3A^2$, becomes 0 at two points: $A = \pm\sqrt{\epsilon/3}$ on both sides of the origin $A = 0$. This translates, for function $P(A)$, as a local minimum at $A = -\sqrt{\epsilon/3}$ and a local maximum at $A = +\sqrt{\epsilon/3}$ (see figure 3.14).

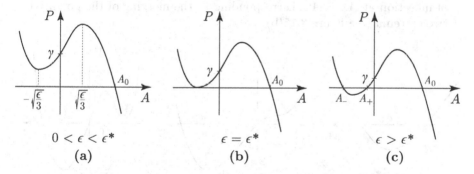

Figure 3.14 – Typical behavior of function P for $\epsilon > 0$.

a) $A > 0$. Taking into consideration the general behavior of curve $P(A)$, the presence of a local maximum at $P = 0$ signals the existence of only one real solution for positive amplitudes $A > 0$, irrespective of the height of this local maximum.

b) $A < 0$. Conversely, for negative amplitudes $A < 0$, equation $P = 0$ can have zero or two solutions, depending on the depth of the local minimum. If the local minimum has a positive value, the curve $P(A)$ does not cross the amplitude axis (x-axis) and so $P(A)$ does not become zero (see figure 3.14(a)). However, if the value of the local minimum is negative, $P(A)$ crosses the amplitude axis at two values, corresponding to two possible solutions where P becomes 0. The critical value of ϵ satisfies $P(-\sqrt{\epsilon/3}) = 0$ and $(dP/d\epsilon)(-\sqrt{\epsilon/3}) = 0$; these conditions define a critical value of parameter ϵ, designated ϵ^*, such that $\epsilon^* = (3\sqrt{3}\nu/2)^{2/3}$.

Using this analysis we thus define a new value ϵ^* based on parameter ϵ to determine the form of the system's potential, which then allows us to reframe the different cases we have just described (see figure 3.14):

1. $\epsilon < \epsilon^*$. V has only one extremum ($P = dV/dA = 0$ has a unique solution). For large amplitudes, the potential is independent of the sign of parameter ϵ because it is dominated by the highest order term with respect to amplitude, where $V \sim +(A^4/4)$ (see equation (3.19)). As the amplitudes tend toward infinity ($A \to \pm\infty$), the potential always goes toward positive infinity ($V \to +\infty$): the extremum of the potential is a minimum, with abscissa A_0 (see figure 3.15(a)).

2. $\epsilon > \epsilon^*$. The potential V has three extrema at three points: A_0 ($A_0 > 0$) corresponding to a minimum, and A_- and A_+ (with $A_- < A_+ < 0$) corresponding to a minimum and maximum, i.e. stable and unstable solutions, respectively (see figure 3.15(c)).

3. $\epsilon = \epsilon^*$. The potential V has two extrema, a minimum at A_0 and a point of inflection at $A_+ = A_-$, corresponding to the merging of the potential's two extrema (see figure 3.15(b)).

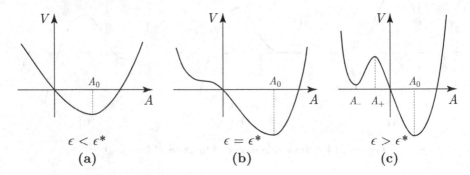

Figure 3.15 – Generic behavior of $V(A)$ for different values of ϵ.

We can now draw the portrait of the equilibrium solutions, also called a bifurcation diagram, and represent the evolution of amplitudes A_0, A_- (of a stable equilibrium, represented by the solid line in figure 3.16) and A_+ (of an unstable equilibrium, represented by the broken line in figure 3.16) as a function of parameter ϵ. This portrait of equilibrium solutions illustrates the difference in behavior between positive and negative amplitudes: the two solutions $A > 0$ and $A < 0$ are not equivalent. The system is characterized by an extrinsic (i.e. due to the existence of an imperfection in the system) symmetry breaking $A \to -A$: we thus refer to this bifurcation as *imperfect*.

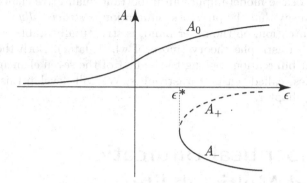

Figure 3.16 – Imperfect bifurcation diagram.

To conclude, we note that the presence of a positive force ($f > 0$), i.e. a force oriented in the direction of the positive amplitude (x) axis in the system creates a bias toward the equilibrium solutions of A_+. If we opt for a negative force ($f < 0$), i.e. a force oriented toward the negative amplitude (x) axis, the same bifurcation is obtained apart from a simple change in the amplitude variable, that is, from A to $-A$ (see figure 3.17).

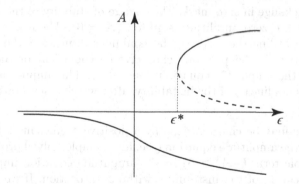

Figure 3.17 – Imperfect bifurcation diagram after changing the sign of the imperfection term.

3.2.1. A Taste of Catastrophe Theory

For real systems, the symmetry $A \to -A$ is an unreachable ideal. The pitchfork bifurcation undergoes an important qualitative change (compare figure 2.4 to 3.16) whatever the amplitude of imperfection term ν, even if arbitrarily small; the bifurcation is thus extremely sensitive to an external perturbation. In this sense, the pitchfork bifurcation may be considered as a "fragile" representation of the model: it is not robust because the smallest perturbation of the model implies an important qualitative change. For this reason we can say that the pitchfork bifurcation is *structurally unstable*. The imperfect bifurcation, on the other hand, is structurally stable as defined in the frame of catastrophe theory (more of which later). Both the pitchfork and imperfect bifurcations belong to the first of the seven elementary catastrophe families, called "cusp" catastrophes. We will develop this further in the following chapter.

3.3. Subcritical Bifurcation and Multistability

In this section we return to models of systems with a perfect left/right symmetry[7] (equivalence between A and $-A$ in the amplitude equation) so that we need only consider the uneven powers of amplitude A in the amplitude equation.

For the pitchfork bifurcation, the sign of the linear term in the amplitude equation (2.26) is given directly by the distance to the instability threshold, ϵ: below the threshold, the sign of ϵ is negative, and above the threshold, ϵ is positive. The change in sign marks the change of stability of the equilibrium solution, here the zero amplitude solution ($A = 0$). There is, however, no simple rule regarding the signs of the nonlinear terms. For the mechanical system described in the previous chapter, the coefficient of the nonlinear term (A^3) in the amplitude equation is negative. This implies a saturation (referred to as nonlinear) of the instability, after which a new stationary state is reached.

Other systems may be characterized by a positive coefficient in front of the cubic term in the amplitude equation (see, for example, solved problem 3.9.2); a positive cubic term ($+A^3 > 0$) in the amplitude equation implies a nonlinear amplification of the instability, without saturation. If we assume the cubic coefficient to be positive, this means that the amplitude equation reads $dA/d\tau = \epsilon A + A^3$, and has fixed points $A = 0$ and $A = \pm\sqrt{-\epsilon}$; the latter two solutions only exist for $\epsilon < 0$. It is simple to verify[8] that for systems

---

7. Even if the perfect bifurcation is structurally unstable, let us first consider, for reasons of simplicity, "perfect" models, and then continue with an introduction of the imperfection term, as in the preceding section.

8. It suffices to analyze the associated potential, $V(A) = -\epsilon A^2/2 - A^4/4$; recall that the maximum (and respectively, the minimum) of potential $V(A)$ corresponds to an unstable state (or stable, respectively).

characterized by a positive coefficient in front of the cubic term, $A = 0$ is stable for $\epsilon < 0$ and unstable for $\epsilon > 0$, while $A = \pm\sqrt{-\epsilon}$ is always unstable (see bifurcation diagram in figure 3.18). There is no stable branch that corresponds to an amplitude $A \neq 0$; thus, to cubic order, the system cannot reach a new stable stationary state. The usual strategy consists of analyzing the problem to higher order and pursuing the nonlinear expansion of terms in the power of amplitude A with the hope of finding terms of a higher order to balance against the linear and cubic[9] terms.

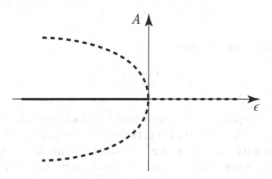

Figure 3.18 – The diagram of an inverted pitchfork bifurcation. Note that, contrary to the case of figure 2.4, the branches emerging from solution $A = 0$ are unstable. We sometimes refer to this bifurcation as the dual of that represented in figure 2.4.

Since the system is assumed to be perfect (equivalence between A and $-A$), the next higher order nonlinear term (i.e. following the cubic term) in the amplitude equation is the fifth order term (A^5); if we assume that it slows down the growth of the instability (nonlinear saturation, see the example of the mechanical system developed in solved problem 3.9.2), then the coefficient of the fifth order nonlinear term is negative and the amplitude equation is written as[10]

$$\frac{dA}{d\tau} = \epsilon A + A^3 - A^5 = -\frac{dV}{dA}. \qquad (3.21)$$

---

9. Many dynamical systems are in practice characterized by nonlinear terms (A^3, A^5, A^7) in the amplitude equation, terms which slow down the temporally exponential linear growth of the instabilities. There exist nonetheless several models of systems for which all the nonlinear terms amplify the instability [52]; these models are symptomatic of a *finite time singularity*.

10. For full generality we could have written $dA/d\tau = \epsilon_1 A + \epsilon_2 A^3 - \epsilon_3 A^5$. Dividing both sides of the equation by ϵ_3 we have $dA/d\tau' = (\epsilon_1/\epsilon_3)A + (\epsilon_2/\epsilon_3)A^3 - A^5$, with $\tau' = \tau\epsilon_3$. We define $A = A'(\epsilon_2/\epsilon_3)^{1/2}$, with $\epsilon = (\epsilon_1/\epsilon_3)(\epsilon_2/\epsilon_3)^{-2}$ and $\tau'' = \tau'(\epsilon_2/\epsilon_3)^2$ and the equation becomes $dA'/d\tau'' = \epsilon A' + A'^3 - A'^5$. In other words, we can restrict our study to a single parameter (assuming the sign of ϵ_2 and ϵ_3 to be fixed, as we have done).

After integrating, we obtain the following potential $V(A)$:

$$V(A) = -\frac{\epsilon A^2}{2} - \frac{A^4}{4} + \frac{A^6}{6}. \tag{3.22}$$

The stationary solutions of the system correspond to the extrema of polynomial V. We find again the null amplitude solution $A_0 = 0$, and the other fixed points satisfy equation

$$A^4 - A^2 - \epsilon = 0, \tag{3.23}$$

which has the following solution:

$$A^2 = \frac{1 \pm \sqrt{1 + 4\epsilon}}{2}. \tag{3.24}$$

Since the square amplitude (A^2) is a real number, parameter ϵ must obey the following condition: $\epsilon > -1/4$. Furthermore, the condition $A^2 > 0$ means that the solution with the minus sign is possible only if $\epsilon < 0$. A fifth order equation has always one, three, or five real solutions, giving us, for different domains, the following stationary solutions for the amplitude equation:

$$\begin{cases} \epsilon < -\frac{1}{4}, & A_0 = 0 \\ -\frac{1}{4} < \epsilon < 0, & A_0 = 0, \quad A_{1\pm} = \pm \left(\frac{1 - \sqrt{1 + 4\epsilon}}{2}\right)^{1/2}, \\ & \text{and } A_{2\pm} = \pm \left(\frac{1 + \sqrt{1 + 4\epsilon}}{2}\right)^{1/2}, \\ \epsilon > 0, & A_0 = 0, \quad A_{2\pm} = \pm \left(\frac{1 + \sqrt{1 + 4\epsilon}}{2}\right)^{1/2}. \end{cases} \tag{3.25}$$

Note that when $\epsilon = -1/4$, $A_1 = A_2$.

The behavior of potential $V(A)$ as a function of amplitude A, given by equation (3.22), depends on the different possible values of parameter ϵ (see figure 3.19). We can deduce the stability of each solution by looking at the nature of the extrema of potential $V(A)$ (minima are stable solutions, maxima are unstable solutions, see figure 3.19). This allows us to plot the bifurcation diagram as a function of parameter ϵ (see figure 3.20 to visualize the stable (continuous lines) and the unstable (broken lines) branches). This bifurcation diagram is an example from the family of "butterfly catastrophes" (see chapter 4), also referred to as *subcritical bifurcations*. This bifurcation is characterized by the existence of two new stable solutions, A_{2+} and A_{2-}, in the regime where parameter ϵ is still negative, that is, before the zero amplitude solution ($A = 0$) loses stability, which occurs at the critical value $\epsilon = 0$.

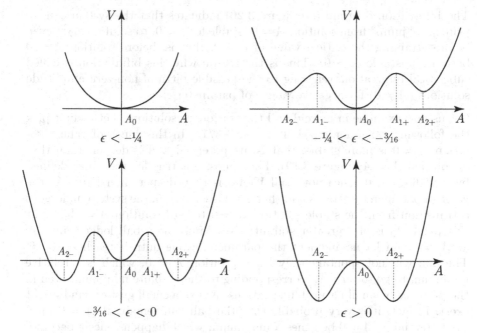

Figure 3.19 – The shape of the potential for a subcritical bifurcation at different values of ϵ. The value $\epsilon = -3/16 \equiv \epsilon'$ corresponds to the situation where the two solutions, A_0 and A_2, have the same stability (see text).

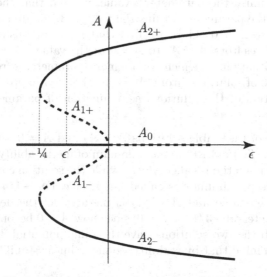

Figure 3.20 – Diagram of a subcritical bifurcation.

The bifurcation diagram (see figure 3.20) indicates that the system can, in principle, "jump" from solution $A = 0$, stable for $\epsilon < 0$, to solution A_2[11] even before attaining the critical value of $\epsilon = 0$, that is, before solution $A = 0$ becomes unstable ($\epsilon > 0$). This is the reason why this bifurcation is called subcritical bifurcation, referring to the possible birth of non-zero amplitude solutions ($A \neq 0$) for negative values of parameter ϵ.

Let us now examine in more detail the stationary solutions. Let us first pick the following range for ϵ: $-1/4 < \epsilon < -3/16$. In this range of values, solution $A = 0$ is more stable, that is, its potential well is deeper, than that of solution $A = A_2$ (figure 3.19). The opposite is true for the range defined by $-3/16 < \epsilon < 0$, since potential $V(A_2)$ is then deeper than $V(0)$. As the value of parameter ϵ attains and then surpasses $-3/16$, the system undergoes a transition from the stable solution $A = 0$ toward solution $A = A_2$, as at this point A_2 is of a greater stability. Now, imagine a ball lodged initially at $A = 0$ and let us augment the parameter value until it reaches $-3/16$. The ball will not spontaneously leave position $A = 0$, which is a relative minimum, to go toward A_2, corresponding to the absolute minimum, even in the presence of small natural fluctuations such as a small gust of wind might exert. In fact, it is very probable that the ball will remain at $A = 0$ until the potential at $A = 0$ becomes a maximum, which happens once ϵ becomes positive. For those values, solution $A = 0$ becomes unstable and the ball falls toward one of the potential's two minima, $A = A_{2+}$ or A_{2-}, where it remains for all values of $\epsilon > 0$. If we instead pick a positive initial value of ϵ and progressively diminish the parameter's value, we find that the ball remains at $A = A_2$ until the parameter attains the value $-3/16$, after which the ball falls back again to $A = 0$. Thus, we find that for a diminishing value of ϵ, the solution changes from $A = A_2$ to $A = 0$ at the value of $\epsilon = -3/16$, while for the reverse situation, in which we augment parameter ϵ from a negative value, the change of solutions (from $A = 0$ to $A = A_2$) happens when $\epsilon = 0$. Thus, the evolution of the solutions as a function of parameter ϵ is not a reversible process.

This coexistence of the stable solutions $A = A_2$ and $A = 0$ for the values of $-1/4 < \epsilon < 0$ (see figure 3.19) define a domain of metastability. The solution with the deepest potential is called *stable*, while the solution corresponding to the other (less deep) minimum is called *metastable*. For $-1/4 < \epsilon < -3/16$, solution $A = 0$ is stable and $A = A_2$ is metastable; the designations are reversed for the interval $-3/16 < \epsilon < 0$, since now $A = 0$ becomes metastable and A_2, stable. If the two solutions have the same potential, they are called bistable. Keep in mind that by stability we mean linear stability. This means

---

11. Take note that $A_{2+} = -A_{2-}$; this means the two solutions are equivalent, so as not to have to distinguish between them, we will globally reference this pair of solutions as A_2.

that in the presence of a perturbation that is strong enough, such as a kick to the ball (to elaborate on the last analogy), the ball can easily be dislodged from the metastable state toward the stable state. To do the reverse, that is, to project the ball from the stable to the metastable state, one needs to furnish a larger amount of energy, which is to say, give a much harder kick. The stable state is more resistant than the metastable state.

Let us now determine the conditions for which we have a bistability of solutions $A = 0$ and $A = A_2$. By definition, the solutions are bistable if:

$$V(A = A_2) = V(A = A_0 = 0). \qquad (3.26)$$

Looking at equation (3.22), we see that the potential vanishes at point A_0, and condition (3.26) implies that the potential must vanish at $A = A_2$ as well ($V(0) = 0 = V(A_2)$), which allows us to express the fourth order term A_2^4 as a function of the lower order terms ($A_2^4 = 3\epsilon + 3/2A_2^2$) and, applying the constraints for a stationary solution (equation (3.23)), we obtain the following bistability condition:

$$A_2^2 = -4\epsilon > 0. \qquad (3.27)$$

Making use of the stationary solution A_2 as a function of parameter ϵ (see equation (3.25)) we find that solutions A_0 and A_2 are bistable for a critical value of ϵ, which we will designate as ϵ', and its numerical value is $\epsilon' = -3/16$. In the domain of $-1/4 < \epsilon < \epsilon'$, the solution A_0 is stable and A_2 metastable; the opposite is true in the domain $\epsilon' < \epsilon < 0$. In the solution to solved problem 3.9.1 we will further develop the geometric interpretation represented by the straight line $\epsilon = \epsilon'$ drawn in the bifurcation diagram and labeled "Maxwell Plateau" (see figure 3.29) by analogy with the concept invented by Maxwell (1831–1879) in the case of certain phase transitions (for example, the liquid-gas transition).

3.3.1. Metastability and the Role of Fluctuations

The phase transitions of a thermodynamic system (like crystallization), called first order transitions, are analogous to subcritical bifurcations. The transitions are catalyzed by thermal fluctuations or impurities in the material; without these perturbative elements, the system would have difficulty spontaneously transitioning from the metastable to the stable state.

For example, pure liquid water, placed into a freezer some degrees below $0°C$, will with difficulty turn into ice[12]. In other words, in the ideal conditions of material purity and without fluctuations of any sort, the liquid water that is exposed to negative temperatures (those below $0°C$) stays in a

---

12. Although even a small impurity, such as that found in potable water, is enough for the formation of ice cubes in the ice tray of a refrigerator!

liquid state even if the thermodynamic potential[13] of the water as ice is lower than that of the water as liquid; the thermodynamic system can just remain in its metastable state. Here, and this is a fundamental difference between microscopic and macroscopic mechanical systems, the thermodynamical fluctuations play a central role in the problem of phase transition, engendering phenomena that operate on the atomic or molecular scales through the bias of collective microscopic movements[14]. At the microscopic scale, all fluctuations, particularly thermal ones, necessarily play a major role. The scenario, valid in the thermodynamic domain, is thus fundamentally different from the scenario of macroscopic bifurcations analyzed in this book. For example, within the spring system studied at the beginning of chapter 2, microscopic fluctuations such as the thermal ones just discussed play no role in the process of transition – or bifurcation – from one state to another. In the same vein, consider a ball placed at the bottom of a fairly deep potential well and in proximity to even deeper troughs. The ball, subject only to natural fluctuations (thermal fluctuations, wind, a banging door, etc.) has no chance of suddenly finding itself in the neighboring pit, even if it is a deeper one (see figure 3.21).

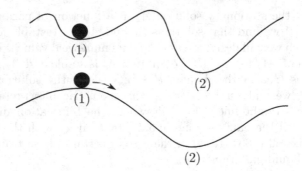

Figure 3.21 – In the top figure, a ball at point (1) has nearly no chance of finding itself at point (2) solely due to natural fluctuations. For this transition, a strong exterior perturbation is needed. By varying a control parameter, the landscape of the potential may change the valley in (1) to a summit (lower figure) and the ball is now in an unstable position. In this case, a small fluctuation is enough to force the transition from (1) toward (2).

---

13. There exist different thermodynamical potentials, such as internal energy, free energy, enthalpy, or Gibbs free energy; they have the dimensions of an energy and are analogous to potential energy in mechanics as they represent the potential energy of the system and they are extremal when in equilibrium.

14. Brownian motion affects a grain of pollen, or a molecule, but not a tennis ball submerged in water.

In order to induce the transition from the metastable to the stable state, there has to be an important fluctuation, such as, in the last example, a kick that is strong enough, i.e. a movement that surpasses the natural and weaker fluctuations. Without this sort of external fluctuation the system is unlikely to transition from the metastable to the stable state. Only once the metastable state becomes unstable – the minimum of the potential thus becoming a maximum – is the path of the ball toward the neighboring trough favored (see figure 3.21).

3.3.2. Hysteresis Cycles

Let us return briefly to the irreversible aspect of the stationary solutions' evolution as a function of parameter ϵ in the case of a subcritical bifurcation (section 3.3). Let us look at the domain of positive amplitudes for negative values of parameter ϵ in the bifurcation diagram (see figure 3.22)[15].

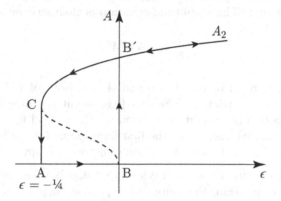

Figure 3.22 – Bifurcation diagram with a hysteresis cycle.

Consider four special points, designated as A, B, B' and C, in figure 3.22. Beginning from the null amplitude solution ($A_0 = 0$) where $\epsilon = -1/4$ at point A, the progressive increase of parameter ϵ changes nothing until ϵ becomes 0 at point B, after which the system evolves from solution A_0 to solution A_2 and ends at point B'; the evolution of the system has followed the path ABB' in the bifurcation diagram. Conversely, if we begin at B' and decrease the value of parameter ϵ, the system remains in solution A_2 until the parameter ϵ reaches a value of $-1/4$, i.e. the critical point C where solution A_2 ceases to exist. Below that value, the system has no choice but to "jump" from branch A_2 to branch $A_0 = 0$ in order to regain point A. This time the system has evolved through the return path of B'CA, which is not

---

15. We exclude, in the following section, the existence of strong perturbations.

equivalent to ABB′ in the bifurcation diagram. This asymmetry between the upward and downward paths describing the system's evolution in the form of a cycle (or a loop) in the bifurcation diagram is called a hysteresis cycle or hysteresis[16].

3.4. Transcritical Bifurcation

Certain nonlinear systems are characterized, independently of any external symmetry breaking evoked earlier, by an intrinsic absence of "left/right" symmetry; in this case, the evolution equation is not invariant under transformation $A \rightarrow -A$ (see chapter 13). Consequently, we must introduce a quadratic nonlinearity (a term in A^2) to the evolution equation for amplitude. If the quadratic term is large enough to saturate the linear growth (that is, if the sign in front of it is negative), we do not need to expand the equation any further. The amplitude equation is then written as follows:

$$\frac{dA}{d\tau} = \epsilon A - A^2. \tag{3.28}$$

This equation has fixed points $A = 0$ and $A = \epsilon$, both of which exist without any parametric restriction. The first fixed point is stable for $\epsilon < 0$ and unstable for $\epsilon > 0$. The stability condition for the second fixed point $A = \epsilon$ is just the inverse with respect to the first fixed point ($A = \epsilon$ is unstable for $\epsilon < 0$ and stable for $\epsilon > 0$, see bifurcation diagram in figure 3.23).

If the next order term A^3 of the Taylor expansion is taken into account in the evolutionary equation, the amplitude equation then takes the following form[17]:

$$\frac{dA}{d\tau} = \epsilon A + A^2 - A^3. \tag{3.29}$$

Repeating the analysis, we find that the evolution equation (3.29) has fixed point $A_0 = 0$ for $\epsilon < -1/4$, and three fixed points for $\epsilon > -1/4$, given by $A_0 = 0$ and $A_{\pm} = (1 \pm \sqrt{1 + 4\epsilon})/2$ (see bifurcation diagram in figure 3.24). This is a *transcritical bifurcation*, characterized by the existence of two

---

16. It is useful to mention that hysteresis is also observed in certain phase transitions of thermodynamical systems despite the microscopic fluctuations, such as thermal ones, that don't easily breach the metastability barrier. It remains the case that the interval in ϵ (i.e. the length of the cycle) of hysteresis of a thermodynamical system undergoing a phase transition is often smaller than that of macroscopic systems we study, for which the domain of hysteresis is almost as large as that of metastability.

17. Note that in the presence of a cubic term, the sign of A^2 is of no importance. Substituting $A \rightarrow -A$ in the amplitude equation and then multiplying the equation by -1, only the sign in front of A^2 changes, and the rest of the terms remain intact.

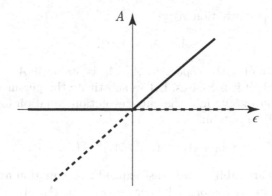

Figure 3.23 – Bifurcation diagram of a quadratic amplitude equation.

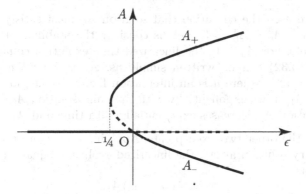

Figure 3.24 – Transcritical bifurcation diagram of a cubic amplitude equation.

branches on either side of the two solution branches ($A = 0$ and $A \neq 0$). Like the pitchfork and imperfect bifurcations, the transcritical bifurcation belongs to the cusp catastrophe family (see next chapter). In all cases, the stability of the fixed points can easily be deduced from the potential $V(A)$ associated with the amplitude equation (3.29), since $dA/d\tau = -dV/dA$.

3.4.1. Linear Stability of Fixed Points

Until now, we have studied systems for which the evolution equation derives from a potential ($dA/d\tau = -\partial V/\partial A$). We have used this property in order to study the linear stability of the pitchfork bifurcation (see chapter 2). However, nonlinear evolution equations need not necessarily derive from a potential. In order to generalize fully, our study of the linear stability of stationary solutions must rely on a different approach, such as we will now develop for the case of the transcritical bifurcation using the third order evolution equation. To this end, we start with the stationary solution (fixed point) A_0

and superpose a perturbation A_1:

$$A = A_0 + A_1(t), \tag{3.30}$$

where $A_1(t)$ is small with respect to A_0. A_0 is stable if A_1 decreases with time, and unstable if it increases. Let us substitute the perturbed stationary solution (equation (3.30)) into the cubic evolution equation for transcritical bifurcation (3.29) to obtain:

$$\dot{A}_1 = \epsilon(A_0 + A_1) + (A_0 + A_1)^2 - (A_0 + A_1)^3. \tag{3.31}$$

To study the linear stability, we must expand this equation and retain only the linear terms in A_1 (ignoring higher order terms A_1^2, A_1^3). This leaves us with the following:

$$\dot{A}_1 = \epsilon A_1 + 2A_0 A_1 - 3A_0^2 A_1, \tag{3.32}$$

where we have used the condition that solution A_0 must satisfy stationary condition $\epsilon A_0 + A_0^2 - A_0^3 = 0$. Let us consider the stability of each fixed point separately. For $A_0 = 0$, the linearized time evolution equation of the perturbation (3.32) can be written simply as: $\dot{A}_1 = \epsilon A_1$; the solution is given by $A_1 = a e^{\epsilon t}$ (where a is an integration factor). Thus, for $\epsilon > 0$, the perturbation A_1 grows exponentially with time and solution A_0 is unstable; for $\epsilon < 0$, solution A_1 decreases exponentially with time and A_0 is stable.

Now consider the other two fixed points. Applying once more the condition for a stationary solution at A_0, the linearized evolution equation (3.32) can be rewritten as:

$$\dot{A}_1 = -(2\epsilon + A_0)A_1. \tag{3.33}$$

The stability of solution A_0 depends on the sign in front of factor A_1 (that is, the sign of $-(2\epsilon + A_0)$). For the solution $A_0 = A_+$, we have $-(2\epsilon + A_+) = -(1 + 4\epsilon + \sqrt{1 + 4\epsilon})/2 < 0$. The evolution equation of the perturbation (3.33) tells us that solution A_+ is always stable. For solution $A_0 = A_-$, we have $-(2\epsilon + A_-) = (-(1 + 4\epsilon) + \sqrt{1 + 4\epsilon})/2$. This quantity is positive (unstable solution) for $-1/4 < \epsilon < 0$ and negative (stable solution) for $\epsilon > 0$ (see the different stationary branches in the bifurcation diagram, figure 3.24).

3.5. Definition of a Bifurcation

So, what is the definition of a bifurcation? From a historical perspective, this term is ancient and in its most literal sense, it signifies a crossing of paths and thus a change from the initial trajectory. The scientist Henri Pioncaré (1854–1912) was the first, it seems, to describe the properties of a bifurcation from a mathematical perspective. In this context, the term "bifurcation" indicates a qualitative change in the behavior of a system; the qualitative change, by definition, must alter the number and/or the nature, stable or unstable, of the solutions with the change of a parameter, which in our examples we have denoted by ϵ. At a bifurcation point the branches of solutions are annihilated, created, or lose or gain their stability.

We must emphasize that behavioral changes of a system do not necessarily create bifurcation points. The change must be "qualitative" in the sense we have previously defined in order to introduce bifurcations in the solutions. Let us illustrate with the example of a system that undergoes a "non-qualitative" change in the behavior of its solutions and for which the existence of nonlinear terms (and a fixed point) does not introduce a bifurcation point. The system with the following evolution equation:

$$\dot{A} = \epsilon - A^5 \qquad\qquad (3.34)$$

represents such an example. At first glance, the evolution equation (3.34) has strong similarities to that of previously studied bifurcations (compare it with the equation for the saddle-node bifurcation, equation (3.9)). Furthermore, the evolution equation (3.34) presents a fixed point, solution $A_0 = \epsilon^{1/5}$, but a quick analysis reveals that the fixed point is unique and exists for all situations! Thus, the variation of parameter ϵ does not change either the number of solutions or the stability, and thus does not modify the behavior of the system in any qualitative fashion (see figure 3.25): the system does not present a bifurcation.

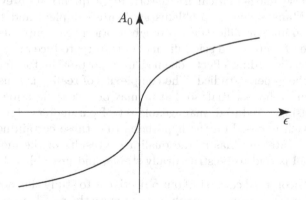

Figure 3.25 – Fixed point branch of equation (3.34).

With the help of this example, we perceive that the meaning of "qualititive change in solutions" covered by the term "bifurcation" is in fact very precise. Accordingly, a formal framework for bifurcation theory has been developed: catastrophe theory, briefly discussed in the following section, and in more detail in chapter 4.

3.6. Catastrophe Theory and Structural Stability

In 1972, mathematician René Frédéric Thom (1923–2002), founder of catastrophe theory, published a work designated for the informed layman. This treatise, entitled *Stabilité structurelle et morphogenèse* (Structural Stability and Morphogenesis, see [101]), has nonetheless become a classic in notably scientific libraries. It presents, in a very intuitive manner, a mathematical framework that allows one to associate all important morphogenetic appearances with a dynamical system that can produce them. The fundamental theorem of this theory states the existence of seven fundamental catastrophes. Thus, the "natural" forms (i.e. forms and patterns encountered in nature, which can be quite complex) can generally be reconstructed through a combination of these elementary catastrophes. Catastrophe theory constitutes the first attempt at formalizing, in the domain of so-called differential topology (a topic which is beyond the scope of this book), the notion of structural stability (see below) for all morphologies. It has, as its goal, to construct the simplest dynamical model (say a set of simple enough nonlinear equations) that can engender, for example, a certain morphology.

What is *structural stability* of a model? This notion is in fact implicitly a natural element in any theory (for example in physics), but it is uncommon to use a precise mathematical framework for a quantitative evaluation of this notion. Many scientific problems are quite complex, and the method consists of making simplifications through a priori reasonable assumptions. The objective of a proposed scientific model is to reproduce an aspect of real systems while discarding effects deemed unimportant in the framework of the specific phenomena studied. The complexity of reality is thus effectively whittled down to the essentials so that we may be able to describe the studied phenomena using a reduced, manageable set of parameters. The legitimacy of this approach is based on the hypothesis that those certain negligible effects do not create any bias in the qualitative results of the model. In this case the model is said to be structurally stable, and unstable otherwise.

Thus, in the theoretical context, if our objective is to study the movement of a planet, say Earth, we can reasonably approximate the problem as a two-body problem, Earth and Sun, and neglect the influence of other bodies in the solar system in order to obtain an elliptical orbit of the Earth around the Sun. If we took into account the interaction between Earth and other bodies in the solar system, there would be a minor modification of the Earth's trajectory on a quantitative scale, but not qualitative: the model is thus said to be *structurally stable.*

The structural stability of models is a requirement analogous to that of an experiment: the scientist expects to obtain similar results in all situations where the experimental conditions are more or less identical. Without this stability in scientific results, no experimental study could be meaningful.

3.6.1. The Difference Between Structural Stability and Dynamical Stability

The study of structural stability in a model is fundamentally different from the study of the dynamical (or temporal) stability of a system's solutions.

The first case is a study of the stability of the model itself and comprises finding the response to this question: would a small modification of the elements constituting the model give the same qualitative results, that is, would the modified model possess the same types of solutions (if they existed)? If the answer is yes, the model is structurally stable.

The second case comprises determining the influence of a small perturbation on the stability of a solution of that model. If the solution persists over time, the perturbation is decreasing over time and we have a solution we call linearly stable. On the other hand, if the solution moves away from the initial non-perturbed state, it is because the perturbation is growing with time, and the solution is unstable.

3.6.2. Structural Stability and Subjectivity

In the nonlinear sciences, the study of a system is illustrated with the help of a bifurcation diagram representing solutions A of the dynamical system's amplitude equation as a function of parameter ϵ (see for example the diagram of the pitchfork bifurcation represented in figure 2.4). The study of a bifurcation's structural stability comprises evaluating the consequences of an infinitesimal perturbation of the model (correspondingly represented in the amplitude equation) from the shape of the bifurcation diagram. Thus, in our study of the imperfect bifurcation in section 3.2, we are testing the structural stability of the pitchfork bifurcation. Indeed, the imperfect bifurcation (figure 3.16) represents the ensemble of solutions for the modified pitchfork bifurcation's amplitude equation (2.4) (modified to become (3.21)), modified by the introduction of an imperfection term ν (with ν arbitrarily small). The pitchfork and imperfect bifurcation diagrams (figures 2.4 and 3.16) are qualitatively different; therefore, the model of the dynamical system (here, a spring) leading to the pitchfork bifurcation is structurally unstable. From a mathematical point of view, there is no doubt about this claim, as the qualitative change is described in a precise manner.

While the notion of structural stability is perfectly defined and non-ambiguous in a mathematical sense, it calls for some clarification when evoked in an experimental context. For example, the bifurcation diagram of an experimental model (see figure 3.26(c)) is representative of a real experiment that is invariably skewed by measurement errors or some weak bias. The theoretical models of the pitchfork bifurcation or of the imperfect bifurcation (see figure 3.26(a,b)) would both be a priori good candidates for the experimental result so long as the criteria for distinguishing between the two is not at our disposal. Thus, the choice of a model to describe the experimentally obtained bifurcation appears to be somewhat subjective. This indeterminacy that comes from the experimental context is something inveterate to the experimental method; nonetheless it should not eliminate the need for a rigorous classification of bifurcations. This classification is found in the framework of catastrophe theory.

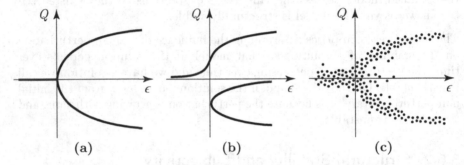

Figure 3.26 – (a) Pitchfork bifurcation; (b) imperfect bifurcation; (c) schematic representation of experimental results and their uncertainty.

3.7. What Does Catastrophe Theory Consist Of?

Although we will undertake this topic in greater detail in the following chapter, we will first introduce and illustrate catastrophe theory with the help of an already familiar example. In contrast to the study of bifurcations in which the evolution of solutions are represented in a bifurcation diagram (for example, see the imperfect bifurcation in figure 3.16 representing the evolution of an internal variable of amplitude A as a function of external control parameter ϵ), catastrophe theory studies the qualitative change in a solution's behavior solely within the parameter space of the control parameters. Furthermore, catastrophe theory relies on the notion of structural stability to determine the generic forms of different types of changes in behavior. We will illustrate the spirit of catastrophe theory with a concrete example in the next section.

We must specify, however, that this theory uses a basic premise that may seem, at first glance, restrictive: it assumes the dynamics of the system are of a potential nature; in other words, it assumes the temporal evolution of the studied variable derives from a potential. This condition is expressed by evolution equations of a specific type: $\dot{A} = -\partial V(A)/\partial A$, where V is the potential. This is actually less restrictive than it seems; even if the dynamics of a system do not derive entirely from a potential, it is possible to describe the behavior of a system's most interesting (descriptive) variable with dynamics derived from a potential. We will discover several examples throughout this book.

3.7.1. Illustration of Catastrophe Theory

We shall now revisit a problem we previously solved using bifurcation theory, this time using catastrophe theory to elaborate.

In our study of bifurcations in subsection 2.3.5, we saw that the evolution equation (2.26) derives from a potential $V = -\epsilon A^2/2 + A^4/4$. To rewrite this in the form conventional to catastrophe theory, we eliminate the constants through an appropriate change of variables ($A \rightarrow \sqrt{2}A$ and $\epsilon \rightarrow -\epsilon$):

$$V(A) = \epsilon A^2 + A^4. \tag{3.35}$$

For what follows, let us adopt the following notations: $\partial V/\partial A \equiv V'$ and $\partial^2 V/\partial A^2 \equiv V''$. The fixed points are, by definition, reached when the solution no longer evolves in time. For an amplitude equation deriving from a potential $V(A)$, the fixed points correspond to the potential's maxima, when $V' = 0$. Using the precedent equation (3.35), we obtain the following stationary solutions: $A_0 = 0$ and for $\epsilon > 0$, $A_\pm = \pm\sqrt{\epsilon/2}$. Once the parameter ϵ reaches 0, the three fixed points become identical, giving us a triple degenerate solution (A_0, A_+, A_- are all 0). A loss of stability of a fixed point (which defines a bifurcation point) corresponds to a change in concavity of the potential. This means that the second derivative of the potential $V'' \equiv \Delta$ changes sign. The demand that V'' vanishes is satisfied when $V'' = \Delta = 12A^2 + 2\epsilon = 0$.

Now let us add an imperfection term, νA, to the potential (equation (3.35)):

$$V = \nu A + \epsilon A^2 + A^4, \tag{3.36}$$

where ν is the imperfection parameter. The fixed points of this expression must satisfy the following condition:

$$V' = \nu + 2A\epsilon + 4A^3 = 0. \tag{3.37}$$

This potential (equation (3.36)), introduced earlier during our study of the imperfect bifurcation, is clearly very similar to the potential of the pitchfork bifurcation (equation (3.35)); and it becomes more and more similar as the imperfection variable ν becomes smaller. Nonetheless, as has already been emphasized, these two potentials are qualitatively different (see figure 3.27).

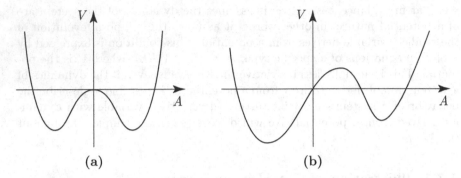

Figure 3.27 – (a) Potential $V = \epsilon A^2 + A^4$; (b) potential $V = \nu A + \epsilon A^2 + A^4$. Here $\epsilon < 0$ and $\nu > 0$. In the neighborhood of $A = 0$, the two potentials are qualitatively different.

At the critical point where $V' = 0$, the concavity of the modified potential V'' does not vanish for $\epsilon = 0$, in contrast to that of the perfect potential (3.35). The modified potential, which includes this imperfection term νA (with $\nu \neq 0$) is thus more robust; indeed, the concavity of the modified potential is expressed by:

$$V'' = \Delta = 2\epsilon + 12A^2 \tag{3.38}$$

and does not necessarily go to 0 when ϵ becomes 0.

So, in the case where we are confronted with a potential $V(A)$ (for example, potential equation (3.35)) where the second derivative becomes zero ($\Delta = 0$) at a given fixed point $V' = 0$, we can always make this potential arbitrarily robust. That is, we can find functions that are arbitrarily close to this potential (i.e. find a modified potential like the one described by equation (3.36) above) and yet have non-zero values for the second derivative ($\Delta \neq 0$) at the fixed point $V' = 0$. In other words, for a "fragile" potential $V(A)$ (for example, potential equation (3.35)) whose second derivative becomes zero ($\Delta = 0$) at a given fixed point $V' = 0$, we can always write a corresponding robust potential (i.e. the modified equation (3.36) in the preceding example) by adding an arbitrarily small term (the imperfection term νA in the preceding example).

However, and this is the crucial point of the study, even if this potential seems robust (as given by definition $\Delta \neq 0$ for $\epsilon = 0$), it is not invincible! There is always a point of balance between the parameters (here ϵ and ν) for which the two conditions $V'' = 0$ and $V' = 0$ are satisfied[18]. One of the two equations determines the value of the fixed point $V' = 0$, and the other equation $V'' = 0$ determines the relationship between ϵ and ν. Together they define a frontier, here a line, outlining two domains in parameter space (ϵ, ν), as seen below. In each domain, the solutions are qualitatively different; the crossing of this frontier corresponds to a catastrophe. In other words, crossing this frontier is equivalent to a sudden change in the number or nature of solutions. Each catastrophe thus has its own geometric figure associated to the frontiers that outline the different domains in parameter space; this geometric figure is called the "signature" or "portrait" of a catastrophe.

In this way, the catastrophe portrait in parameter space defines the frontiers that delimit regions of qualitative change in the solutions of a model. That brings us to the second important point of this study: the number of "independent control parameters" (external variables) for a system is what ultimately determines the number of possible catastrophes, and this result is independent of the number of internal variables (degrees of freedom of the system), which might even be infinite. For example, when the number of independent control parameters is less than or equal to four (see next chapter), only seven distinct catastrophes are possible: these are the seven *elementary catastrophes*; beyond four control parameters, the problem becomes more complex and its treatment is outside the scope of this book.

Let us now, as an example, specify the mathematical form of the frontier for the catastrophe associated with the modified potential (equation (3.36)) we have dealt with up to this point. The two conditions $V' = 0$ (equation (3.37)) and $V'' = 0$ (equation (3.37)) lead to the following expression:

$$\Delta \equiv 8\epsilon^3 + 27\nu^2 = 0. \tag{3.39}$$

The cubic equation (3.37) possesses one or three real solutions, depending on the sign of the concavity for potential $V'' = \Delta$. If the concavity is negative ($\Delta \leqslant 0$), there are three real solutions; if the concavity is positive ($\Delta > 0$) there exists only one real solution, and when the concavity vanishes, two of the three solutions merge[19]. The condition for zero concavity ($\Delta = 0$) defines a frontier of the catastrophe in parameter space (see figure 3.28); this frontier delineates the different domains associated with the different forms of the potential.

---

18. Note that the greater the number of parameters, the easier it is to have this interplay of parameters.

19. In the particular case where two control parameters are 0, $\epsilon = \nu = 0$, the solution becomes triple degenerate.

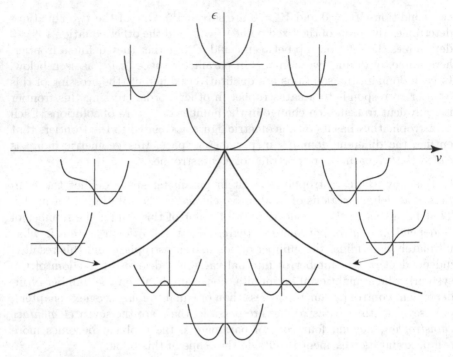

Figure 3.28 – The *cusp* catastrophe and the shape of the potential
$V = \nu A + \epsilon A^2 + A^4$ in each regime.

When traced in the parameter space (see figure 3.28) the catastrophe line of
the modified potential (equation (3.36)) reveals a recognizable shape: a cusp
(angular point) at $(\nu = 0,\ \epsilon = 0)$. This is why this catastrophe has been
donned *cusp* catastrophe. The portrait of the cusp catastrophe outlines two
zones (see figure 3.28). The first is situated above the line that traces the frontier of the singularity, and there potential V has one extremum (a minimum),
while below the line it has three extrema (two minima, and one maximum).
At the *cusp* ($\nu = 0$ and $\epsilon = 0$) the three extrema merge. At this point, the
potential is of the form $V = A^4$ (see equation (3.36)) and the null amplitude
solution $A = 0$ is thus triple degenerate since the derivative of the potential is
simply $V' = 4A^3$. Note that, furthermore, the potential defines an inflection
point for solution $A = 0$ ($V'' = 12A^2 = 0$) and that the third derivative is
also zero at this point $V''' = 24A = 0$. In other words, at the cusp point, the
potential is more "fragile" and thus more structurally unstable. This indicates
that catastrophes in the parameter space define precisely the places where
the potential loses its structural stability. The *cusp catastrophe* is likely to
occur in many nonlinear models. Indeed, whatever the associated potential,
the model is potentially a candidate for a *cusp catastrophe* as soon as it has
at least two control parameters.

3.8. Can One Accept Structural Stability with an Infinite Number of Exceptions?

Having got to this point in our study, we might feel we have reached a certain paradox. We had first reached the conclusion that the pertinent description of a phenomenon rests on the necessity of elaborating a structurally stable model. For a phenomenon that can be approximated by a model deriving from a potential, this necessity is fulfilled by a potential that has no change in curvature ($V'' \neq 0$).

However, in the previous section we have seen that it is actually impossible to avoid such situations (i.e. to avoid $V'' = 0$) as soon as the model introduces control parameters, since we have proved that one can annul the first and second derivatives of the potential through an adequate combination of the control parameters (condition (3.39) provides the relation between the parameters in order to have a zero concavity). This conclusion might lead one to assert – a bit hesitantly – that there is thus no structurally stable model appropriate to nonlinear phenomena.

In order to escape this apparent paradox, it is enough to add a nuance to our formulation of a nonlinear model by renouncing the necessity for structural stability in an absolute sense. Thus the model of the modified potential (equation (3.36)) is structurally stable *except* for points that are along the line tracing the *cusp* catastrophe (see figure 3.28). The model is thus "almost" structurally stable. This qualification of "almost stable" signifies, in this context, "with an eventually infinite number of exceptions", which are found exactly at the points (or curves or surfaces) of catastrophe: frontiers characterized by a null curvature of the potential $V'' = 0$. In other words, we admit that there can be an infinite number of exceptions to the structural stability of a model without completely putting into question the model's structural stability[20]. The presence of catastrophes is universal in systems described by nonlinear equations which contain control parameters. A strategy for attempting to escape these catastrophes, and thus to guarantee a greater

---

20. To better understand the meaning of the term "almost", let us use an analogy and consider points on a plane (x–y). They are an infinite number; the same way that the Ox-axis contains an infinite number of points; but the relationship between the number of points on the axis and the one on the plan is very small! Let us posit that a variable such as temperature has a value of $100°C$ at every point except for those that are along the Ox-axis where the temperature is zero. We would be able to state that the plane is globally "hot" without being really incorrect. The horizontal axis, though it comprises of an infinite number of points, is not representative of the state of the whole plane (x–y).

stability for a model, is to reduce the domain of variation of the parameters as much as possible. Of course, a poor knowledge of the system, or the absence of a quantitative model (as, for example, is the case with most biological systems) sometimes makes this strategy difficult to execute.

The appearance of these sudden changes, in the context of models that are described by an a priori smooth potential (see equations (3.35) and (3.36) which describe a potential using an analytic function, there a polynomial, free of singularities), leads to a remarkable result, mysterious the first time one encounters it: occurrence of catastrophes is ubiquitous for any nonlinear system since through a combination of control parameters it is always possible to induce a catastrophe (i.e. $V'' = 0$). These brusque changes in behavior obtained by the variation of parameters occur through the line (or surface) of catastrophes. Having studied the example of the *cusp* catastrophe in detail, we are now equipped to present a systematic classification of elementary catastrophes: this is the goal of the next chapter.

3.9. Solved Problems

3.9.1. Maxwell Plateau

Show that the line $\epsilon = \epsilon'$ in the bifurcation diagram below corresponds to the Maxwell plateau found in phase transition problems in the $(\epsilon - A^2)$ plane.

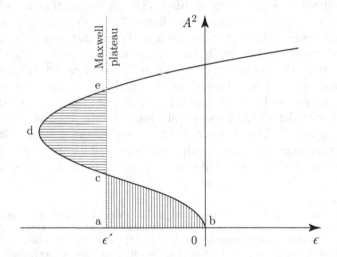

Figure 3.29 – Maxwell construction: the straight line $\epsilon = \epsilon^*$ cuts the curve of the stationary solution's amplitudes into two regions that have the same area (the area filled in with vertical lines is equal to the area of the one denoted by horizontal lines).

Solution. We start from $V(A_2) = V(A_0)$, which corresponds to the coexistence of stable and metastable solutions (bistable solutions) where A_0 and A_2 represent the amplitudes of the stable and metastable solutions, respectively, in coexistence. We can rewrite this equality as $V(A_2) - V(A_0) = 0$ and reinterpret this as the solution to an integration of the total derivative of the potential dV between the states 0 and 2 corresponding to stationary solutions A_0 and A_2, e.g.:

$$\int_{(0)}^{(2)} dV = 0. \tag{3.40}$$

The expression (see equation (3.22)) for the potential $V(A, \epsilon)$ is a function of two variables. Therefore we can express the derivative of the potential dV as a function of two partial derivatives: $dV = (\partial V/\partial A)dA + (\partial V/\partial \epsilon)d\epsilon$. Since dV is a total derivative, we can choose the integration path at will. We opt here for the path composed of four sections, (a,b), (b,c), (c,d), and (d,e), all situated on the curve of fixed points (see figure 3.29), along which, by definition, the partial derivative of the potential with respect to the amplitude is 0 ($\partial V/\partial A = 0$). Now, the potential is linear in ϵ (see equation (3.22)), and the partial derivative with respect to ϵ is equal to amplitude A ($\partial V/\partial \epsilon = -A^2/2$). Along the section (a,b), the amplitude A is zero everywhere ($A = 0$). With this information we rearrange the integral of the potential derivative (see equation (3.40)) to obtain the following:

$$-\int_b^c A d\epsilon = \int_c^d A d\epsilon + \int_d^e A d\epsilon. \tag{3.41}$$

The left-hand side of this equation corresponds to the area enclosed by the straight line $\epsilon = \epsilon'$, the axis of null amplitudes, and the curve of amplitude A_1 (vertically shaded area in figure 3.29). The term to the right of the equality refers to the area delimited by the curve of amplitude A_1 and the line $\epsilon = \epsilon'$ (horizontally shaded area in figure 3.29). We thus uncover Maxwell's equal area rule, seen in phase transition problems (e.g. liquid-gas transition).

3.9.2. A Mechanical Device Giving Rise to a Subcritical Bifurcation

We will now describe an example of a mechanical system in which we find a subcritical bifurcation[21]. Let us consider a solid mass attached between two horizontal springs, with fixed extremities; the joint between the two springs allows the mass to be displaced vertically (see figure 3.30). Let us posit k as the springs' constant, ℓ_0 as the rest length of the springs, ℓ the length imposed on the springs when they are lying along the horizontal (the situation $\ell > \ell_0$

---

21. I warmly thank Olivier Pierre-Louis for suggesting this setup.

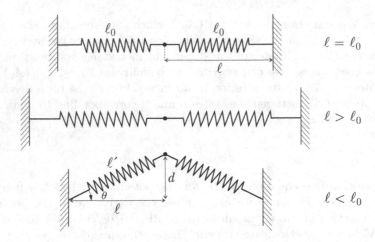

Figure 3.30 – Schema of the example system. Top: spring at rest; middle: spring stretched, forcing the mass to stay on the horizontal axis; bottom: compressed spring, allowing the mass to move up or down to release some of the compression.

corresponds to an extension of the springs) and ℓ' the length of the springs in the configuration they find themselves in once the mass is no longer on the horizontal axis ($\ell' = \ell$ when the mass is on the horizontal axis). The potential energy of the system is given by the sum of elastic potential energy of both springs ($k(\ell' - \ell_0)^2$) and the curvature energy of the joint between the two springs ($K \sin^2 \theta / 2$), where K represents[22] the rotational rigidity of the joint and θ the angle between the horizontal axis and the axis of the springs in any configuration (see figure 3.30); this energy is zero when $\theta = 0$. The total energy of the system, E_T, is thus written as:

$$E_T = \frac{K}{2} \sin^2(\theta) + k(\ell' - \ell_0)^2. \tag{3.42}$$

with a maximum at $\theta = \pi/2$.

Defining $x = d/\ell$, with d the distance with respect to the horizontal axis (see figure 3.30), we can express the total energy as:

$$E_T = \frac{K}{2} \frac{x^2}{1 + x^2} + k \left(\ell \sqrt{1 + x^2} - \ell_0 \right)^2. \tag{3.43}$$

We constrain our study of this function to the neighborhood near the bifurcation. In anticipation of $x = 0$ being a fixed point of the system, we make the appropriate Taylor expansion around it, and expanding the total energy

---

22. Note the dimension of the rigidity constant K, where $[K] = L^2 M T^{-2}$, which is different from the spring constant k, where $[k] = M T^{-2}$.

to fourth order, we obtain:

$$E_T = [K + 2k\ell(\ell - \ell_0)]\frac{x^2}{2} + [k\ell\ell_0 - 2K]\frac{x^4}{4}. \tag{3.44}$$

We have omitted the constant $k(\ell - \ell_0)^2$, which is independent of x, since ℓ and ℓ_0 are parameters of the system. At $x = 0$, the sign of the curvature of the total energy is given by $d^2 E_T/dx^2 = K + 2k\ell(\ell - \ell_0) = 2k[K/2k + \ell(\ell - \ell_0)] = 2k\Delta$ with

$$\Delta = \alpha + \ell^2 - \ell\ell_0, \quad \alpha = \frac{K}{2k}, \tag{3.45}$$

where the parameter α measures the ratio of rotational rigidity over spring constant.

If Δ is positive, the total energy E_T has a minimum at $x = 0$ corresponding to a stable fixed point. This point becomes unstable when Δ is negative. The condition $\Delta = 0$ is satisfied for critical value ℓ_c of horizontal length ℓ of the spring such that:

$$\ell_c = \frac{\ell_0 + \sqrt{\ell_0^2 - 4\alpha}}{2}. \tag{3.46}$$

There exists a second critical value ℓ_c': $\ell_c' = (\ell_0 - \sqrt{\ell_0^2 - 4\alpha})/2 < \ell_c$.

Suppose that at some initial time the spring is stretched ($\ell > \ell_0$), let us follow the evolution of the system's total energy E_T as a function of the decay of parameter ℓ until the parameter attains critical value $\ell = \ell_c$, corresponding to the energy's first change in curvature.

To ensure that length ℓ_c is a real number, we impose the following condition: $\alpha < \ell_0^2/4$. Physically this means that the cost in curvature energy cannot be so large that the fixed point $x = 0$ become unstable. This condition can be understood intuitively in the following way: if the energy cost for bending is too large, the mass does not deviate from the horizontal.

The energy curvature is negative, $\Delta < 0$, when the length of the spring projected to the horizontal ℓ is smaller than the critical value ℓ_c ($\ell < \ell_c$). In this case, the energy curvature changes (the minimum becomes the maximum): the mass deviates from the horizontal axis. To determine the new equilibrium position of the mass, we must study the effect of the x^4 term. In the neighborhood of the critical point ($\ell \simeq \ell_c$), the coefficient of x^4, noted as C_4, is given by:

$$C_4 = \frac{k\ell_0^2}{2}\left(1 + \sqrt{1 - \bar{\alpha}} - 2\bar{\alpha}\right), \tag{3.47}$$

with $\bar{\alpha} \equiv 4\alpha/\ell_0^2$. A look at the different possible signs of coefficient C_4 allows us to define different domains for parameter $\bar{\alpha}$:

$$\begin{cases} C_4 > 0, & 0 < \bar{\alpha} < \frac{3}{4}, \\ C_4 < 0, & \frac{3}{4} < \bar{\alpha} < 1. \end{cases} \tag{3.48}$$

When the coefficient of C_4 is positive, the x^4 term found in the Taylor expansion of the total energy around bifurcation $x = 0$ (see equation (3.44)) leads to a saturation of the instability: the mass moves toward a new equilibrium position, in this case a pitchfork bifurcation (also known as a supercritical bifurcation). Note that there is a value of $\bar{\alpha}$ which annuls coefficient C_4 ($\bar{\alpha} = 3/4$); this point is sometimes called the *Lifschitz point* (a point where the bifurcation turns from supercritical to subcritical).

When coefficient C_4 is negative, the term in x^4 reinforces the instability, which defines the situation of a subcritical bifurcation. To determine the state of the system, we must continue the expansion of total energy E_T to the next higher order, that is, the sixth order, giving us coefficient C_6 with term x^6. If we do this near the critical parameter $\ell \simeq \ell_c$, we obtain:

$$C_6 = \frac{k\ell_0^2}{8}\left(2\bar{\alpha} - \frac{1 + \sqrt{1 - \bar{\alpha}}}{2}\right), \tag{3.49}$$

where ℓ has been replaced with ℓ_c. An analysis of the sign of coefficient C_6 establishes:

$$\begin{cases} C_6 < 0, & 0 < \bar{\alpha} < \frac{7}{16}, \\ C_6 > 0, & \frac{7}{16} < \bar{\alpha} < 1. \end{cases} \tag{3.50}$$

The conditions on parameter $\bar{\alpha}$ corresponding to a negative C_4 coefficient are such that C^6 is positive (i.e. when $C_4 < 0$ we have $C_6 > 0$). The x^6 term acts against the x^2 and x^4 terms, ensuring an eventual saturation of the instability (see figure 3.31 which illustrates the shape of the system's potential energy constrained to a subcritical bifurcation (i.e. when $C_4 < 0$) for different valuesof ℓ).

If we continue to decrease parameter ℓ, we reach critical value ℓ_c'; the quantity Δ changes sign and solution $x = 0$ becomes stable once again. Physically, this result can be interpreted in the following way: when ℓ becomes smaller, and for a fixed vertical change in the mass's position with respect to $x = 0$, the corresponding inflection angle becomes larger. It thus requires more and more energy for the mass to move from the horizontal axis. Eventually, when ℓ becomes too small, the cost in curvature energy is too large and the mass remains on the horizontal axis (stability of $x = 0$).

3.9.3. The Euler Problem of a Soap Film

Consider a film of soap stretched by two rings (see figure 3.32). We invite the reader to consult the famous article by Taylor and Michael on this subject [100]. In this section we will endeavor to find the equilibrium form adapted by the soap film. Constrained by the presence of two rings at the extremities, this form can be found by minimizing the surface energy $F = \int \gamma \, dS$ where γ is the energy density of the surface (energy per surface unit) and dS is the elementary surface unit.

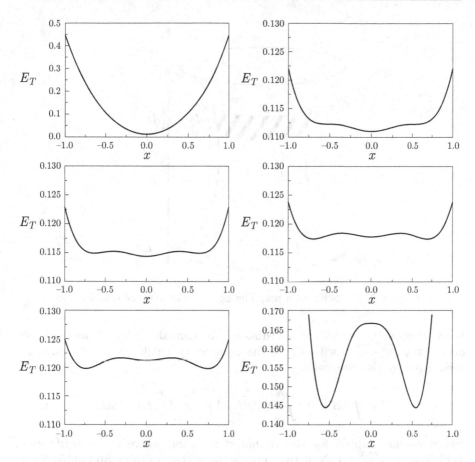

Figure 3.31 – The total energy (equation(3.43)), E_T (in units of $k\ell_0^2$) as a function of x for different values of $\ell - \ell_c$ (proportional to ϵ, and measuring distance to the linear instability threshold). From left to right, we have chosen the following values for $\ell - \ell_c$: 0.5; 0.055; 0.05; 0.045; 0.04; −0.02. We fix $\ell_0 = 1$ and $\bar{\alpha} = 0.95$.

Assuming that the energy density of the surface is constant, the expression for energy F is:

$$F = \gamma \int dS. \tag{3.51}$$

Thus, minimizing the surface energy F is equivalent to minimizing the film soap's area $S = \int dS$

In curvilinear coordinates, where r is the radius of the circle at height z and ds is the curvilinear abscissa such as indicated in figure 3.32, the area of elementary surface element dS (hatched area) is given by $dS = 2\pi r\, ds$.

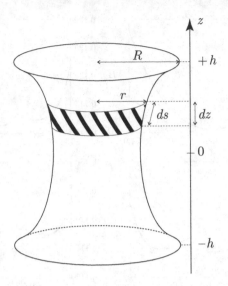

Figure 3.32 – Schema: a soap film between two rings of radius R.

If we change to cylindrical coordinates, we can also express area dS as: $dS = 2\pi r\sqrt{dr^2 + dz^2}$ with dz as the elementary height. The total area is then given by the integral:

$$S = \int dS = 2\pi \int_{-h}^{h} r\sqrt{r'^2 + 1}\, dz \equiv \int_{-h}^{h} f(r, r')dz, \qquad (3.52)$$

where we have opted for the simplified notation where $r' = dr/dz$ and $f(r, r') = 2\pi r\sqrt{r'^2 + 1}$. Now to minimize the surface of the soap bubble S we must find the functions $r(z)$ that define the extrema of surface S.

The area of soap film S is a functional, since it is a function of the function $r(z)$. The search for a minimal area S, or more generally for the minimum of any functional F, is a problem analogous to the search for the least action in Lagrangian mechanics[23]. Generally, we can find the extremum of energy $F = \int f(r, r', z)\, dz$ (for any functional $f(r, r', z)$) by using the Euler–Lagrange

---

23. The Lagrangian $L = T - V$ of a dynamical system, where T is the kinetic energy of the system and V is the potential energy, is a function of dynamic variables. It allows us to describe the equations of motion for the (Lagrangian mechanical) system using the principle of least action, the action being a functional defined as the integral of the Lagrangian over all variables. Its name comes Joseph Louis Lagrange (1736–1813), who established the principles of this process.

equation[24]:

$$\frac{\partial f}{\partial r} - \frac{d}{dz}\frac{\partial f}{\partial r'} = 0. \tag{3.53}$$

The function f is a function of two mutually independent variables r and r'. For example, if $r(z) = z^2$, r' is independent of r, since $r' = dr/dz = 2z$. Furthermore, if the function f is a linear composition of the two variables $f = ar + br'$ with a and b two arbitrary parameters, and we minimize f for any value of z, both parameters must be null $a = b = 0$, thus confirming that r and r' are two independent quantities.

Since f does not explicitly depend on variable z in our example, we can rewrite the Euler–Lagrange equation[25] (equation (3.53)) as follows:

$$f - r'\frac{\partial f}{\partial r'} = C, \tag{3.54}$$

C is an integration constant. Using the expression of function $f(r,r')$ (see equation (3.52), $f = 2\pi r\sqrt{1 + r'^2}$) we have explicitly:

$$r = a\sqrt{1 + r'^2}, \tag{3.55}$$

where we define $a = C/(2\pi)$.

---

24. In this mathematical footnote we will describe the principal steps for obtaining the Euler–Lagrange equation starting with the search for an extremum of a functional (variational problem).
Say $I = \int F(x, y(x), y'(x))dx$, a functional of function $y(x)$, with $y'(x) = dy/dx$. Let us look for the extremum of functional I, that is, let us find the function $y(x)$ which makes functional I an extremum, with boundary conditions such that $y(x)$ has fixed values at the boundaries.
The variation of I with respect to a variation of function $y(x)$ is written $\delta I = \int dx[(\partial F/\partial y)\delta y + (\partial F/\partial y')\delta y']$, with $\delta y' = \delta(dy/dx)$. Be careful not to confuse the effects of the differential d/dx with respect to variable x and the action of δ, which refers to a variation with respect to function $y(x)$ (i.e. $d/d(y(x))$). Under the action of δ, x is simply a fixed parameter of the curve. We can thus invert the two operations, and write $\delta dy/dx = d(\delta y)/dx$.
Let us substitute this last change in the expression δI, and then integrate by parts, to obtain $\delta I = \int dx[\partial F/\partial y - (d/dx)(\partial F/\partial y')]\delta y$ where we have used the condition $\delta y = 0$ at the end of the integration interval since we have assumed that $y(x)$ is fixed at the boundary of the domain. Thus, our search for extremum I comes down to nullifying the variation of δI in favor of a small, but arbitrary, variation in δy. It follows that the function $y(x)$ must satisfy the following equation: $[\partial F/\partial y - (d/dx)(\partial F/\partial y')] = 0$, the variational equation called the Euler–Lagrange equation. The left-hand quantity is called the Euler–Lagrange derivative, or functional derivative, or still Fréchet derivative. It is conventional to use the symbolic notation of $\delta I/\delta y$ for this derivative.
25. Since f does not depend explicitly on z ($f = f(r,r')$) the differential f is written as $df = (\partial f/\partial r)dr + (\partial f/\partial r')dr'$, such that $df/dz = (\partial f/\partial r)r' + (\partial f/\partial r')r''$. If we multiply the Euler–Lagrange equation (3.53) by r', and use the previous expression of df/dz, we have the following equation: $(d/dz)(f - r'(\partial f/\partial r')) = 0$. We integrate this to find the simplified Euler–Lagrange equation in the case where f is independent of z: $f - r'(\partial f/\partial r') = C$, with C a constant.

Since $r' = dr/dz$, we have:

$$\frac{dr}{\sqrt{\frac{r^2}{a^2} - 1}} = dz, \tag{3.56}$$

where we recognize the derivative of a hyperbolic cosine function, allowing us to finally obtain the expression for function $r(z)$ for which the area of S is an extremum:

$$r = a \cosh\left(\frac{z - b}{a}\right), \tag{3.57}$$

where b is another integration constant. This equation describes the form, known as catenary, of the hyperbolic cosine shape which a string, suspended by two extremities and subject to its own weight, takes. The form of the surface obtained by the rotation of this catenary curve $(r(z))$ around the axis of Oz is called a catenoid (see the schematic representation in figure 3.32).

In choosing the origin of the systems to be the center of the catenoid between $z = -h$ and $z = h$, we get rid of constant b ($b = 0$). Furthermore, the constant a is potentially determined from the catenary equation (3.57) with satisfaction of the boundary conditions as fixed by the two rings situated at the extremities of the soap film: radius r of the film at height h is equal to radius R of the ring. Using this as our boundary condition (that is $r(z = h) = R$) that we then use in our catenary equation (3.57), we obtain the following:

$$\frac{R}{a} = \cosh\left(\frac{h}{a}\right). \tag{3.58}$$

However such an equation of variable a does not always have a solution. For a given height h, the solution to equation (3.58) only exists while the value of the ring's radius R lies between a certain interval. Thus, the variation of parameter R controls the existence – or lack thereof – of a solution. This is reminiscent of the saddle-node bifurcation.

By making a change of variable $a \to 1/\lambda$ we obtain from the last relation (equation (3.58)) the following equation:

$$\lambda R = \cosh(h\lambda), \tag{3.59}$$

where the solution only exists if the straight line $D(\lambda) = R\lambda$ of slope R and the hyperbolic function $g(\lambda) = \cosh(h\lambda)$ intersect (see figure 3.33 for a graph depiction). Qualitatively, there is no doubt that the two curves cannot cross when R is too small (see figure 3.33(a)). As R grows, we find the critical value R_c where $R\lambda$ becomes tangent to the hyperbolic function $\cosh(h\lambda)$; note that λ_c is the abscissa of the tangent point. When radius R is greater than the critical radius $(R > R_c)$ we have two intersection points and thus two solutions (see figure 3.33(b)); their abscissa are λ_- and λ_+. When R is

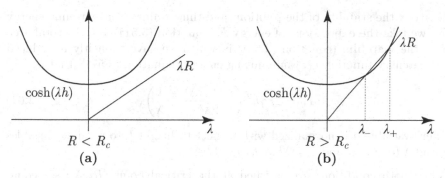

Figure 3.33 – The qualitative resolution of equation $\lambda R = \cosh(h\lambda)$: (a) where R is small enough and there is no intersection; (b) where R is big enough. There are two intersections, typical of the saddle-node bifurcation.

less than this critical radius $(R < R_c)$, the two curves do not intersect, and the number of solutions changes abruptly from two to zero.

The critical point with coordinates (R_c, λ_c) is defined by two conditions. The first is derived from the necessity of an intersection between the straight line $R\lambda$ and the hyperbolic function $\cosh(h\lambda)$, that is:

$$R_c\lambda_c = \cosh(h\lambda_c). \tag{3.60}$$

The second condition derives from the fact that at the critical point, the straight line $R\lambda$ is exactly tangent to the hyperbolic curve $\cosh(h\lambda)$, that is, the slope of the line $dR\lambda/d\lambda = R$ is equal to the derivative of the hyperbolic function $d\cosh(h\lambda)/d\lambda = h\sinh(h\lambda)$:

$$R_c = h\sinh(h\lambda_c). \tag{3.61}$$

We now make a Taylor approximation in the neighborhood of the critical point, $(R - R_c, \lambda - \lambda_c)$, and starting from equation (3.59) we obtain:

$$R - R_c = \frac{h^2 R_c}{2}(\lambda - \lambda_c)^2. \tag{3.62}$$

This equation has no solution when the radius of the ring R is smaller than the critical radius R_c $(R < R_c)$. On the other hand, when the radius of the ring R is greater than the critical radius R_c $(R > R_c)$, we have two solutions:

$$\lambda - \lambda_c = \pm\frac{\sqrt{2}}{h}\sqrt{\frac{R - R_c}{R_c}} = \frac{1}{a} - \frac{1}{a_c}. \tag{3.63}$$

When $R = R_c$, the two solutions merge. We can already guess that one of the two solutions is stable and the other unstable (the stable one has a lower surface than the unstable one). To convince ourselves, we undertake a detailed analysis of the solution's stability.

To study the stability of the solutions and thus evaluate the extremal energy F_0, we take the expressions of energy F (equation (3.51)) and the total area S of the soap film (equation (3.51)) in which we have explicitly introduced the catenary function $r(z)$ minimizing energy (equation (3.57)), that is:

$$F_0 = 2\pi\gamma h^2 \left(\frac{\sinh(2\bar{\lambda})}{2\bar{\lambda}^2} + \frac{1}{\bar{\lambda}}\right), \qquad (3.64)$$

where variable a is normalized with respect to height h to redefine variables $\bar{a}$ and $\bar{\lambda}$ ($\bar{a} = a/h$ and $\bar{\lambda} = \lambda h = h/a = 1/\bar{a}$).

When both conditions are satisfied at the critical point (R_c, λ_c; see equations (3.60) and (3.61)), we obtain an expression of critical variable $\bar{\lambda}_c$:

$$\bar{\lambda}_c \tanh(\bar{\lambda}_c) = 1. \qquad (3.65)$$

If we solve the equation numerically, we find $\lambda_c \simeq 1.199$. Substituting this value into the analytic expression that reflects the existence of the point tangent to the critical point (equation (3.61)), we obtain the normalized value of the critical radius $\bar{R}_c$: $\bar{R}_c = R/h \simeq \sinh(1.199) \simeq 1.507$.

The solutions exist only if radius R is greater than or equal to this critical value (see equation (3.63)). Thus, for $R \simeq 1.6$ we obtain from the conditions for existence of a solution (equation (3.59)) two values for variable $\bar{\lambda}$: $\bar{\lambda}_1 \simeq 1.577$ and $\bar{\lambda}_2 \simeq 0.887$. Let us substitute these values into the expression for energy at the extremum point (equation (3.64)) to obtain:

$$\frac{F_0(\bar{\lambda}_1)}{2\pi\gamma h^2} \simeq 2.985, \qquad \frac{F_0(\bar{\lambda}_2)}{2\pi\gamma h^2} \simeq 2.945. \qquad (3.66)$$

The solution with minimal energy, i.e. the stable solution, is obtained with the smaller value of $\bar{\lambda}$, that is, $\bar{\lambda} = \bar{\lambda}_2$; the second solution, found for $\bar{\lambda} = \bar{\lambda}_1$, corresponds to an unstable solution, since there F_0 is a maximum. We can verify that this result corresponds with our physical intuition of the problem: the stable profile for the soap film is associated with the minimal area, and thus with a minimal value of surface energy (see figure 3.34, showing the stable and unstable profiles of the soap film).

Let us now make a qualitative assessment of the form of the extremal energy[26] F_0 (equation (3.64)) as a function of variable $\bar{\lambda}$ (and $\bar{a}$, see figure 3.35). The minimum energy is found at the degenerate point, that is, where the normalized radius $\bar{R}$ of the ring is equal to the normalized critical radius ($\bar{R} = \bar{R}_c$ and for the critical value of $\bar{\lambda}$, $\bar{\lambda} = \bar{\lambda}_c$). At the critical point ($\lambda_c, R = R_c$),

---

26. Remember, extremal energy F_0, called the relaxation energy, is found when the catenary function $r(z)$ which minimizes energy (equation (3.57)) is introduced into the expression of total area S of the film soap (equation (3.52)) taken from the expression for energy (equation (3.51)).

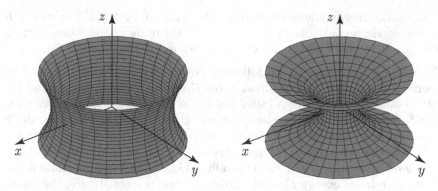

Figure 3.34 – The form of the two Euler solutions: (a) (stable form) has a smaller area than (b) (unstable form). Here $\bar{R} = 1.9$, $\bar{\lambda} \simeq 0.636$ (a) and $\bar{\lambda} \simeq 2.023$ (b).

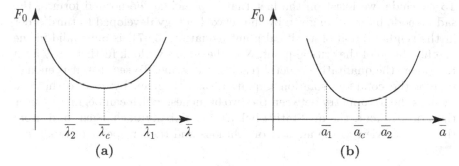

Figure 3.35 – The relaxed energy as a function of $\bar{\lambda}$ (left) and of $\bar{a}$ (right).

the energy is: $F_{0c} = 2\pi\gamma(R_c^2 + h^2)/(h\lambda_c)$. For $R < R_c$ there is no minimal form, and thus no solution. For $R > R_c$, there are two solutions represented by λ_1 and λ_2 (see figure 3.35); we have seen that λ_2 corresponds to a lesser energy with respect to λ_1.

We can recover the results detailing the stability of the solutions through a slightly more formal analysis if we expand the energy F_0 (equation (3.64)) to third order around critical value $\delta\bar{a} = \bar{a} - \bar{a}_c$:

$$\frac{F_0}{2\pi\gamma h^2} = \frac{\bar{R}_c^2 + 1}{\bar{\lambda}_c} + \frac{\bar{R}_c^2}{\bar{a}_c^3}\delta\bar{a}^2 - \frac{2}{3}\frac{5\bar{R}_c^2 + 1}{\bar{a}_c^4}\delta\bar{a}^3, \tag{3.67}$$

where we have used the two constraints found at a critical point (R_c, λ_c) (see equations (3.60) and (3.61)).

Taking into account the expression of the energy in the neighborhood of the critical point (equation (3.67)), only the third order term (in $\delta\bar{a}$) allows us to classify the stability of the solutions. Since the coefficient of the $\delta\bar{a}^3$ term is negative, the energy minimum corresponds to $\delta\bar{a} > 0$ ($\bar{a} > \bar{a}_c$). The energy F is asymmetric: for variable $\bar{a}$ such that $\delta\bar{a} > 0$, the branch of possible solutions has a small energy with respect to the energy in $-\bar{a}$ (see figure 3.35).

We bring our attention to the following point: we would have obtained an incorrect result if we had undertaken our third order expansion around the critical value using variable $\bar{\lambda}$ instead of $\bar{a}$. The asymmetry of the relaxed energy F_0 is only seen when developed around $\bar{\lambda}$: the deviation from a parabolic form in terms of variable $\bar{\lambda}$ is important even when it is a weak variation of $\delta\bar{a}$. On the other hand, a bigger deviation $\delta\bar{\lambda}$ is necessary to reflect the same level of asymmetry in F_0. Typically, $\delta\bar{\lambda}/\bar{\lambda}_c \sim 30\%$ creates an asymmetry in the relaxed energy F_0 of the order of a few percent. Hence, the choice of the variable for our analysis is essential for deciding which order to expand the equation to. It is a good idea to check, in each case, whether or not the following order in the expansion would affect the result.

To conclude, we insist on the fact that in principle the normal form of the saddle-node bifurcation given by the relaxed energy developed to third order in the neighborhood of a critical point (equation (3.67)) is only valid in the neighborhood of the critical point. Nonetheless, as we look further away from the point, the qualitative general result is the same. To see this, it is enough to plot the complete solution (equation (3.59)), given by $\bar{R} = \cosh(\bar{\lambda})/\lambda$. Besides the asymmetry between the two branches, which can begin even near the critical point, the qualitative behavior is globally unchanged, both from the point of view of the number of solutions and their respective stabilities.

3.9.4. A Subcritical Bifurcation with a Saddle-Node Branch

Show that for a subcritical bifurcation, the amplitude equation in the neighborhood of a turning point T is characterized by a saddle-node bifurcation (see figure 3.36).

Solution. Let $A^{(T)}$ and $\epsilon^{(T)}$ be the coordinates of the turning point (see figure 3.36). Let $A = A^{(T)} + B$ and $\epsilon = \epsilon^{(T)} + \mu$. We expand the amplitude equation (3.21) around the inflection point to the highest order in B and μ; we obtain

$$\dot{B} = \mu A^{(T)} + \epsilon^{(T)} A^{(T)} + A^{(T)3} - A^{(T)5}$$

$$+ B\left(\epsilon^{(T)} + 3A^{(T)2} - 5A^{(T)4}\right) + B^2\left(3A^{(T)} - 10A^{(T)3}\right). \quad (3.68)$$

Since the turning point is on the branch of fixed points, we have $\epsilon^{(T)}A^{(T)} + A^{(T)3} - A^{(T)5} = 0$. For the precise value of ϵ at the turning point, we have $\epsilon^{(T)} = -1/4$, which determines amplitude A at this point: $A^{(T)} = 1/\sqrt{2}$ (see equation (3.25)). The factor for the linear term in B is canceled out, and the

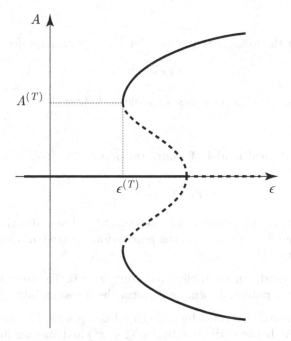

Figure 3.36 – The subcritical bifurcation has the shape of a saddle-node bifurcation in the vicinity of the turning point of coordinates $(\epsilon^{(T)}, A^{(T)})$.

equation for amplitude B around A^T is written as

$$\frac{1}{\sqrt{2}}\dot{B} = \frac{\mu}{2} - B^2. \tag{3.69}$$

With a minor change of variables, this expression becomes the normal form of the saddle-node bifurcation (equation (3.9)).

3.10. Exercises

3.1

Plot the bifurcation diagram for each of the following dynamical equations (as a function of ϵ, which is the control parameter):

$$\dot{x} = -\epsilon x + x^2 + \nu \tag{3.70}$$

$$\dot{x} = \epsilon x - x^2 - x^3 + \nu \tag{3.71}$$

$$\dot{x} = \epsilon x + x^2 + \epsilon^2 - \nu; \tag{3.72}$$

ϵ and ν can be positive or negative. Discuss all possible cases as a function of ϵ and ν. Specify in each case the type of bifurcation.

3.2

Let us consider the following equation giving rise to an imperfect bifurcation:

$$\dot{x} = \epsilon x - x^3 + \nu. \tag{3.73}$$

Show that this equation possesses a saddle-node branch.

3.3

Consider a biological model of budworm population [65], which takes the form:

$$\dot{x} = rx\left(1 - \frac{x}{k}\right) - \frac{x^2}{1 + x^2}, \tag{3.74}$$

where r and k are two positive parameters and x designates the population variable (the number of insects in the population, written in some dimensionless form, at time t).

1. Show there exists an unstable fixed point $x = 0$. To study the stability, either adopt a potential form or perform the linear stability analysis.

2. Write the equation verified by the other fixed points. Express it in terms of an equality between the function $x/(1 + x^2)$ and another function to be specified.

3. With the help of a graphical representation, show that the equality determined above has either one, two, or three solutions. Specify qualitative conditions of their existence.

4. On the basis that $x = 0$ is unstable, can you deduce the stability of the other fixed points (when they exist). Plot the bifurcation diagram as a function of r for a given k. Which type of bifurcation do we have?

3.4

Analyze the bifurcation diagrams associated with the following nonlinear equations, and specify in each case the type of bifurcation and the stability and instability of each branch:

$$\dot{x} = \epsilon x - \ln(1 + x) \tag{3.75}$$
$$\dot{x} = x - \epsilon x(1 - x) \tag{3.76}$$
$$\dot{x} = x(\epsilon - e^x). \tag{3.77}$$

3.5

We propose to study a problem associated with a laser system (see [38]). Let N denote the number of excited atoms and n that of photons emitted by the laser. It is known that these two quantities are linked by the following equations [38]:

$$\dot{n} = GnN - kn \tag{3.78}$$

$$\dot{N} = -GnN - fN + p \tag{3.79}$$

where G is the so-called gain for stimulated emission, k the attenuation rate of the photons (due to diffusion, etc.), f is the attenuation rate of the spontaneous emission, and p is the amplitude of the optical pumping. All these parameters are positive except p which can have both signs.

1. Assume that N relaxes very quickly toward its quasi-static values, i.e. N is such that $\dot{N} \simeq 0$. Express N as a function of n (this is the so-called adiabatic elimination we have already encountered) and deduce the equation satisfied by n.

2. Show that the fixed point $n_0 = 0$ becomes unstable for $p > p_c$. Determine p_c.

3. Which type of bifurcation do we have at $p = p_c$ (this is the critical value required for laser emission)? Plot the bifurcation diagram.

3.6

Consider the following nonlinear equation:

$$\ddot{x} + \mu\dot{x} = c\left(\cos\left(\frac{x}{2}\right) - a\right)\sin\frac{x}{2} - b\sin(x). \tag{3.80}$$

a, b, c and μ (friction coefficient) are positive numbers.

1. Can you imagine a simple mechanical device (involving a particle attached to a spring) described by this equation? Find expressions for a, b and c in terms of the original physical parameters. Hint: the mass is bound to move along a simple curved trajectory (to be guessed); we assume $a < 1$.

2. Find the number of fixed points and their stability as a function of the values of the parameters.

3. Which type of bifurcation do we have?

3.7

Consider the imperfect bifurcation studied in this chapter. The corresponding amplitude equation is given by equation (3.21). Show that $A \sim \nu^{1/3}$ for small ϵ. Indicate the amplitude of ϵ for which this is valid.

3.8

Consider the following amplitude equation in which the introduction of an imperfection variable ν breaks the left/right symmetry of the system:

$$\frac{dA}{d\tau} = \epsilon A + A^3 - A^5 + \nu. \qquad (3.81)$$

Discuss the form of the bifurcation diagram as a function of the parameter ϵ. Consider the two possible signs of the imperfection term $\nu > 0$ and $\nu < 0$ (refer to figure 3.37, which shows partial results and may guide your analysis). We consider ν to be small.

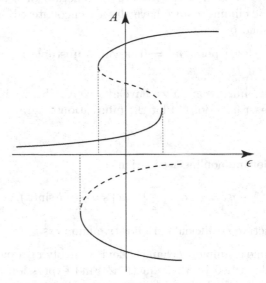

Figure 3.37 – Diagram of supercritical imperfect bifurcation with respect to a given imperfection term, ν (to be determined).

Chapter 4

Classification of the Seven Elementary Catastrophes

Abstract *This chapter describes the classification of seven elementary catastrophes and introduces the basic terminology native to catastrophe theory. We will see that for a nonlinear system of four (or fewer) (dimensionless) control parameters the solutions can undergo only seven different types of qualitative change (catastrophes). Many systems found in nature can be classified in a general way within this framework irrespective of our knowledge of the evolution equations governing the systems. Even if we cannot imagine the form of their equations, we can make the classification and see its results hold true.*

This is a more formal chapter, although the mathematical notions remain relatively uncomplicated. Nonetheless it is strongly advised that the reader review the previous chapters before beginning this one. For instance, the previous chapter outlines a simple introduction to catastrophes which may be particularly useful for better comprehension of this chapter. The reader may also find a pedagogical study by M. Demazure useful [27].

In the preceding chapters, we studied several bifurcations and classified each within a particular family of catastrophes without further elaboration. We learned that the pitchfork bifurcation – both the perfect and imperfect forms (chapter 2 and section 3.2, respectively) – and the transcritical bifurcation (section 3.4) belong to the family of "cusp" catastrophes; that the subcritical bifurcation belongs to the "butterfly" family (section 3.3); and that the saddle-node bifurcation belongs to the "fold" family (section 3.1). How many different families of catastrophe exist? One may think at first that the answer to this general question must depend on the specific system. In reality, as we will see in this chapter, our answer depends exclusively on the number of control parameters of the system, whatever the system may be, and not

© Springer Science+Business Media B.V. 2017
C. Misbah, *Complex Dynamics and Morphogenesis*,
DOI 10.1007/978-94-024-1020-4_4

its degrees of freedom. If the number of control parameters is less than or equal to four, we can observe up to seven elementary catastrophes. This is an important result, especially because one need not know the equations describing the system in order to state it; we can apply this to any arbitrarily complex (or even mathematically irreducible) systems.

Let us begin our classification of the different catastrophes through the study of a system with only one degree of freedom – such as the mechanical spring system described in chapter 2 – corresponding to dynamic variable A (in the spring example, A corresponds to the position of the mass suspended by the spring). A general assumption of catastrophe theory is that the dynamics of the system has a potential form:

$$\dot{A} = F(A, \{\mu\}) = -\frac{\partial V}{\partial A}, \tag{4.1}$$

where V is the potential and F (which would be the force in a mechanical analogy) is an analytic function[1] of variable A and of $\{\mu\}$, the ensemble of control parameters. Let us suppose that a bifurcation point exists at $A = 0$[2], by definition characterized by a stationary solution's[3] qualitative change in response to the variation of a control parameter. Hypothesizing F as an analytic function is justified by our choice, independent of the system and its model, to a priori rule out singular functions, since singularity will emerge naturally from catastrophes. In choosing an analytic function F, the initial model has no apparent singularity, and yet, singularities will emerge.

In order to assess the eventual existence of a qualitative change in solutions, we shall first expand function F to some order in A (say A^n, with n some positive integer; this means that the highest power in the potential behaves as A^{n+1}).

---

1. Which is to say it can be expanded in powers of A.
2. We have assumed that the bifurcation happens when $A = 0$ in order to simplify and compress the notation of any further expansions in the power of A without affecting the results. We could just as well have adopted any finite value A_0 at the bifurcation point, making the expansion around that value A_0 instead.
3. Also called an equilibrium solution or fixed point.

4.1. Fold or Turning Point Catastrophe

Expanding the dynamical equation of the system (4.1) to second order, we have:

$$F = \epsilon + \alpha A + \beta A^2 = \beta \left(\frac{\epsilon}{\beta} + \left(\frac{\alpha}{\beta} \right) A + A^2 \right),$$

or

$$\frac{F}{\beta} = \frac{\epsilon}{\beta} + \left(\frac{\alpha}{\beta} \right) A + A^2.$$

We redefine function F such that $F/\beta \to F$, and parameters ϵ and α such that $\epsilon/\beta \to \epsilon$ and $\alpha/\beta \to \alpha$ to obtain:

$$F = \epsilon + \alpha A + A^2, \tag{4.2}$$

where parameters ϵ and α are the only two independent parameters of the problem. The potential is cubic and we shall see that this potential is structurally unstable (in the sense defined in chapter 3). Let us superpose this function F (equation (4.2)) with a polynomial of the same order[4] in A and with the arbitrarily small coefficients ν_i:

$$F = \epsilon + \alpha A + A^2 + \nu_0 + \nu_1 A + \nu_2 A^2 = (\epsilon + \nu_0) + (\alpha + \nu_1)A + (1 + \nu_2)A^2. \tag{4.3}$$

This is equivalent to making an arbitrarily small change to each coefficient without changing the qualitative form of the potential. Note that the linear term in A in the expression of function F (equation (4.2)) is unimportant since it can be eliminated through the simple translation $A \to A - \alpha/2$; the generic form of the dynamical equation is therefore given by:

$$\dot{A} = \epsilon + A^2. \tag{4.4}$$

The fixed points, solutions of the stationary equation $\dot{A} = 0 = \epsilon + A^2$, are given by:

$$A = \pm\sqrt{-\epsilon} \tag{4.5}$$

for negative ϵ ($\epsilon < 0$); there are no fixed points when parameter ϵ is positive. The point $\epsilon = 0$ is therefore a critical point that marks the qualitative change of solutions (see the bifurcation diagram figure 4.1). This dynamical equation (4.4) is characteristic of the saddle-node bifurcation (see section 3.1).

---

4. Superposing F with a higher order polynomial would not change the result, see subsection 4.1.2.

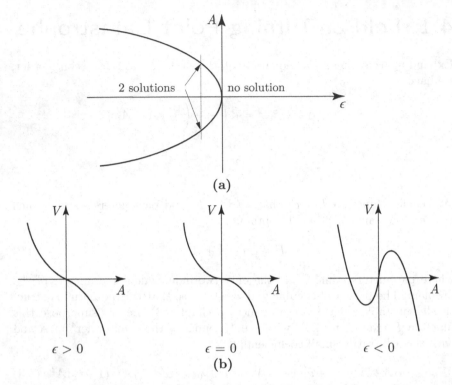

Figure 4.1 – (a) The fold catastrophe. (b) The different forms of the potential $V = -\epsilon A - A^3/3$ for different values of ϵ. [© LiPhy]

So, in the case of having a single parameter ϵ, the catastrophe frontier is defined by a critical point ($\epsilon = 0$). Representing this critical point in the parameter space, equivalent to plotting the single point $\epsilon = 0$ on the axis of parameter ϵ, is a relatively poor geometric representation. It is more illuminating to represent the catastrophe in a bifurcation diagram; indeed the geometric figure traced out in that space (A, ϵ) is what inspired the name: "fold" catastrophe (sometimes also called *turning point* catastrophe).

4.1.1. Unfolding Singularities, in Simple Terms

Before going any further, let us introduce some useful concepts for the classification of catastrophes. The first concept is that of the unfolding of singularities.

First of all we note that the function $F = A^2$ (or the associated potential $V \sim A^3$) has a structurally unstable form. In practice, simply adding a linear term νA to the associated potential V will introduce a qualitative change to the solutions at $A = 0$. The potential[5] $V = A^3$, which has a horizontal slope at $A = 0$, displays, upon addition of a linear term, a slope of 1 at $A = 0$. The unique fixed point $A = 0$ exists as a solution of $F = V' = dV/dA = 0$. Actually, it is more precise to say that there are two fixed points at $A = 0$, for it is a double solution. Expressed in terms of the potential $V(A)$, this degeneracy signifies not only that the first derivative of the potential is zero at $A = 0$ ($V' = 0$), but also that the second derivative becomes zero ($V'' = 0$). This kind of potential is called "singular"[6]. Remember that when the first and second derivatives of the potential are zero for all values of the control parameter, the potential is structurally unstable (see subsection 3.7.1).

The potential $V = A^3$ has an inflection point at $A = 0$ (see figure 4.2(a)). Adding a linear term of type ϵA (with ϵ arbitrarily small and negative), we obtain a new potential that is qualitatively different (figure 4.2(b)) and possesses two extrema: a minimum and a maximum.

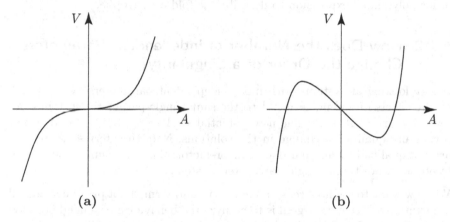

(a)   (b)

Figure 4.2 – (a) The shape of the perfect potential $V = A^3$. (b) The shape of the perturbed potential $V = A^3 + \epsilon A$, with $\epsilon < 0$. [© LiPhy]

On the other hand, if we also add a quadratic term νA^2 (with ν a small parameter) there is no qualitative change in the solutions. In fact, if the

---

5. In this chapter, we do not pay special attention to numerical prefactors.
6. For a function of two or more variables, the determinant (called the discriminant) of the Hessian matrix plays the role of the second derivative V''. By way of example, the Hessian matrix for a two variable function is the matrix comprised of the function's four second derivatives. A matrix with a determinant of zero (analogous to $V'' = 0$ for a single variable) is not invertible; the matrix is thus called singular. The point A that satisfies this condition (of having a Hessian matrix of determinant zero) is a "singular point".

potential has the form

$$V = A^3 + \nu A^2 + \epsilon A, \tag{4.6}$$

the quadratic term may be eliminated with a change of variables $A = B - \nu/3$. In terms of the new variable B, the potential V contains terms in B^3, B, and an unimportant constant. We may thus conclude that it is the addition of the linear term $-\epsilon A$ that reveals the hidden structure embedded in the singular potential $V(A) \sim A^3$: this is how one *unfolds the singularity* A^3 (the potential). The addition of a minuscule linear perturbation unfolds the hidden structure (the two extrema) at the heart of potential A^3: to quote from Timothy Saunders' book dedicated to catastrophe theory, *it is like the unfolding of a bud to reveal a flower* (see [91]). For an illustration of unfolding, compare the two figures 4.2(a) and (b). The potential $A^3 + \epsilon A$ represents the *universal unfolding* of singularity A^3. Here, the word universal refers to the fact that this is the most concise and general (after the elimination of A^2 through the aforementioned change of variable) third order polynomial expression to describe the fold catastrophe.

4.1.2. How Does the Number of Independent Parameters Change the Order of a Singularity?

Before looking at systems with more independent parameters, we will look at the effects of adding term A^4 to the same single-parameter system's potential $V(A) = \epsilon A + A^3$. The new potential $V(A) = \epsilon A + A^3 + A^4$ does not reveal any qualitative change in the solutions. Note that terms A^3 and A^4 are not qualified by any parameter or prefactor; this is a result of our initial hypothesis, positing a single-parameter[7] system.

We now want to see how reasonable it is to add term A^4, a priori considered a perturbative term. Our goal is to analyze the behavior in the neighborhood of the fixed points in order to determine any eventual qualitative change in the solutions. In the neighborhood of $A = 0$, the term A^4 is small with respect to the other terms; however, as we move into neighborhoods of other potential fixed points with values other than $A = 0$, the effect of the A^4 term may be felt or even become dominant. We can check. The following fixed points come from the second order dynamical equation (4.4): $A = \pm\sqrt{-\epsilon}$. In the neighborhood of these fixed points, the two first terms of the potential $V = \epsilon A + A^3 + A^4$ are of order $\epsilon^{3/2}$, while the A^4 term is of order ϵ^2. Thus, the A^4 term is negligible with respect to the first two terms, justifying our use of it as a simply perturbative term.

---

7. Even if we were to speculate that the term in A^3 had, as a prefactor, a function of ϵ $(g(\epsilon))$, we could always simplify to the initial form as follows: the potential V is defined by $V = \epsilon A + g A^3 = g((\epsilon/g)A + A^3)$. Redefining parameter $\epsilon \to \epsilon/g$ and thus potential $V \to V/g$, we have: $V(A) = \epsilon A + A^3$.

Apart from this, we can look at the perturbed model's fixed point equation

$$\epsilon + 3A^2 + 4A^3 = 0. \tag{4.7}$$

The presence of A^3 signals the possible existence of other fixed points ($V' = 0$). In the non-perturbed model there is a single fixed point $A = \pm\sqrt{-\epsilon}$, for which $V' = 0$ (condition for a fixed point) and $V'' = 0$ when $\epsilon = 0$ (condition for a singular point). However, and this is key to our argument, the new fixed point, though satisfying $V' = 0$, cannot also be a catastrophe point (as would be characterized by satisfaction of the two conditions $V' = 0$ and $V'' = 0$). With only a single parameter, ϵ, the potential V cannot be adjusted to simultaneously satisfy both conditions $V' = V'' = 0$ at two separate points.

Now suppose that the placement of the new fixed point, induced by A^3, was very far away from the old fixed point $A = \sqrt{-\epsilon}$, which was close to the origin ($A = 0$) like in the non-perturbed model. The new fixed point can always be brought back into the neighborhood of the origin by employing the appropriate change of variable when expanding the potential V around the new fixed point. This change of variable brings us back to a fixed point close to zero. We arrive at the following conclusion: in the case of a single parameter, the potential

$$V = \epsilon A + A^3 \tag{4.8}$$

is the most generic representation of the catastrophe.

We will now take a look at an everyday situation in which we see this catastrophe: the flow of water inside a garden hose made of relatively stiff material. The movement of incompressible fluids is described by the Navier–Stokes equations, well known in the field of hydrodynamics (for interested readers, see [42] and chapter 9 of this book, in which we solve a hydrodynamical problem). The movement of an incompressible fluid is characterized by a single control parameter which relates the inertial and viscous forces. This is called the Reynolds number:

$$Re = \frac{RV_0}{\rho\eta}, \tag{4.9}$$

where R is the radius of the hose (more generally, R is the characteristic length of whatever geometric object contains the flow), V_0 is the characteristic velocity of the flow, ρ the density of the fluid, and η the fluid's dynamic viscosity coefficient. It is known that for small enough Reynolds number Re (a condition that is satisfied if the flow velocity is very small), there exists a unique stable solution which is called laminar flow (flow occurring in parallel layers; no cross-currents perpendicular to the direction of flow, nor eddies nor swirls of fluids). Beyond a certain critical value of Re, this solution coexists with a pair of solutions that represent a non-laminar regime, one stable and the other unstable: this is a fold catastrophe (see figure 4.3), also known as a saddle-node bifurcation.

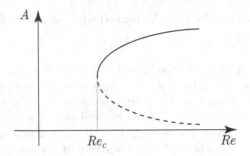

Figure 4.3 – A represents the amplitude of the perturbation movement as a function of the Reynolds number Re. For $Re < Re_c$ the movement is laminar. For $Re > Re_c$ the laminar movement coexists with two non-laminar solution branches, constituting a fold catastrophe. [© LiPhy]

4.1.3. Co-dimensions

Space is commonly conceived of in three linear dimensions. In Cartesian co-ordinates, the surface of a sphere of radius R is described by a single equation of the form: $x^2 + y^2 + z^2 = R^2$. Generally speaking, the surface of a three dimensional object (i.e. the sphere) is determined by a single equation. In the same physical space, a one dimensional object (e.g. a curve, a straight line) requires two equations. A curve may effectively be thought of as an intersection of two surfaces, each described by one equation. In general, the number of equations needed to represent an object is equal to the difference between the dimension of the space in which the object is embedded and the dimension of the object itself; this quantity is called the *co-dimension*. Thus, in a space of three dimensions, the co-dimension of a surface is equal to 1 and that of a curve is 2; in a space of two dimensions (such as a plane), the co-dimension of the curve is 1. Whatever the dimension of the space in consideration, only an object of co-dimension 1 can divide space into two distinct regions. Thus, a three dimensional space (respectively, bidimensional) is divided into two distinct regions by a surface (respectively, a curve), and a curve is divided into two half-curves by a point.

Having thus defined the co-dimension, let us return to catastrophe theory and determine the role of co-dimensions in that context. With the goal of identifying the region of a catastrophe, our first step is to determine the fixed points A_0 (points at which the first derivative of the potential becomes zero, $V' = \partial V / \partial A = F(A_0, \mu) = 0$). The second step is to look at the parameter space and search, amongst the fixed points, for the condition that further derivatives (second, third...) also go to zero. If the potential is such that only the second derivative can be made zero: $V''(A_0, \mu) = \partial^2 V / \partial A^2 = 0$, this corresponds to a single condition that requires a single tuning (or control) parameter. So, if we have only a single parameter (the parameter space has dimension equal to one) only a point can divide that space into two distinct regions. The co-dimension of the separating object is equal to one and this point defines the catastrophe.

By way of example, consider the potential $V = \epsilon A + A^3$, the generic form for the "fold" catastrophe; this catastrophe is thus said to be of co-dimension 1. The fixed points are given by $\pm\sqrt{-\epsilon}$ and the second derivative at these points is equal to $\pm 2\sqrt{-\epsilon}$ which vanishes at $\epsilon = 0$. In this case the third derivative never vanishes, while higher derivatives vanish automatically. Since we have a single equation to satisfy, namely $V'' = 0$, we say that the catastrophe is of co-dimension 1. Be careful not to confuse between the dimension of the separating object, which is always of co-dimension 1, and the number of equations needed to satisfy the process of singularity unfolding (see section 4.1.1), which turns out to be one in this example.

If the potential is of higher order (i.e. contains a term in A^4), then second as well as third derivatives can be made equal to zero provided we have at least two control parameters. This is the case for the *cusp* catastrophe (see chapter 3) where the separation between regions of different qualitative solutions correspond to a line (still a co-dimension 1 object). However, since the catastrophe is a line it implicitly means that we have two control parameters, and for this reason the catastrophe is called a co-dimension 2 catastrophe[8]. Indeed, with two parameters we can make both second and third derivatives be zero, hence two equations. We can continue with this reasoning; if we simultaneously annul the first n derivatives of the potential, we will find a catastrophe of co-dimension $(n - 1)$, which would require the existence of $(n - 1)$ independent parameters. We conclude that in catastrophe theory, the co-dimension represents the number of control parameters needed to produce a catastrophe of a certain type.

4.2. Cusp Catastrophe

Let us now take a model with two control parameters and expand the potential V to fourth order – or, equivalently, expand function F to third order – as follows: $F = \epsilon + \alpha A + \beta A^2 + A^3$, with ϵ, α and β as our three parameters. The form of function F is structurally stable. The quadratic term in the new equation F (corresponding to the cubic term of potential V) can be eliminated through the transformation $A \to A = A - \beta/3$. We can redefine new parameters, ν and ϵ in order to rewrite function F as: $F = \nu + \epsilon A + A^3$.

---

8. Actually, in the presence of two parameters we can perform singularity unfolding for both A^3 and A^4 by annulling the second and third derivatives. We thus have two equations to define unfolding, and this is a more appropriate reason why this is called a co-dimension 2 catastrophe.

This defines our dynamical equation:

$$\dot{A} = F = \nu + \epsilon A + A^3. \tag{4.10}$$

We recognize the normal form of the imperfect bifurcation studied previously in section 3.2. Adopting the typical notation used in catastrophe theory, where u and v denote the two independent control parameters, we rewrite the potential as:

$$V = vA + uA^2 + A^4. \tag{4.11}$$

As mentioned at the end of the last chapter, this imperfect bifurcation belongs to the family of *cusp* catastrophes.

Using the potential of the imperfect bifurcation above (equation (4.11)), we find the fixed points using the cubic equation

$$\dot{A} = -\frac{\partial V}{\partial A} = v + 2uA + 4A^3 = 0. \tag{4.12}$$

We can use the determinant

$$\Delta = 8u^3 + 27v^2 \tag{4.13}$$

to find the number of solutions of the equation. If the discriminant is positive, equation (4.12) has one real solution; if the discriminant is negative, we have three real solutions (see [1]). The number of solutions changes abruptly from one to three when the discriminant passes through zero. The condition for a zero value discriminant thus describes the necessary condition for a *cusp* catastrophe[9]. When the discriminant is zero, two possible scenarios emerge: (1) at least one of the two control parameters is non-zero ($u \neq 0$ or $v \neq 0$), and so two of the three solutions are identical; (2) the control parameters are all zero simultaneously ($u = v = 0$) and we have three coinciding solutions. This triple point is called the cusp point, and traces the shape of a cusp in the parameter space, lending the catastrophe its name; see figure 4.4 showing the frontier defined by $\Delta = 0$ (equation (4.13)).

The fixed point equation (4.12) represents the universal unfolding of the singularity A^4 (see section 4.1.1 in which we described the concept of unfolding singularities). Since this is a double singularity, characterized by annulled second and third derivatives of the potential ($V'' = 0$ and $V''' = 0$), we have two equations with which to determine the catastrophe, and as such the co-dimension of this catastrophe is 2. The fixed point equation allows us to determine the amplitude of the variable $A(u, v)$ as a function of the control

---

9. In the previous chapter, we found this equation by simultaneously annulling the first and second derivatives of the potential and by eliminating variable A. This was done using the two equations ($V' = 0$) and ($V'' = 0$) to find the relation between the two independent control parameters (see equation (3.39)).

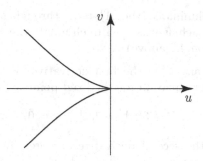

Figure 4.4 – Qualitative representation of the *cusp* catastrophe. [© LiPhy]

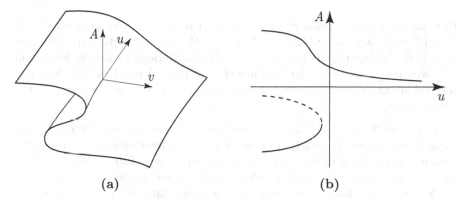

(a)   (b)

Figure 4.5 – (a) Qualitative representation of the fixed point as a function of parameters u and v. (b) Projection on the plane $v =$ some negative constant, where we recognize the diagram of imperfect bifurcation. [© LiPhy]

parameters u and v (see figure 4.5, which portrays the surface described by variable A in the space (A, u, v)). Cutting this surface with an axis $v = v_0$, where v_0 is a constant, reveals the shape of an imperfect bifurcation. This bifurcation thus belongs to the family of *cusp* catastrophes.

4.3. Swallowtail Catastrophe

Now, for the case of three control parameters, denoted by (u, v, w), we expand the potential V to fifth order (corresponding to a fourth order expansion of F). The generic potential is written as:

$$V = A^5 + uA^3 + vA^2 + wA. \tag{4.14}$$

Note that we have eliminated the A^4 term through a change of variable (using the same approach as was used to eliminate the A^3 term for the *cusp* catastrophe, see section 4.2 above).

We find fixed points any time the first derivative becomes zero ($V' = 0$). Accordingly, this is our equation for the fixed points:

$$5A^4 + 3uA^2 + 2vA + w = 0. \tag{4.15}$$

For singular points, the second derivative of potential V must go to zero ($V'' = 0$). This condition is expressed as follows:

$$20A^3 + 6uA + 2v = 0. \tag{4.16}$$

In principle we may eliminate the variable A using the two equations $V' = 0$ and $V'' = 0$ (equations (4.15) and (4.16)), thereby extracting the relationship between the parameters (u, v, w). The analytic expressions derived in this way are a little complex and are thus of limited practical use. We may instead adopt another approach to construct, step by step, the shape of the catastrophe.

To begin, we fix the value of parameter u and confine A to some range $[-a, a]$ (a is a real value). For each value of A we then determine the value of the second parameter v using the equation for singular points (4.16), and then we can substitute our result of value v into the fixed point equation (4.15) to determine the third parameter, w. For every value of u, we have an ensemble of pairs (v, w), and can plot what will be the geometric shape of a swallowtail[10] in the plane of parameters (see figure 4.6). The operation is repeated for different values of parameter u, and we can thus construct the three dimensional image of the catastrophe in the space (u, v, w) (see figure 4.6, right). For $u > 0$ the parameter w is of a single value, while for $u < 0$ it has several values. The swallowtail catastrophe is of co-dimension 3. Indeed, the singularity A^5 is triple degenerate at $A = 0$, because the second, third, and fourth derivatives go to zero at that point ($V'' = V''' = V'''' = 0$, at $A = 0$). The universal unfolding of this singularity requires three independent control parameters.

Let us now analyze the extrema of potential V. To begin, we consider the case where parameter v is zero ($v = 0$) and the parameter u is negative ($u < 0$). The solution of the second order (A^2) fixed point equation (4.15) is given by:

$$A^2 = \frac{1}{10}\left(-3u \pm \sqrt{9u^2 - 20w}\right) \tag{4.17}$$

10. The term "swallow" refers to the bird; today it is commonly used in the expression "swallowtail" to describe the form of an object similar to the shape of a swallow's tail, that is, where one extremity is larger than the other.

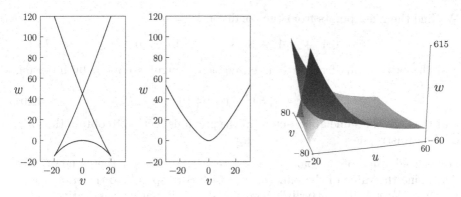

Figure 4.6 – Left: a section of the swallowtail catastrophe for $u < 0$ (here $u = -10$). Center: the same, for $u = 2$. Right: the swallowtail catastrophe in three dimensions. There are no fixed points in the upper region, one in the region below, and two in the interior region of the swallowtail. [© LiPhy]

for $w < 9u^2/20$. For $w > 0$ (respectively, $w < 0$) there are four (respectively, two) fixed points; when $w > 9u^2/20$, the fixed point equation has no solution.

Thus, for a model with three control parameters and a potential which we have expanded to fifth order, three cases will arise:

1. $w - 9u^2/20 > 0$: the fixed point equation (4.17) has no solution, the potential V has no extrema;

2. $0 < w < 9u^2/20$: there exist four solutions:
$$A = \pm[-3u \pm \sqrt{9u^2 - 20w}]^{1/2}/\sqrt{10}.$$
The potential V thus has four critical points, two minima and two maxima;

3. $w < 0$: there exist two solutions:
$$A = \pm[-3u + \sqrt{9u^2 - 20w}]^{1/2}/\sqrt{10} > 0.$$
The potential V has two extrema, a minimum and a maximum.

4.4. Butterfly Catastrophe

We continue with the case of a system with four parameters, written (u, v, w, t) but still with just one dimension in the model (i.e. one degree of freedom, in the form of a single variable A). We expand potential V to sixth order to reveal the universal unfolding of the butterfly catastrophe:

$$V = A^6 + tA^4 + uA^3 + vA^2 + wA. \qquad (4.18)$$

We have again eliminated the term A^5 by the appropriate change of variable.

We find the fixed points from the condition $V' = 0$:

$$6A^5 + 4tA^3 + 3uA^2 + 2vA + w = 0 \qquad (4.19)$$

and the catastrophe frontier (the ensemble of singular points) by imposing $V'' = 0$:

$$30A^4 + 12tA^2 + 6uA + 2v = 0, \qquad (4.20)$$

and then use the same approach as in the previous section to obtain the form of the catastrophe.

For a fixed value of the pair of parameters (t, u) we vary the value of A to determine the value of the pair (v, w), which corresponds to the catastrophe frontier. We call it "butterfly" from the shape drawn from negative values of parameter t. When t is positive, the curve is in the shape of a cusp (see figure 4.7).

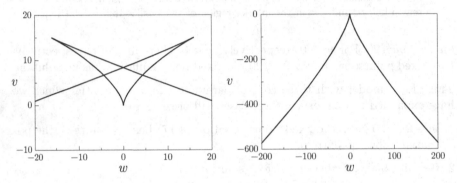

Figure 4.7 – The butterfly catastrophe. Left: the case when $u = 0$ and $t = -5$. Right: $u = 0$ and $t = 5$. [© LiPhy]

4.5. What Happens If We Have Many Degrees of Freedom?

Throughout our exploration of classifications in catastrophe theory, we have kept to systems which have only one degree of freedom: each model invariably had a single variable that we call A. In general, however, a real system will have many degrees of freedom. For example, to describe the movement of a bed of sand caressed by wind one must know, at each instant, what happens at each point in this bed of sand. The variables that are indispensable in our description of these dynamics, at each point in the bed of sand, are of order n, n being a very large number (even infinite for a continuous description of a bed of sand).

What happens to the classification of catastrophes in a situation such as when the degrees of freedom become numerous or sometimes infinite? The response may be surprising: the number of degrees of freedom is irrelevant; our only concern is the number of control parameters. We can use the classification provided by catastrophe theory, without any use of the evolution equation, purely by identifying the number of control parameters. Beyond four control parameters, the dynamics through the approach of catastrophe theory become very complex. However, if we retain only four independent control parameters, the relevant dynamics as described by catastrophe theory involve only two degrees of freedom at most; and in the case of two degrees of freedom, there exist only three possible independent forms of the potential. In other words, only three further distinct types of catastrophe are possible. Adding these to the first four catastrophes we have already discussed (fold, cusp, swallowtail, and butterfly), we count a total of seven elementary catastrophes.

Let us provide now the general reason why only the number of control parameters matters, regardless of the number of degrees of freedom. As we have seen, the frontier of a catastrophe is found by annulling the concavity, Δ, of a potential. For a system with one degree of freedom, A, the concavity of the potential, is equal to the second derivative of the potential ($V'' = d^2V/dA^2$). For a system with two degrees of freedom, A and B, the concavity is a function of the second order partial derivatives, that is:

$$\Delta = (\partial^2 V/\partial A^2)\,(\partial^2 V/\partial B^2) - (\partial^2 V/\partial A \partial B)^2.$$

Setting $V_{AA} = \partial^2 V/\partial A^2$, $V_{BB} = \partial^2 V/\partial B^2$ and $V_{AB} = \partial^2 V/\partial A \partial B$, we define the catastrophe frontier with the following condition:

$$\Delta = V_{AA}V_{BB} - V_{AB}^2 = 0. \tag{4.21}$$

A simple way of bringing the concavity of V to zero is to impose the simultaneous annulment of the three second derivatives. In this case, the potential V is degenerate along the two axes associated with variables A and B, since $V_{AA} = V_{BB} = 0$. In the terminology of catastrophe theory, variables A and B are called "essential" variables. If only condition $V_{AA} = 0$ is satisfied, then variable A is the only essential variable, and we are back to having a one dimensional problem as before. Note that in order to simultaneously satisfy the three conditions ($V_{AA} = V_{BB} = V_{AB} = 0$), three parameter sets must be adjusted independently in order to fulfill the three conditions separately. Thus, in the case of three or more degrees of freedom (say, A, B, and C), four parameters are not enough to simultaneously annul the six second derivatives (V_{AA}, V_{BB}, V_{CC}, and the three cross partial derivatives). Accordingly, for a system of four control parameters there can be no catastrophe that has more than two essential variables, regardless of how many degrees of freedom were present in the initial problem. With these preliminary remarks, we go on to present the last three catastrophes.

4.6. Hyperbolic Umbilic Catastrophe

As we have already emphasized for systems of a single degree of freedom, the weakest nonlinear term of the potential V is the cubic term. Making a simple generalization from the unidimensional case, we can write the two dimension singularity as $A^3 + B^3$. As was the case in the one dimensional problem, the quadratic terms A^2 and B^2 can be eliminated through a change of variables. However, since the term AB cannot be eliminated simultaneously and since the linear terms A and B must remain in the equation, as in the unidimensional case, the generic form of the potential is:

$$V = A^3 + B^3 + wAB - uA - vB. \qquad (4.22)$$

Since we will need three parameters for the universal unfolding of $A^3 + B^3$, we can say that this catastrophe under consideration is of co-dimension 3. We have three conditions: $V_A = V_B = 0$ and the Hessian $\Delta = 0$. We have five unknowns: A, B, u, v and w. The two first conditions permit us to eliminate A and B in favor of (u, v, w), and the third condition provides the relationship between these three parameters from which we can plot the surface of the catastrophe which we call *hyperbolic umbilic*[11] (see figure 4.8 for plots in three dimensions and on the plane $(u$–$v)$).

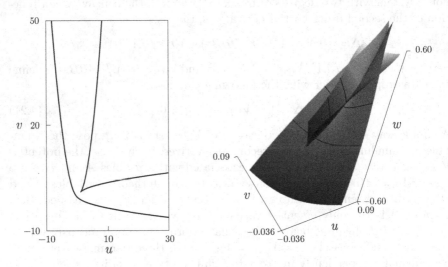

Figure 4.8 – Hyperbolic umbilic catastrophe. Left: a section corresponding to $w = 2$. Right: a representation of the catastrophe in three dimensions. [© LiPhy]

---

11. The word umbilic signifies a point of a curved surface where all the sections of a normal plane have the same radius of curvature. This catastrophe is sometimes called the "wave" catastrophe.

4.7. Elliptic Umbilic Catastrophe

In the last section, we studied the potential containing the cubic singularity of type $A^3 + B^3$. If we now include the crossed cubic terms A^2B or B^2A in the potential, we find a new catastrophe. The generic potential associated with this new category is written as:

$$V = A^3 - AB^2 + w(A^2 + B^2) + uA + vB. \qquad (4.23)$$

Only the cubic term in A^3 is included in the generic expression of the potential because, using the two identities $A^3 = A(A^2 - B^2) + B(AB)$ and $B^3 = -B(A^2 - B^2) + A(AB)$, the B^3 term can always be rewritten as a combination of the other terms already present in the expression for the potential $(A(A^2 - B^2)$ and $A^2 + B^2)$. Note that the inclusion of cross term AB^2 in the expression of the potential (equation (4.23)) necessarily implies the presence of a quadratic term in A^2 or B^2. (To convince yourself, consider the following identities: $(A^2 + B^2) + (A^2 - B^2) = 2A^2$ and $(A^2 + B^2) - (A^2 - B^2) = 2B^2$.) Since it is common practice to highlight symmetries when writing the potential (as long as we are not affecting the qualitative behavior of the solutions), the quadratic term left in the generic expression of the potential is written as the sum of these two quadratic terms $(w(A^2 + B^2))$.

Why is the potential with cubic terms $A^3 + B^3$ (equation (4.22)) not equivalent to the potential with the crossed cubic terms A^2B or B^2A (equation (4.23))? A first, simple answer is the following. If we consider the particular case of all three parameters being zero $(u = v = w = 0)$, the first potential (equation (4.22), $V = A^3 + B^3$) will have one inflection point at $A = B = 0$ $(\partial^2V/\partial A^2 = \partial^2V/\partial B^2 = 0)$ while the second potential (equation (4.23), that is $V = A^3 - AB^2$) will show that, if there exists an inflection point for A variable, the potential for B direction follows parabolic behavior; there is thus a qualitative difference between the two potentials. For a second answer, we may also verify that it is not possible to apply any change of variable in order to transform one potential to the other, making them qualitatively inequivalent.

The same approach as previously used for the hyperbolic umbilic catastrophe gives us the shape of this three dimensional catastrophe, or its projection in the parameter space (see figure 4.9), known as the *elliptic umbilic* catastrophe and sometimes nicknamed "hair" catastrophe.

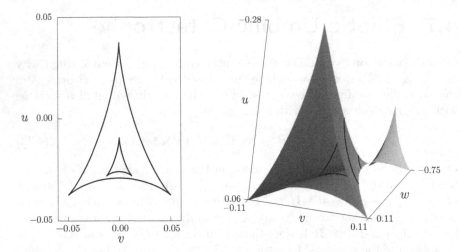

Figure 4.9 – The elliptic umbilic catastrophe. Left: the intersection of plane
(u–v) with $w = 0.5$ (external curve) and $w = 0.25$ (internal curve). Right:
a representation in three dimensions. [© LiPhy]

4.8. Parabolic Umbilic Catastrophe

We now come to the seventh and final of the elementary catastrophes. Consider a higher order polynomial for our potential V, through one of the two variables A or B. The presence of a singularity in A^4 implies not only the annulment of the second partial derivative of the potential at $A = 0$ ($V_{AA}(A = 0) = 0$) but also the annulment of the third partial derivative at the same point ($V_{AAA}(A = 0) = 0$). This gives us four conditions that must be satisfied (three conditions on the second derivatives with respect to A and B and a fourth condition on the third derivative with respect to A), meaning we must consider four independent control parameters for the universal unfolding of the singularity. Note that if we were to include an additional singularity around B^4 in the potential, we would be defining an additional condition to satisfy ($V_{BBB}(B = 0) = 0$), calling for a fifth control parameter, surpassing the limit of the four control parameters we will study in this book. With the inclusion of this one higher order polynomial, we can, at last, write the generic form of the potential associated with the seventh catastrophe:

$$V = A^4 + A^2B + wA^2 + tB^2 - uA - vB, \qquad (4.24)$$

where the cubic term in A^3 is eliminated through a change of variables. This potential defines the parabolic umbilic catastrophe, sometimes also called the "mushroom" catastrophe (see figure 4.10 for plots in three dimensional space and on a plane).

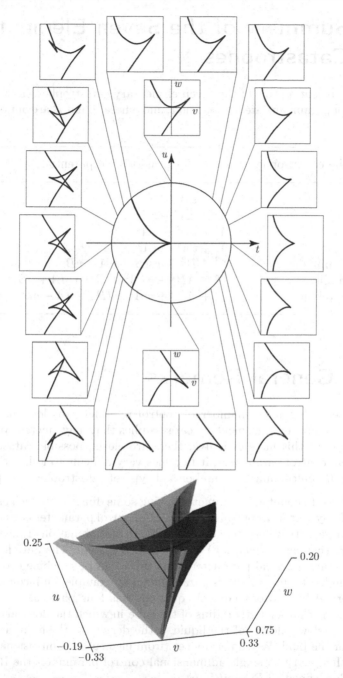

Figure 4.10 – The parabolic umbilic catastrophe. Top: a projection for different values of u and t. Bottom: a representation in three dimensions. Note that, unlike for other catastrophes, there is no simple way of representing this catastrophe as a projection. [© LiPhy]

4.9. Summary of the Seven Elementary Catastrophes

Below we present a table of the seven elementary catastrophes, in which the four control parameters are u, v, w and t, and where A and B are our dynamic variables.

Name of the catastrophe	Form of the associated potential
Fold	$A^3 + uA$
Cusp	$A^4 + uA^2 + vA$
Swallowtail	$A^5 + uA^3 + vA^2 + wA$
Butterfly	$A^6 + tA^4 + uA^3 + vA^2 + wA$
Hyperbolic umbilic	$A^3 + B^3 + wAB - uA - vB$
Elliptic umbilic	$A^3 - AB^2 + w(A^2 + B^2) + uA + vB$
Parabolic umbilic	$A^4 + A^2B + wA^2 + tB^2 - uA - vB$

4.10. General Remarks

We have seen that if the number of control parameters is less than or up to four, there are seven elementary catastrophes that may emerge. We could show that once this number is five, there are eleven possible catastrophes. Beyond five control parameters, it becomes very difficult to follow the same procedure in determining the number and type of catastrophes (see [91]).

Be careful not to count any parameter having some dimension (such as meter, second) of a system as a control parameter. A control parameter is an adimensional parameter that corresponds, generally, to a combination of the physical parameters (having a dimension) of the studied system. The initial problem may have many physical parameters, but these can be combined to make a smaller number of dimensionless parameters. For example, an incompressible fluid governed by hydrodynamical equations has four physical parameters, namely R, the characteristic radius of the tube in which the flow takes place, V_0 the mean flow velocity of the liquid, ρ the density of the fluid, and η the viscosity of the fluid. We can easily see (from elementary dimensional analysis) that there exists a single adimensional control parameter, the Reynolds number Re, defined as $Re = \rho R V_0 / \eta$.

We have classified the seven catastrophes by the number of parameters and not by the number of degrees of freedom of the system. In other words, and this is an essential point, this classification is done without having to write the equations governing the systems. One need not even bother about whether the equations of the system are known. In other words, the classification is universal and may be applied to any system. As René Thom remarked: *That we can construct an abstract, purely geometrical theory of morphogenesis, independent of the substrate of forms and the nature of the forces that create them, might seem difficult to believe, especially to the seasoned experimentalist used to working with living matter and always struggling with an elusive reality...* [101]. This idea was already expressed in the classic treatise *On Growth and Form*, published for the first time in 1917 and written by the biologist and mathematician Sir D'Arcy Wentworth Thompson (1860–1948) (see [102]).

René Thom continues this reflection and asks himself: *This general point of view raises the following obvious question: if according to our basic hypothesis, the only stable singularities of all morphogenesis are determined solely by the dimension of the ambient space, why do not all phenomena of our three-dimensional world have the same morphology? Why do clouds and mountains not have the same shape, and why is the form of crystals different from that of living beings? To this I reply that the model attempts only to classify local accidents of morphogenesis, which we will call elementary catastrophes, whereas the global macroscopic appearance, the form in the usual sense of the word, is the result of the accumulation of many of these local accidents and the correlations governing their appearance in the course of a given process are determined by the topological structure of their internal dynamics, but the integration of all these accidents into a global structure would require, if we wanted to pursue the application of the model, a consideration of catastrophes on spaces of many more dimensions that the normal three. It is the topological richness of the internal dynamics that finally explains the boundless diversity of the external world and perhaps even the distinction between life and inert matter.*

A basic postulate in the classification of catastrophes argues that the dynamics of the system are of a variational nature. In other words, the evolution equation can be written in the form of a gradient of potential V. With a single degree of freedom, it is obvious that all evolution equations of form $\dot{A} = F(A)$ can be written as $\dot{A} = -\partial V/\partial A$ where $V(A) = \int F(A)dA$. For higher dimensions, this is not generally possible.

However, we will show, through specific examples, that if the bifurcation is stationary (non-oscillatory), the evolution equation may still be written in the form of a potential in the neighborhood of the bifurcation. For an oscillating bifurcation (see the following chapter), the evolution equation cannot be written in potential form, in which case we have a so-called *general catastrophe*. The problem becomes more sophisticated, and we no longer have a simple classification to rely on. Nonetheless, adopting the formal framework of bifurcations, we can procure descriptions founded on the existing symmetries of the problem.

4.11. Solved Problem

4.11.1. Umbilic Hyperbolic Catastrophe

Write the three equations which determine the umbilic hyperbolic catastrophe in vector space (u, v, w).

Solution. Starting with potential $V(A, B)$ (equation (4.22)), we obtain the fixed point equations $\partial V/\partial A = \partial V/\partial B = 0$. We take the determinant of the Hessian matrix to zero, expressing the annulment of the potential's concavity, resulting with $\Delta = V_{AA}V_{BB} - (V_{AB})^2 = 0$ (see equation (4.21)). These conditions give us the following three relations:

$$3A^2 + wB - u = 0,$$
$$3B^2 + wA - v = 0,$$
$$36AB - w^2 = 0. \tag{4.25}$$

Suppose that parameter w is not zero ($w \neq 0$). The above equations give us the parametric equations for the two parameters u and v:

$$u = 3A^2 + \frac{w^3}{36A},$$
$$v = \frac{3w^4}{36^2 A^2} + wA. \tag{4.26}$$

We fix the value of parameter w, and vary the values for A (for example, between -2 and 2). We can calculate an ensemble of pairs (u, v) for each value of A which would allow us to trace a curve in the plane $(u–v)$ (see figure 4.8). This curve is the projection onto the plane $(u–v)$ of the umbilic catastrophe. If w is in fact zero ($w = 0$), the third condition of annulment for the potential's concavity (equations (4.25)) implies that either variable A or B is annulled: (i) if $A = 0$, the first condition of the same system (equations (4.25)) implies $u = 0$, while the second implies $v > 0$; (ii) if $B = 0$, we find $v = 0$ and $u > 0$.

To conclude, if $w = 0$, the curve of the catastrophe is the ensemble of axes u and v. Finally, note that for small enough A, v is positive regardless of the sign of A, because the first term in $1/A^2$ of the parametric equation for v (equation (4.26)) is dominant over the second term in A, while u changes sign with A. This results in a discontinuous curve in the plane of $(u–v)$; the curve is composed of two disjoint sections.

4.12. Exercises

4.1

Consider the following nonlinear equation

$$\dot{x} = ux + vx^2 - x^3, \tag{4.27}$$

where u and v are two control parameters (having arbitrary signs). Let us set $v = 0$. Which type of bifurcation do we have? Is this bifurcation structurally stable?

1. Show qualitatively that when $v \neq 0$, but is arbitrarily small, the bifurcation diagram changes qualitatively.

2. By varying v (from positive to negative by passing through zero), plot the bifurcation diagrams.

3. Provide a summary of the different solutions in the plane $(u\text{–}v)$ where you represent the lines of catastrophes. Which type of catastrophe do we have? You can follow the same type of analysis presented in the chapter.

4.2

Let us examine the van der Waals equation of state and show that the liquid-gas transition (described by that theory) corresponds to a cusp catastrophe. This equation of state reads:

$$\left(P + \frac{a}{V^2}\right)(V - b) = RT, \tag{4.28}$$

where P is the pressure, V the volume, T the temperature, and a, b and R are positive constants. The equation is cubic in V.

1. Show that the three roots of the cubic equation in V coincide when $V = V_c$, $P = P_c$, $T = T_c$, where $P_c = a/(27b^2)$, $T_c = 8a/(27bR)$ and $V_c = 3b$.

2. Show the qualitative behavior of the function $P(V)$ for $T > T_c$, $T = T_c$ and $T < T_c$. For which values of T is the function $P(V)$ univalued and for which values is it not?

3. Let us introduce the following dimensionless parameters $p = P/P_c - 1$, $\phi = V_c/V - 1$ and $\epsilon = T/T_c - 1$. Rewrite the equation of state as $F(\epsilon, p, \phi) = 0$, where F has the form:

$$F = \phi^3 + \frac{1}{3}(8\epsilon + p)\phi + \frac{2}{3}(4\epsilon - p). \tag{4.29}$$

Show that there is a double solution to $F = 0$ if the following condition is fulfilled:

$$81(4\epsilon - p)^2 = -(8\epsilon + p)^3. \tag{4.30}$$

Plot this in the $(p\text{–}\epsilon)$-plane. Which type of catastrophe do we have? For which values of p and ϵ is the catastrophe less structurally stable? Justify your answer.

4.3

Consider the following nonlinear equation:

$$\dot{x} = \epsilon x - x^2 + \nu. \tag{4.31}$$

If $\nu = 0$, which type of bifurcation do we have? The parameter ν plays the role of imperfection.

1. Plot the bifurcation diagram for $\nu > 0$ and $\nu < 0$. What is your observation on the qualitative aspect? Is the bifurcation for $\nu = 0$ structurally stable? Why?

2. Plot in the $(\epsilon\text{–}\nu)$-plane the lines of catastrophes. What type of catastrophe do we have?

3. In each distinct region obtained in question 2, plot the potential associated to equation (4.31).

4.4

Consider the same problem as the previous one but by considering the following equation:

$$\dot{x} = \epsilon x + x^2 - x^3 + \nu. \tag{4.32}$$

4.5

We would like to analyze the notion of caustics encountered in optics and show that it corresponds to a catastrophe. The general definition of caustics designates an envelope of optical rays issued from a point located either at a finite or infinite distance after having undergone a reflection on a surface. However, the term "caustic" often refers to a very bright optical spot resulting from a strong light focalization, due either to reflection or refraction. A quite common example is shown in figure 4.11 where a cup of coffee is illuminated in an oblique way.

1. Consider a cup having a circular section with radius unity and a light ray originating from a point D having $-d$ as ordinates (figure 4.12). This ray is reflected at a point M. Show that if that point has y as ordinate, then its abscissa is given by $\sqrt{1 - y^2} - 1/2$.

2. The light ray then follows the path toward a point having (X, Y) as coordinates as shown on the figure. For any point having (X, Y) as coordinates inside the circle we have a family of possible optical paths originating from point M (which has its coordinates parameterized by y, as seen above).

Figure 4.11 – A caustic which is visible on the coffee surface, illuminated in an oblique direction.

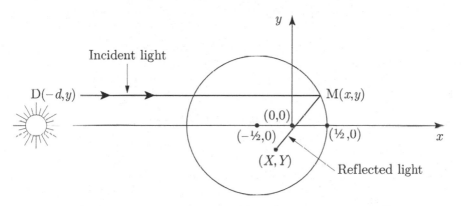

Figure 4.12 – Schematic view of the optical path (see text).

Between two points, the source and the target (the latter having (X,Y) as coordinates), several optical paths are possible. The Fermat principle (contemporary viewpoint) stipulates that the light describes the optimal path with respect to variation of paths parameterized by y. For each vertical position y where reflection occurs, there is a corresponding new path. Following the Fermat principle, we have to find the optimal length of optical rays. Show that the optical length from point D to the point of coordinate (X,Y) is given by:

$$L = d - \frac{1}{2} + \sqrt{1-y^2} + \left[\left(X + \frac{1}{2} - \sqrt{1-y^2}\right)^2 + (Y-y)^2\right]^{1/2} . \quad (4.33)$$

3. Show that, to order four, in y, and to order one in X and Y, the length L takes the form:

$$L = -\left(2X + \frac{1}{4}\right) y^4 + 2Y y^3 + 2X y^2 - 2Y y + (d + 1 - X). \qquad (4.34)$$

Because our wish is to obtain the optimal length, L may be viewed as a potential in a mechanical analogy from which we compute the derivative with respect to a variable (here y is the variable of interest). Which type of catastrophe do we have (i.e. represented by L)? Which quantity takes the role of control parameter? Explain your answer.

4. Plot the lines of catastrophes in the plane of control parameters.

5. What are the coordinates of the point where the singularity of the catastrophe is the strongest? That point also corresponds to the brightest spot of the light focalization.

Chapter 5

Hopf Bifurcation

Abstract *This chapter introduces the Hopf bifurcation (a time oscillating solution) through various examples found in electronics, population dynamics, and chemical reactions, to illustrate the diversity of this bifurcation's manifestations in nature.*

Up until now we have studied stationary bifurcations corresponding to changes in stationary solutions. This chapter describes some examples of nonlinear systems which manifest Hopf bifurcations, named after Eberhard Hopf (1902–1983). The bifurcation is characterized by a transition from a stationary solution toward an oscillatory solution (temporal oscillation).

5.1. Van der Pol Oscillator

To begin, we revisit the linear system of the electronic circuit of a resistor R, an inductor L, and a capacitor C in series (commonly known as an RLC circuit, see figure 5.1). Kirchhoff's second law stipulates that the total sum of currents in a closed circuit adds up to zero, from which we can write $V_L + V_R + V_C = 0$; $V_L = L(\dot{i}) = Ldi/dt$ is the potential difference (voltage) through the inductor with i the electric current, $V_R = Ri$ the voltage felt by the resistor, and $V_C = q/C$ the voltage with q the electric charge of the capacitor. Since the time derivative of charge q is equivalent to electric current i, we can take the time derivative of Kirchhoff's law to obtain the following:

$$(\ddot{i}) + r(\dot{i}) + \omega_0^2 i = 0, \qquad (5.1)$$

where we have defined $r = R/L$ and $\omega_0 = \sqrt{1/LC}$ (ω_0 is the oscillator's resonance frequency). This is the basis for setting up the classic problem of an RLC circuit as band-pass or band-stop bandwidth filter.

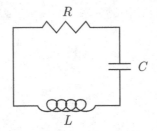

Figure 5.1 – A classic *RLC* circuit.

Without a source of current, this system is dissipative: whatever the initial conditions, over time the final current will become zero through ohmic dissipation (a phenomenon associated with the Joule effect and responsible for the heat felt from conductors). If, on the other hand, the system includes a current generator, the case is that of a forced oscillator and the flow of current is conserved. Furthermore, if the generator creates an alternating voltage (a temporal oscillation in $\sin(\omega t)$, with ω such that $\omega = 2\pi f$, where f is frequency), the linearity of the system ensures that the current in the circuit will follow the same sinusoidal behavior.

However, in most devices, there are "nonlinear" elements at the terminals and we see that the voltage is not proportional to either the current or its derivative[1]. Let us consider the circuit shown in figure 5.2 where a tunnel diode is inserted into the circuit. Unlike the classical (and ideal) diode where the current is blocked in one direction, the tunnel diode allows the current to flow in both directions.

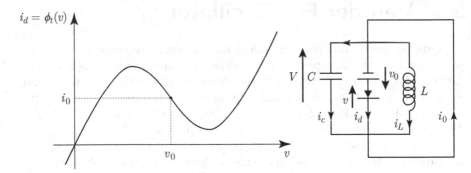

Figure 5.2 – The current against voltage curve for a circuit featuring a tunnel diode.

---

1.   Rogowski terminals are coils placed around a conductor in order that the magnetic fields produced by current induce a voltage in the conductor proportional to the derivative of the current.

This diode has a special characteristic that distinguishes it from the classical diode. The relation between the current i_d and the diode voltage v (in the normal condition of an electrical polarity) has the shape presented in figure 5.2 and is denoted as $i_d = \phi_t(v)$. An outstanding feature is the existence of a region with a negative slope in the current-voltage curve that plays an important role in what follows. Using the current conservation law, we see from figure 5.2 that we have $i_0 = i_c + i_d + i_L$ where i_c is the current across the capacitor, i_L the current across the inductance and i_d the current across the diode. Using the well-known relations $i_c = CdV/dt$, and $di_L/dt = V/L$ (with $V = v - v_0$; v_0 is the generator voltage), we obtain from the relation $i_0 = i_c + i_d + i_L$, after differentiation with respect to time, the following equation

$$C\frac{d^2V}{dt^2} + \frac{1}{L}V + \frac{di_d}{dt} = 0 \qquad (5.2)$$

with $i_d = \phi_t(v) = \phi(v - v_0) - i_0$ where $\phi_t(v)$ is the current across the diode, as represented in figure 5.2. Close to $(v = v_0, i_d = i_0)$, v is a linear function of i_d but with a negative slope, so that we write it as $i_d = -(v - v_0)/R$, where $-R$ plays the role of a local resistance and is negative (we take R to be positive). To next order we have (by assuming antisymmetry around $v = v_0$) $i_d = -(v - v_0)/R + \alpha(v - v_0)^3$ (with α a positive constant). This constitutes a good approximation to the curve presented in figure 5.2 provided that i_d and v are not very far from i_0 and v_0. Equation (5.2) becomes

$$\ddot{V} + \omega_0^2 V - \left(\frac{1}{RC} - \frac{3\alpha}{C}V^2\right)\dot{V} = 0. \qquad (5.3)$$

Setting $V = I\sqrt{\omega_0 C/(3\alpha)}$, $t\omega_0 = t'$, and $\epsilon = 1/(\omega_0 CR)$ yields (where dots will now refer to derivatives with respect to t')

$$\ddot{I} - (\epsilon - I^2)\dot{I} + I = 0. \qquad (5.4)$$

This is called the van der Pol equation. A positive value (or respectively, negative) of the control parameter ϵ corresponds to a negative resistance (or respectively, positive). In what follows we will call I a current (not to be confused by the currents defined above). It is proportional to the potential but has the dimension of a current. In addition, when we refer to a small current value, it must be remembered that we mean that v is close to v_0 and i close to i_0.

5.1.1. Qualitative Discussion

For a positive value of control parameter ϵ (remember that by convention $\epsilon > 0$ corresponds to a negative resistance $R < 0$), and considering only the linear term of the evolution equation $-\epsilon\dot{I}$ (see equation (5.4)), and further-more supposing that the current is very weak, we can write the following linear equation: $\ddot{I} - \epsilon\dot{I} + I = 0$ for which the temporal solution is an oscillating current with an exponentially increasing amplitude. For short timescales, while the amplitude of the current I is still weak, the nonlinear term may be ignored. At longer timescales, this approximation ceases to be accurate and we must look at the nonlinear term $(-(\epsilon - I^2)\dot{I})$. In the growing phase of current I, as soon as term I^2 overtakes parameter ϵ (see the second term of equation (5.4)), the coefficient of the current's time derivative $\dot{I}$ becomes positive. This term takes the role of a positive resistance which brings about a decrease in the value of I. If the term I^2 becomes weaker than parameter ϵ, the current I grows again, and so on and so forth. There is thus a compromise between term $-\epsilon\dot{I}$, which calls for an increasing I, and the antagonist term $+I^2\dot{I}$, which gives rise to a self-sustaining oscillation I, for which the amplitude is of order $\sqrt{\epsilon}$ (a result foreshadowed by the condition of changing the sign of factor $\epsilon - I^2$ in equation (5.4)). We recover the behavior previously described of an RLC circuit endowed with a tunnel diode, in which the current grows with time until the attainment of values in the region where the effective resistance $(I^2 - \epsilon)$ is positive.

5.1.2. Study of the Nonlinear Dynamics

In the third order expansion of the evolution equation (5.4), the effective resistance $(I^2 - \epsilon)$ changes sign for $I = \sqrt{\epsilon}$, indicating that current I remains small (of order $\sqrt{\epsilon}$). It makes sense for us then to search for a solution based on a power series of that same parameter, $\sqrt{\epsilon}$. For a small enough value of the parameter, the series may a priori be cut off at a certain power: the regime in which the device works is then considered weakly nonlinear. For a weakly nonlinear regime, the solution should look like a small deviation from the linear regime's oscillatory solution. We thereby put forward the hypothesis that the weakly nonlinear regime's solution is indeed also of an oscillatory nature. If our choice of solution proves to be incorrect, we will find a contradiction at one or another stage of our analysis. This type of approach, in which a hypothesis is asserted and must stand the test of self-consistency (i.e. functioning without contradiction) as a solution, is relatively common in physics and allows the physicist to proceed in elaborating a model without rigorously justifying each choice. With that said, let us proceed in our search for a solution using the following form:

$$I \simeq \rho\sin(t). \tag{5.5}$$

Note that if the oscillation is self-sustaining, the average of the dissipated power must equal zero over the period of one oscillation. Since the total energy E of the oscillator is given by:

$$E = \frac{1}{2}(\dot{I}^2 + I^2), \tag{5.6}$$

the instantaneous value for dissipated power, equal to the time derivative of energy, is written as[2]:

$$\dot{E} = \frac{1}{2}(2I\dot{I} + 2I\dot{I}) = (\epsilon - I^2)\dot{I}^2, \tag{5.7}$$

where we have used the third order expansion of the evolution equation (5.4) to obtain the second equality. The average power over a period of time T is given by the weighted integration of the instantaneous power over this interval of time, that is: $\langle \dot{E} \rangle \equiv \frac{1}{T} \int_0^T \dot{E} dt$. Now, with the help of our guessed solution for current (equation (5.5)), we can find the value of the two averages[3] $\langle \dot{I}^2 \rangle$ and $\langle I^2\dot{I}^2 \rangle$:

$$\langle \dot{I}^2 \rangle = \frac{1}{2}\rho^2 \quad \text{and} \quad \langle I^2\dot{I}^2 \rangle = \frac{1}{8}\rho^4. \tag{5.8}$$

Using these values in the expression for average dissipated power and positing that $\langle \dot{E} \rangle = 0$, given that the average dissipated power is zero, we can finally determine the value of amplitude ρ of current I:

$$\rho \simeq \pm 2\sqrt{\epsilon}. \tag{5.9}$$

5.1.3. Hopf Bifurcation

We can now state that the solutions[4] for the third order expansion for the current's evolution equation (5.4) are:

$$\begin{cases} \epsilon < 0, & I = 0, \\ \epsilon > 0, & I = \pm 2\sqrt{\epsilon}\sin(t). \end{cases} \tag{5.10}$$

Indeed, when the control parameter ϵ is negative, the effective resistance is positive and current I decreases over time until it reaches zero; when the control parameter is positive, the current oscillates with amplitude $2\sqrt{\epsilon}$.

The threshold value $\epsilon = 0$ of this control parameter thus marks the frontier where we see a qualitative change in solutions. It represents a bifurcation point where the stationary solution $I = 0$ bifurcates into two branches, each a symmetric solution that oscillates in time. In actuality, all solutions with $\sin(t+\theta)$ (with θ representing one phase) are also valid; the solutions $\pm\sin(t)$ are for the particular case of $\theta = 0, \pi$. We can thus say that there exists a

---

2. Note that, by analogy, we recognize in the previous expression a form similar to that for mechanical energy, equal to the sum of kinetic and potential energies.

3. Rigorously speaking, the averages evaluated through this method are not exact, because the oscillation period is modified by nonlinear effects. We could take the displacements caused by the nonlinear effects into account by including higher order terms in our calculation of the averages.

4. Note that we are interested in the solutions valid over long timescales, ignoring transitory solutions.

continuous family of solutions parameterized by phase θ. This is our first example of a *Hopf bifurcation*.

The zero current solution $I = 0$ is stable when control parameter ϵ is negative ($\epsilon < 0$). If we linearize the third order expansion of the evolution equation (5.4) we find: $\ddot{I} - \epsilon \dot{I} + I = 0$, with a solution of exponential form $e^{\epsilon t}$ (and multiplied by an oscillating function). For a negative parameter ϵ (characterizing a stable solution), the current decreases over time, approaching zero, but for positive ϵ (characterizing an unstable solution), the current tends toward infinity (in the mathematical sense). The loss of stability of solution $I = 0$ coincides with the emergence of oscillating solutions (see figure 5.3, the bifurcation diagram plotting the behavior of amplitude ρ as a function of parameter ϵ).

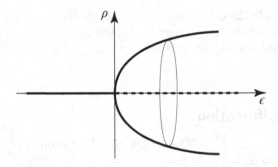

Figure 5.3 – Qualitative bifurcation diagram for the Hopf instability. We show the stable and unstable branches. The amplitude ρ of the limit cycle grows as $\sqrt{\epsilon}$. The phase θ grows with time, due to the oscillation of I. To highlight this aspect (and especially to highlight that we are working with a problem that has two degrees of freedom), we plot a cycle in the third direction in order to indicate that I is a periodic function of time.

5.1.4. Phase Space and Introduction of the Notion of a Limit Cycle

It is often useful to represent the system's dynamics using *phase space*. For example, in mechanics, for a point particle of one dimension, we only need two initial conditions, velocity and position, to entirely determine the particle's trajectory. We can choose to represent the evolution of the system by writing position and velocity as functions of time t, $x(t)$ and $\dot{x}(t)$ respectively, but if instead we decide to eliminate the temporal variable t altogether between position and velocity, we obtain a direct relationship between the position x and velocity $\dot{x}$. We can then use them as the coordinate bases to make a plane and plot the trajectory in the space (called phase space) of $(x, \dot{x})$.

Newton's equation ($m\ddot{x} - F(x, \dot{x}) = 0$, with mass m, position variable x, and force F, a function of position x and velocity $\dot{x}$) is, like the van der Pol equation (5.4), a second order differential equation with respect to time. As such, the phase space is constructed from a function x and its derivative, $\dot{x}$. To determine a solution for a dynamical system described by a differential equation of order n, there must be n initial conditions. In other words, representing a trajectory in phase space reduces to tracing a trajectory in an n dimensional space, represented by variable x and its $(n-1)$ derivatives. If the dynamics are given directly by n first order equations, the phase space is composed of the n variables appearing in the system of n equations.

For our consideration of the van der Pol oscillator, we posit $X = I$ and $Y = \dot{I}$. The third order evolution equation for current (5.4) can be rewritten as two first order equations with respect to time, defining the pair (X, Y):

$$\dot{X} = Y \tag{5.11}$$

$$\dot{Y} = (\epsilon - X^2)Y - X. \tag{5.12}$$

Eliminating temporal variable t between X and Y allows us to plot a trajectory in the phase space (X, Y). For a small value of control parameter ϵ, we have seen in this chapter (equations (5.5) and (5.9)) that the solution of the van der Pol system[5] is of the form:

$$X = I = 2\sqrt{\epsilon}\sin(t). \tag{5.13}$$

Therefore we can express variable Y, which is equal to the time derivative of variable X, as:

$$Y = \dot{I} = 2\sqrt{\epsilon}\cos(t). \tag{5.14}$$

In combining the two last equations to eliminate the time variable, we can write the following:

$$I^2 + \dot{I}^2 = X^2 + Y^2 = 4\epsilon. \tag{5.15}$$

This equation describes a circle of radius $2\sqrt{\epsilon}$. As time tends to infinity, the phase space trajectory tends toward this circle irrespective of initial conditions (see figure 5.4). It is as if the dynamics are pulled into this trajectory: this is called a *limit cycle*[6], which is a special attractor of dynamics. For an initial condition corresponding to a point within the circle, amplitude of X grows until reaching the circle's edge. If, inversely, the initial condition refers us to a point outside the circle, amplitude X decreases until it reaches a point on the circle (see figure 5.4 for a schematic representation of the trajectory in the phase space).

---

5. We keep the positive value solution here; the other solution brings us to the exact same conclusion.

6. In the most general terms, a limit cycle is any curve closed on itself; a circle constitutes one particular case of a limit cycle.

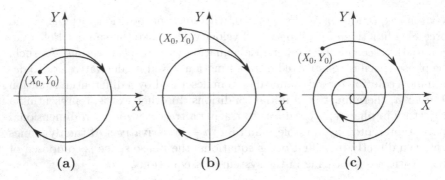

Figure 5.4 – (a) and (b): $\epsilon > 0$, giving way to a stable limit cycle. The evolution toward the limit cycle is represented for two different initial conditions, denoted by (X_0, Y_0). (c): $\epsilon < 0$, there exists one stable fixed point at $(0,0)$.

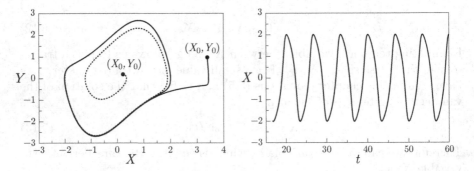

Figure 5.5 – Left: limit cycle for $\epsilon = 1$ with two different initial conditions. Whatever the initial condition (solid or dotted line), the solution tends toward the limit cycle. Right: the behavior of $X(t)$ showing a clear deviation from the sinusoidal function.

We have seen that for a negative value of the control parameter ($\epsilon < 0$), and for any initial amplitude, variables X and Y will decrease over time until they reach the origin $(0,0)$. Here the origin acts as an attractor, or stable fixed point.

What happens when ϵ is not small enough? In that case, either (i) we must expand the equation one order more with respect to ϵ, or (ii) use a numerical calculation. The numerical results are shown in figure 5.5. We see that if ϵ is big enough, the current $I(t)$ is no longer a function of the type $\sin(t)$. The nonlinearities generate harmonics of a higher order ($\sin(2t)\dots$).

5.2. The Prey-Predator Model, an Example of Population Dynamics

The famous prey-predator model of Lotka–Volterra (LV) (see [64, 108]), is a nonlinear model describing the interaction between two biological species where one, the predator species, grows at the expense of the other, the prey species. The LV model is described by the following system of equations:

$$\dot{N} = a_1 N (1 - b_1 M)$$
$$\dot{M} = -a_2 M (1 - b_2 N) \qquad (5.16)$$

where M and N represent, respectively, the number of predators and preys at a given time (the variables $M(t)$ and $N(t)$ are time-dependent), and a_1, a_2, b_1, and b_2 are positive constants. In the absence of a predator ($M = 0$), the prey population (first equation above) grows with time; the characteristic time of this evolution is equal to $1/a_1$.

The first term of the first equation ($a_1 N$) describes the increase, with rate $1/a_1$, in the number of preys for the situation of no predators (i.e. $M = 0$). The second term ($-a_1 b_1 NM$) describes the decrease in the prey population in the presence of predators. Similarly, the first term of the second equation expresses the decline, with rate $1/a_2$, of the predator population in the absence of preys; the second term $a_2 b_2 NM$ represents the survival of predators as may be ascertained by the available preys. In spite of this simple formulation, this model shows interesting behavior, which we will explore in the following pages.

5.2.1. Essential Results of the Lotka–Volterra Model

The system of equations (5.16) for the LV prey-predator model admits two fixed points defined by the condition of a zero first derivative $\dot{M} = \dot{N} = 0$. The first fixed point is trivial ($M = N = 0$) and the second fixed point, denoted M_0 and N_0, is given by:

$$M_0 = \frac{1}{b_1}, \quad N_0 = \frac{1}{b_2}. \qquad (5.17)$$

We begin our study of linear stability by introducing small perturbations M_1 and N_1 around fixed points M_0 and N_0, respectively, as follows:

$$M(t) = M_0 + M_1(t), \quad N(t) = N_0 + N_1(t). \qquad (5.18)$$

We assume that perturbations M_1 and N_1 are small enough that we may exclude any nonlinear terms. Thus, including only the linear contributions from M_1 and N_1 found in the dynamical equation (5.16), we may write:

$$\dot{N}_1 = a_1 N_1 - a_1 b_1 M_0 N_1 - a_1 b_1 N_0 M_1,$$
$$\dot{M}_1 = -a_2 M_1 - a_2 b_2 N_0 M_1 N_1. \tag{5.19}$$

The resulting system of linear differential equations with constant coefficients is called autonomous (in that the coefficients are independent of time). The form of the general solution for the preceding equation (5.19) is as follows:

$$M_1(t) = Ae^{\omega t}, \quad N_1(t) = Be^{\omega t}, \tag{5.20}$$

with constants A and B, and parameter ω, a priori a complex number, to be determined. If parameter ω has a real positive part, the term $e^{\omega t}$ grows with time and the solution (equation (5.20)) is unstable.

Substituting the general solution (equation (5.20)) into the linear system of equations (5.19), we obtain a homogeneous system with two algebraic linear equations of constants A and B. For the non-trivial solution (i.e. $(A, B) \neq (0, 0)$), the determinant of the system must be zero. The condition of a null determinant defines the equation we call the dispersion equation, relating parameter ω to the physical parameters. Applied at the non-trivial fixed point M_0 and N_0 (equation (5.17)), the dispersion equation is written as:

$$\omega^2 = -a_1 a_2 < 0. \tag{5.21}$$

Since the constants a_1 and a_2 are positive, the parameter $\omega = \pm i\sqrt{a_1 a_2}$ is a purely imaginary number. When the real part of ω equals zero, the solution (M_0, N_0) is neither stable nor unstable; it is neutral. Perturbation $e^{\omega t}$ neither grows nor diminishes over time, and the solution simply oscillates, keeping the same initial amplitude. In the neighborhood of the fixed points, we obtain the following solutions for $M = M_0 + M_1$ and $N = N_0 + N_1$:

$$M = M_0 + Ae^{i\sqrt{a_1 a_2}t} + c.c.$$
$$N = N_0 + Be^{i\sqrt{a_1 a_2}t} + c.c., \tag{5.22}$$

where the abbreviation $c.c.$ stands for "complex conjugate" (M and N are real quantities). Using the system of equations solved with constants A and B, we can determine the constants' ratio: $A/B = \pm i b_2/(a_1 b_1) \equiv \pm iC$, with $C = b_2/(a_1 b_1)$.

This reveals that the numbers of preys N and predators M oscillate one fourth out of phase (i.e. they evolve with a phase difference equal to $\pi/2$). In the linear regime, the amplitudes A and B cannot be determined (since the linear equations are invariant when parameters M and N are multiplied by arbitrary constants).

If we define $B = \nu$ (and thus $A = \pm i\nu C$), we can rewrite the solutions in the neighborhood of the fixed point (see equation (5.22)) as:

$$M = M_0 \pm 2C\nu \sin(\sqrt{a_1 a_2} t),$$

$$N = N_0 + 2\nu \cos(\sqrt{a_1 a_2} t). \qquad (5.23)$$

From this we can easily find the relationship between M and N:

$$(M - M_0)^2/C^2 + (N - N_0)^2 = 4\nu^2.$$

This describes an ellipse (which becomes a circle if $C = 1$; in this case the radius of the circle is 2ν and its center is found at (M_0, N_0) in the plane $(M\text{–}N)$).

The result brings to mind the van der Pol system (which also has an oscillating solution), but actually the situation is drastically different. Here, the final trajectory in phase space (equal to radius of the circle, when $C = 1$) depends on the initial conditions. As we shall see, when the initial conditions change, the closed trajectory changes. There are as many closed trajectories as initial conditions. In other words, there exists a continuum of possible cycles. This assertion can be inferred from the fact that the eigenvalue ω is purely imaginary, but we shall give a more general proof below. On the other hand, our study of the van der Pol system showed its limit cycle to be an intrinsic property of the system, determined by the physical parameters characterizing the system and not by any independent initial condition. Thus, unlike the LV system, any initial condition will result in the same limit cycle for the van der Pol oscillator (see figure 5.5). We can show (see next chapter) that the eigenvalue of the van der Pol oscillator has both real and imaginary parts, and this is the source of the main difference between this and the LV model. The presence of a non-zero real part is related to dissipation. It is equally of interest to note that the LV prey-predator system does not have any bifurcation, while for the van der Pol system, a dissipative damped system, the trivial solution exists for any control parameter ϵ and becomes unstable for positive values of ϵ ($\epsilon > 0$).

The problem described by the LV model is similar to that of a frictionless oscillator (a conservative system). Conservative systems display non-dissipative oscillations whose amplitudes depend on initial conditions. By way of example, the frictionless pendulum system conserves its energy, and the maximum angle with respect to the vertical (maximum amplitude) is equal to the initial angle from the vertical (if we let go of the mass with zero initial velocity). Since the LV model also has a conservation law, we can observe the same type of phenomena. To see this, let us rewrite the initial system of equations (5.16) of the LV model by taking out the temporal variable, as follows:

$$\frac{\dot{M}}{\dot{N}} = \frac{dM}{dN} = -\frac{a_2 M}{a_1 N} \frac{1 - b_2 N}{1 - b_1 M}, \qquad (5.24)$$

wherein the solution is given by:

$$a_1 \ln(M) + a_2 \ln(N) - b_1 a_1 M - a_2 b_2 N = K, \qquad (5.25)$$

where K is an integration factor (analogous to a constant of motion, such as the energy for the perfect pendulum).

If we expand the equation around the fixed point (M_0, N_0), equation (5.25) reduces to $a_1 b_1^2 (M - M_0)^2 + a_2 b_2^2 (N - N_0)^2 = Cte$. This is the equation of an ellipse, in agreement with the linear theory delineated above. If we plot the trajectories in the phase space determined from the time-independent solution (equation (5.25)), we confirm that the LV model allows for an infinite number of closed orbits corresponding to the infinite possible initial conditions (see figure 5.6). This behavior contrasts strongly with that of the van der Pol system, which admits only a single, unique orbit toward which all the trajectories, with any initial conditions, will eventually fall.

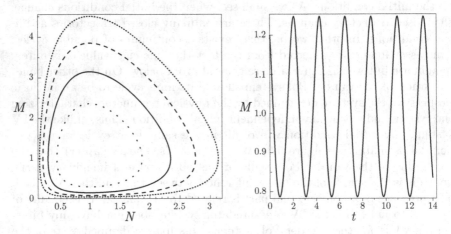

Figure 5.6 – Left: a phase space portrait of the LV model. From the outside toward the interior of the cycle, we have $K = -10, -9, -8, -7$. We have chosen the values for parameters $a_1 = 2$, $a_2 = 4$, $b_1 = b_2 = 1$. Right: $M(t)$ as a function of time. $N(t)$ presents the same qualitative behavior.

5.2.2. A More Realistic Dynamical Model of Populations with a Limit Cycle

The LV model stipulates that in the absence of a predator, the prey population grows indefinitely (equation (5.16) for N, with $M = 0$); this approximation is only valid when the number of preys N is small. Beyond a certain population size, this hypothesis is no longer valid, if only because of the limitation of resources available to the preys for nourishment.

The basic LV model has been correspondingly improved upon in order to create more realistic models. The most popular of these is the "logistic" model, which adds a nonlinear term to the linear term of our base equation (equation (5.16)), creating a regulating contribution of the form $N(1 - N/\bar{N})$. This term is said to be logistic, with $\bar{N}$ the threshold above which the population will decrease. Thus, for small N, such as $N \ll \bar{N}$, the term $N/\bar{N}$ is negligible with respect to the whole and we find the base equation of the LV model associated to a regime of unlimited growth. Beyond the limit $N = \bar{N}$, the term $(1 - N/\bar{N})$ changes sign and the population decreases, and so on. The point characterized by $N = \bar{N}$ is a fixed stable point for the dynamics of the prey population.

There is a second refinement of the LV model in which the logistic term is replaced by a contribution whose form is inspired by field observations (see [43]): $N^2(1 - N/\bar{N})/(N + c)$, with c a constant (for $c = 0$, we have again the logistic model). The system of coupled equations for this prey-predator model is written as follows:

$$\dot{N} = \frac{a_1 N^2(1 - N/\bar{N})}{N + c} - a_1 b_1 N M,$$

$$\dot{M} = -a_2 M(1 - b_2 N). \tag{5.26}$$

Introducing dimensionless variables: $N' = N/\bar{N}$, $M' = Mb_1$, and $t' = a_1 t$, the equations take the following form:

$$\dot{N} = \frac{N^2(1 - N)}{N + n} - NM,$$

$$\dot{M} = -\gamma M(m - N), \tag{5.27}$$

where we have written that $n = c/\bar{N}$, $m = 1/(\bar{N}b_2)$ and $\gamma = a_2 b_2 \bar{N}/a_1$. We have kept the same notation for variables M and N (instead of variables with prime) to avoid complication. The system of equations (5.27) has a fixed point given by:

$$N_0 = m, \quad M_0 = \frac{m(1 - m)}{n + m}, \tag{5.28}$$

with the condition $m > 1$ (the variables N and M are positive). We designate the deviations from points M_0 and N_0 as M_1 and N_1 respectively, and study the linear stability of the fixed points by searching for the new solutions to the linear equations in the form: $M_1 = Ae^{\omega t}$ and $N_1 = Be^{\omega t}$.

By introducing these expressions into the above system of coupled equations (5.27), we can then write in matrix form: $\mathcal{M}X = \omega X$, where $\mathcal{M}$ is a 2×2 matrix and X is the column vector A, B. We obtain the dispersion relation from the condition (in order to avoid the trivial solution $A = B = 0$) of a zero value determinant for the matrix $\mathcal{M} - \omega \mathcal{I} = 0$ ($\mathcal{I}$ is the identity matrix):

$$\omega^2 - Tw + D = 0, \tag{5.29}$$

where T is the trace of matrix $\mathcal{M}$ and D its determinant. We denote the two eigenvalues of the matrix $\mathcal{M}$ as $\omega_\pm$, generally complex numbers, and then use these to express trace $T = \omega_+ + \omega_-$ and the determinant $D = \omega_+\omega_-$. To simplify the algebraic calculations without changing the main results, we restrict our discussion to the case $\gamma = 1$ and explicitly express the trace and the determinant of the matrix as a function of the problem's parameters:

$$T = \frac{nm - 2nm - m^3}{(n+m)^2}, \quad D = \frac{m^2(1-m)}{n+m}. \tag{5.30}$$

Since the condition $m > 1$ must be met to ensure that the number of predators is positive ($M_0 > 0$), the determinant of the matrix must also be positive ($D > 0$). Furthermore, the trace is positive ($T > 0$) for all parameter n such that

$$n > n_c \equiv \frac{m^2}{1 - 2m}. \tag{5.31}$$

When the two quantities, determinant and trace, are positive ($D > 0$ and $T > 0$), the real part of each of the two eigenvalues is positive and the fixed point (M_0, N_0) is unstable. Inversely, if the parameter n is such that $n < n_c$, the real part of both eigenvalues is negative, and the fixed point becomes stable. We therefore determine that the parameter value $n = n_c$ marks the critical threshold at which an instability forms, defining a bifurcation point. Since the two parameters n and m are linked, there is actually a bifurcation line in the plane (n–m). For $n = n_c$, the trace is zero ($T = 0$), and the dispersion relation (equation (5.29)) can be simplified to let us determine the eigenvalue:

$$\omega^2 = -D = -\frac{m^2(1-m)}{n+m} \equiv -\omega_0^2 < 0. \tag{5.32}$$

Thus, at the bifurcation, the real part of the eigenvalues is zero, as we might expect, while the imaginary part is finite. This is a Hopf bifurcation, and the critical parameter $n = n_c$ corresponds to the Hopf bifurcation point. Unlike the LV model, where parameter ω is a purely imaginary number, this improved model involves a complex ω consisting of a non-zero real part, except for at the bifurcation point – again, the characteristic of any Hopf bifurcation.

Plotting the trajectories in phase space reveals two different behaviors, dependent on the position of parameter n with respect to the bifurcation point. When the parameter n is smaller than bifurcation point n_c, the fixed point (M_0, N_0) of the dynamics is stable and the trajectory will trace a spiral motion around the fixed point for any initial condition. The fixed point defines a "spiral sink". When parameter n is greater than n_c, the trajectory explodes or gets away from the fixed point and the fixed point becomes a "spiral source". Will this outward spiral end in a closed limit cycle? One of two general cases is possible: (i) the trajectory falls into a limit cycle, as in the case of the van der Pol oscillator (see figure 5.4); or (ii) no saturation occurs and the spiral continues outward ad infinitum.

In the case of the population dynamics model described here, the behavior of the phase space trajectories depends on the values of m and n. For small values of both parameters, it is scenario (i) that prevails, while scenario (ii) occurs at large values of m and n. This result is derived from a numerical analysis of the model (not treated here), but it is also common to use an analytic approach to analyze the phase space trajectories (see next chapter).

5.3. Chemical Reactions

Found everywhere in nature and used within many domains (pharmaceutical industry, geology, biology, etc.), chemical reactions play an essential role in nonlinear science. Not only do chemical reactions create temporal instabilities (notably, many examples of the Hopf bifurcation), but they are also main actors in spatial organization. Morphogenesis, the birth of spatial structure, is one of the critical aspects in the study of nonlinear phenomena.

Before delving into an in-depth account of the nonlinear chemical reactions' behaviors, let us go over some of the key concepts of their descriptions.

5.3.1. Law of Mass Action

Consider a chemical reaction during which the chemical species, denoted A_i, ($i = 1, \ldots, n$), interact with each other to create new species denoted A_i', ($i = 1, \ldots, m$).

If, after some time, the system made up of species A_i and A_i' ceases to evolve, we say we have reached a state of chemical equilibrium. This equilibrium is only possible if the chemical reaction happens in both directions at comparable rates; that is, if it is a reversible reaction. Accordingly, we may describe the reaction with the following symbolic equation:

$$N_1 A_1 + N_2 A_2 + \cdots \rightleftharpoons N_1' A_1' + N_2' A_2' + \cdots, \tag{5.33}$$

where parameters N_i and N_i', called stoichiometric coefficients, represent the number of moles in species A_i and A_i', and where the symbol $\rightleftharpoons$ refers to the fact that the reaction happens in both directions. Let us denote the molar concentrations of each species as C_i and C_i' respectively.

The law of mass action, a law of thermodynamics, can be applied to a chemical reaction in equilibrium:

$$\frac{C_1'^{N_1'} C_2'^{N_2'} \cdots}{C_1^{N_1} C_2^{N_2} \cdots} = K, \tag{5.34}$$

where K is a function of temperature called the *equilibrium constant*. As an example, let us look at the creation of ammonia in the following chemical reaction:

$$N_2 + 3H_2 \rightleftharpoons 2NH_3. \tag{5.35}$$

The law of mass action for this reaction is written as:

$$\frac{[NH_3]^2}{[N_2][H_2]^3} = K, \tag{5.36}$$

where $[NH_3]$ refers to the concentration of species NH_3, etc.

5.3.2. Reaction Kinetics (also Known as Chemical Kinetics)

For cases in which the reaction is irreversible, which is to say that one direction of the reaction is favored with respect to the other, we replace the symbol $\rightleftharpoons$ by $\rightarrow$ in our notation:

$$N_1 A_1 + N_2 A_2 + \cdots \rightarrow N_1' A_1' + N_2' A_2' + \cdots \tag{5.37}$$

Unlike reactions in equilibrium, characterized by a constant exchange between species, in both directions and at comparable rates, one may distinguish irreversible reactions by uneven directionality and differing rates of exchange between the two species. As such, it is very useful to use reaction kinetics, in which we study the speed of the reaction between initial reactants A_i and the products A_i', to analyze these chemical reactions.

To begin, we note down the number of moles in reactants A_i and A_i' at a given moment as n_i and n_i'. Over the course of the reaction, we can say that for a duration of time dt the number of moles n_i diminishes at a rate of $-dn_i$, while the number of moles n_i' grows as dn_i'. Conservation of mass requires that:

$$-\frac{dn_1}{N_1} = -\frac{dn_2}{N_2} = \cdots = \frac{dn_1'}{N_1'} = \frac{dn_2'}{N_2'} = \cdots \equiv d\lambda, \tag{5.38}$$

where λ is a parameter that quantifies the progress of the reaction.

We use v_i to denote the rate of conversion of species A_i and v_i' for the reaction rate of species A_i' (units of the reaction speed, hereafter called speed, expressed in moles per second[7]), so we have:

$$v_i = -\frac{dn_i}{dt} = N_i \frac{d\lambda}{dt}, \quad v_i' = \frac{dn_i'}{dt} = N_i' \frac{d\lambda}{dt}, \ldots \quad (5.39)$$

We can relate the different rates to each other as follows:

$$\frac{v_1}{N_1} = \frac{v_2}{N_2} = \cdots = \frac{v_i}{N_i} = \cdots = \frac{v_1'}{N_1'} = \frac{v_2'}{N_2'} = \cdots = \frac{v_i'}{N_i'} = \cdots . \quad (5.40)$$

Since each rate is equal to the rest, it is possible to define the rate of conversion, v, independently of the choice of a particular species (this is fairly intuitive, since the reaction only occurs when all the species making up the reaction are reacting together). We obtain the following relationship:

$$v = -\frac{1}{N_1}\frac{dn_1}{dt} = -\frac{1}{N_2}\frac{dn_2}{dt} = \cdots \frac{1}{N_1'}\frac{dn_1'}{dt} = \frac{1}{N_2'}\frac{dn_2'}{dt} = \frac{d\lambda}{dt}, \quad (5.41)$$

where v is equal to the time derivative of the parameter λ defined above.

Furthermore, since concentration C_i of substance A_i is written as $C_i = n_i/V$, where V is the volume in which the reaction takes place, we have:

$$\frac{dC_i}{dt} = -\frac{v_i}{V}. \quad (5.42)$$

Armed with these preliminary expressions, we can now proceed to write the equations for the chemical reactions by the introduction of coupled nonlinear terms in concentrations C_i and C_i'.

5.3.3. Nonlinear Evolution Equations

To clarify some points, we begin with a simple example – the hydrogenation of nitric oxygen[8]:

$$2NO + 2H_2 \longrightarrow N_2 + 2H_2O \quad (5.43)$$

Experiments show that if the partial pressure of nitric monoxide NO is fixed, then varying the partial pressure of molecular hydrogen H_2 will result in a proportional variation of the reaction rate. Say we vary the partial pressure of hydrogen in a successive manner, in increments p_0, $2p_0$, and $3p_0$, the reaction rate will be v, $2v$, and $3v$, respectively. In other words, the reaction

---

7. It is common for the rate to be defined with respect to the variation of the volumetric concentration and not with respect to the number of moles. In this case $v \sim dC_i/dt$; consequently, since the concentration is expressed in moles per volume, the rate is expressed (in international units) as mol m^{-3} s^{-1}.

8. In the absence of a catalyst, the reaction requires a high temperature of order 800°C.

rate increases with the partial pressure. The ideal gas law guarantees a linear relationship between pressure and concentration[9], which means that the reaction grows linearly with the concentration of molecular hydrogen H_2.

If instead the partial pressure of the molecular hydrogen H_2 is kept constant, we can vary the partial pressure of the nitric monoxide NO in the same fashion, but this time we observe the successive values of v, $4v$, and $9v$ for the rate of the reaction. This implies that the reaction rate grows quadratically with the concentration of N_2.

From these observations we may write:

$$v = k\,[H_2]\,[NO]^2, \tag{5.44}$$

where k is a reaction rate constant depending on temperature (note that the rate is not given by quantity $k\,[H_2]^2\,[NO]^2$, as the equation describing equilibrium would imply). This type of reaction is considered complex because the coefficients of the reaction rate are not the same stoichiometric coefficients corresponding to the chemical species in the reaction equation. A simple reaction rate would be:

$$v = kC_1^{N_1}C_2^{N_2}\ldots, \tag{5.45}$$

where k is the reaction constant.

It is also important to note that most chemical reactions happen in various stages. If one had to take each stage into account, the problem would be too complex to solve in any simple manner, but luckily many simplifications are possible. One of the most important is that of adiabatic elimination, in which the fast variables are slaved to the slow variables (i.e. the fast variables adapt quickly), letting us focus our attention purely on the slow reactions. This is a similar concept to the adiabatic elimination already described in subsection 2.3.3.

5.3.4. The Brusselator

In this section we study the theoretical model for a chemical reaction that is simple but exhibits a rich variety of behaviors – the Brusselator. The name, which was communally decided upon, comes from the name of Belgium's capital, Brussels, where chemist Prigogine (1917–2003) and his team, having received the 1977 Nobel Prize in Chemistry for their work on dissipative structures, introduced the model.

---

9. For most cases, a good approximate for the equation of state for the concentration found in volume V of a reaction is that of a perfect gas, $P = nRT/V$, where P is pressure, T is temperature, and R is the ideal gas constant.

First we designate the concentration of the two essential species (the slow variables) as X and Y. The chemical reaction takes place in four stages, as follows:

$$A \to X, \text{ of reaction constant } k_1;$$
$$X \to \text{products, of reaction constant } k_2;$$
$$2X + Y \to 3X, \text{ of reaction constant } k_3;$$
$$B + X \to Y, \text{ of reaction constant } k_4. \tag{5.46}$$

The term "products" in the second stage of equation (5.46) refers to intermediate products in the reaction. We can ignore them because species A and B are the intermediary products playing the pivotal role in the reaction. The evolution equation for variable X is simple: X is produced by A, whose reaction speed is $k_1[A]$, where $[A]$ is the concentration of species A (from now on, we will omit the brackets when designating concentrations). The species X is converted at a rate characterized by reaction constant k_2, giving way to an intermediate product of no importance. This contributes $-k_2X$ to the reaction rate. Afterward, in the third stage, two moles of species X react with species Y to create $3X$; this contributes k_3X^2Y to the reaction rate. Finally, species X produces Y with reaction constant k_4. Adding everything together we have the following differential equation to describe the evolution of the concentration of species X (noted simply as X):

$$\dot{X} = k_1A - k_2X - k_4BX + k_3X^2Y. \tag{5.47}$$

We can determine the evolution equation for concentration Y from the same reasoning:

$$\dot{Y} = k_4BX - k_3X^2Y. \tag{5.48}$$

By using the following change of variables:

$$X = \sqrt{\frac{k_2}{k_3}}\bar{X}, \quad Y = \sqrt{\frac{k_2}{k_3}}\bar{Y}, \quad t = \frac{\bar{t}}{k_2}, \tag{5.49}$$

we obtain the coupled evolution equations for X and Y:

$$\dot{X} = a - X - bX + X^2Y,$$
$$\dot{Y} = bX - X^2Y, \tag{5.50}$$

with the two parameters a and b defined by:

$$a = A\sqrt{k_2k_3}, \quad b = \frac{k_4B}{k_2}. \tag{5.51}$$

Note that we have kept the original notation for X, Y and t (instead of the variables with a bar). The fixed point of these equations is given by:

$$X_0 = a, \quad Y_0 = \frac{b}{a}. \tag{5.52}$$

A study of the fixed point's stability (see examples in previous subsection 5.2.1 of linear stability analyses for procedure) results in the following characteristic equation:

$$\omega^2 - \omega T + D = 0, \tag{5.53}$$

where the quantity ω gives us the values for the stability matrix, the trace of the matrix $T = b - (1 + a^2) < 0$, and the determinant $D = a^2$.

When parameter b is equal to the critical value $b_c = (1 + a^2)$, the eigenvalue of $\omega = \pm ia$ is a purely imaginary number, describing an oscillatory behavior over time for concentrations X and Y. For values of parameter b that are above critical value b_c, the trace of the stability matrix is positive ($T > 0$). Taking into account the fact that determinant D of the matrix is always positive, the two eigenvalues (complex conjugates of one another) each have a positive real part, and so the fixed point loses its stability. For values of parameter b below the critical value b_c ($b < b_c$), the real part of both eigenvalues is negative and the fixed point is stable again. The critical value of parameter $b = b_c$ marks the frontier of a Hopf bifurcation.

5.4. Exercises

5.1

Consider the following system of nonlinear equations:

$$\dot{x} = \mu x - y + xy^2$$
$$\dot{y} = x + \mu y + y^3. \tag{5.54}$$

1. Show that this system presents a Hopf bifurcation. Determine the condition on μ for which there is a Hopf bifurcation.

2. Is this bifurcation subcritical or supercritical?

5.2

Consider a modified version of the van der Pol oscillator:

$$\ddot{x} + \mu(x^2 - 1)\dot{x} + x = a. \tag{5.55}$$

Find, in the $(\mu\text{–}a)$-plane, the location of trajectory where there is Hopf bifurcation.

5.3

Consider the following model system for prey-predator [79]:

$$\dot{x} = x[x(1 - x) - y]$$
$$\dot{y} = y(x - a), \tag{5.56}$$

where $x > 0$ and $y > 0$ represent the number (in a dimensionless form) of preys and predators, respectively, and a is a control parameter, taken to be positive.

1. Show that $(0,0)$, $(1,0)$ and $(a, a − a^2)$ are fixed points.

2. Show there exists a Hopf bifurcation for $a_c = 1/2$.

5.4

Any system undergoing a Hopf bifurcation can be represented at the bifurcation point by the following system (after an adequate change of variables):

$$\dot{x} = −\omega y + f(x, y)$$
$$\dot{y} = \omega x + g(x, y), \tag{5.57}$$

where f and g are nonlinear functions of x and y vanishing at the origin $(0, 0)$.

1. Provide a proof for the above statement.

2. One can show that the bifurcation nature (subcritical or surpercritical) can be determined by examining the sign of the following quantity (see [40]):

$$16a = f_{xxx} + f_{xyy} + g_{xxy} + g_{yyy}$$

$$+ \frac{1}{\omega} [f_{xy}(f_{xx} + f_{yy}) − g_{xy}(g_{xx} + g_{yy}) − f_{xx}g_{xx} + f_{yy}g_{yy}], \tag{5.58}$$

where the subscripts designate derivatives (as we have often adopted in this book). We have (i) a supercritical bifurcation if $a < 0$, and (ii) a subcritical bifurcation if $a > 0$. Determine a for the system $\dot{x} = −y + xy^2$, $\dot{y} = x − x^2$. It is assumed here and below that the bifurcation refers to the point $(x, y) = (0, 0)$. What can you conclude?

3. Which type of bifurcation do we have for the system $\dot{x} = \mu x − y + xy^2$ and $\dot{y} = x + \mu y − x^2$ when $\mu = 0$?

4. The same question for the following systems:

$$\dot{x} = y + \mu x, \qquad \dot{y} = −x + \mu y − x^2 y$$
$$\dot{x} = y + \mu x − x^3, \quad \dot{y} = −x + \mu y + 2y^3$$
$$\dot{x} = y + \mu x − x^2, \quad \dot{y} = −x + \mu y + 2x^2. \tag{5.59}$$

5.5

Consider the following system of nonlinear equations:

$$\dot{x} = −y + \frac{x(1 − x^2 − y^2)}{(x^2 + y^2)^{1/2}}$$

$$\dot{y} = x + \frac{y(1 − x^2 − y^2)}{(x^2 + y^2)^{1/2}}. \tag{5.60}$$

Show that $x = r\cos(t + \theta_0)$ and $y = r\sin(t + \theta_0)$ represent a stable limit cycle.

5.6

Consider the following system:

$$\dot{x} = -y + xf((x^2 + y^2)^{1/2})$$
$$\dot{y} = x + yf((x^2 + y^2)^{1/2}), \tag{5.61}$$

where $f(r) = \sin(1/(r^2 - 1))$ for $r \neq \pm 1$ and $f(\pm 1) = 0$.

1. Show that $x = r_j \cos(t+\theta_0)$ and $y = r_j \sin(t+\theta_0)$, with $j = 0, 1, 2, \ldots$, represent an infinite set of limit cycles, with $r_0 = 1$ and $r_j = (1 + 1/(j\pi))^{1/2}$.

2. Discuss the stability of the limit cycles.

5.7

The following system represents a model for the dynamics of a biomimetic red blood cell (RBC) subject to an external shear flow (see [71]):

$$\begin{cases} \dot{\mathcal{R}} = h\left(1 - 4\dfrac{\mathcal{R}^2}{\Delta}\right)\sin(2\psi), \\[2mm] \dot{\psi} = -\dfrac{1}{2} + \dfrac{h}{2\mathcal{R}}\cos(2\psi), \end{cases} \tag{5.62}$$

where ψ is the inclination angle of the RBC with respect to the flow direction, and $\mathcal{R}$ the deformation amplitude of the RBC; h and Δ are two positive parameters. We assume (without restriction) that $\mathcal{R} > 0$ and $-\pi/2 < \psi < \pi/2$.

1. By using the adiabatic approximation for $\mathcal{R}$ (i.e. $\dot{\mathcal{R}} = 0$), extract the equation satisfied by ψ.

2. Find the fixed points of this equation.

3. Analyze the linear stability of these fixed points. Do we have a bifurcation? If so, which type?

4. By focusing on the vicinity of the bifurcation point, show that ψ obeys an equation characteristic of the saddle-node bifurcation.

5. Consider now the equation of ψ (obtained above) after adiabatic elimination of $\mathcal{R}$, and show that it has the following exact solution:

$$\psi(t) = A\arctan\left[B\tan\left(C(t - t_0)\right)\right], \tag{5.63}$$

where t_0 is a real constant chosen such that $\psi(t = t_0) = 0$. Determine the values of the constants A, B and C. Under which condition does this solution exist? In the case where this solution does not exist, do we have a fixed point solution?

6. We now consider the full system of equations (5.62). Find all the fixed points and determine their stability. One of the fixed points has a complex stability eigenvalue. Does this imply that there is a Hopf bifurcation? Provide explanations to your answer. Does the system have a limit cycle? Explain your answer.

7. Use the following change of variables: $\xi(t) = \mathcal{R}(t)\cos(2\psi(t))$, $\zeta(t) = \mathcal{R}(t)\sin(2\psi(t))$ to show that the system (5.62) becomes

$$\begin{cases} \dot{\xi} = \zeta\left(1 - \dfrac{4h}{\Delta}\xi\right), \\[2mm] \dot{\zeta} = h - \xi - \dfrac{4h}{\Delta}\zeta^2. \end{cases} \qquad (5.64)$$

Introducing the new variable ρ, defined by $\rho = \xi - h/\zeta$, show that ρ satisfies:

$$\dot{\rho} = \rho^2 + 1 - \frac{4h^2}{\Delta}. \qquad (5.65)$$

Show that ρ has the following solutions: for $h > h_c = \sqrt{\Delta}/2$,

$$\rho(t) = -\omega\frac{e^{\omega t} + C_1 e^{-\omega t}}{e^{\omega t} - C_1 e^{-\omega t}}, \qquad (5.66)$$

whereas for $h < h_c$,

$$\rho(t) = \omega\tan\left(\omega t + C_3\right). \qquad (5.67)$$

C_1 and C_3 are integration factors and

$$\omega = \sqrt{\left|1 - \frac{4h^2}{\Delta}\right|}. \qquad (5.68)$$

8. Using the system of equations obeyed by ξ and ζ, and the equation obeyed by ρ, show that $\dot{\zeta} = -\rho\zeta - (4h/\Delta)\zeta^2$.

Show that the solution is given by:

$$\zeta(t) = \frac{\Delta\omega}{4h}\frac{e^{\omega t} - C_1 e^{-\omega t}}{C_2 + e^{\omega t} + C_1 e^{-\omega t}},$$

$$\xi(t) = \frac{\Delta}{4h} + \frac{\omega^2\Delta}{4h}\frac{C_2}{C_2 + e^{\omega t} + C_1 e^{-\omega t}}, \qquad (5.69)$$

for $h > h_c$, and by:

$$\zeta(t) = \frac{\Delta\omega}{4h}\frac{\cos(\omega t + C_3)}{C_4 + \sin(\omega t + C_3)},$$

$$\xi(t) = \frac{\Delta}{4h}\frac{\Gamma + \sin(\omega t + C_3)}{C_4 + \sin(\omega t + C_3)},$$

$$\Gamma = \frac{4h^2}{\Delta}C_4, \qquad (5.70)$$

for $h < h_c$, where C_i are integration factors.

9. Deduce then that ψ and $\mathcal{R}$ are given by:

$$\psi(t) = \frac{1}{2} \arctan\left(\omega \frac{\Delta \left[e^{\omega t} - C_1 e^{-\omega t}\right]}{4h^2 C_2 + \Delta \left[e^{\omega t} + C_1 e^{-\omega t}\right]}\right), \qquad (5.71)$$

and

$$\mathcal{R}^2(t) = \frac{\Delta^2}{16h^2} \frac{\omega^2 (e^{\omega t} - C_1 e^{-\omega t})^2 + (C_2' + e^{\omega t} + C_1 e^{-\omega t})^2}{(C_2 + e^{\omega t} + C_1 e^{-\omega t})^2},$$

$$C_2' = \frac{4h^2}{\Delta} C_2, \qquad (5.72)$$

for $h > h_c$. How do these solutions behave for $t \to \infty$? What can you conclude?

Show that for $h < h_c$, the solution is given by:

$$\psi(t) = \frac{\pi}{4} \frac{\Gamma}{|\Gamma|} \frac{\cos \omega t}{|\cos \omega t|} \left(1 - \frac{\Gamma}{|\Gamma|} \frac{\Gamma + \sin \omega t}{|\Gamma + \sin \omega t|}\right)$$
$$+ \frac{1}{2} \arctan\left(\omega \frac{\cos \omega t}{\Gamma + \sin \omega t}\right). \qquad (5.73)$$

Discuss the solution according to whether $|\Gamma| > 1$ or $|\Gamma| < 1$. Which qualitative differences exist between these two solutions corresponding to $|\Gamma| > 1$ and $|\Gamma| < 1$? The type of motions under question are called "tumbling" and "vacillating-breathing". Can you guess which one is which? Plot $\psi(t)$ in both cases.

Chapter 6

Universal Amplitude Equation in the Neighborhood of a Hopf Bifurcation

Abstract *In this chapter we show that all systems manifesting a Hopf bifurcation can be described by the same universal equation (that is, with an equation independent of the system itself). We first derive the universal equation using a concrete example, and then from symmetries reveal and explain its universal character. We introduce a few more general concepts, such as that of the limit cycle associated with the Hopf bifurcation, and will show that a time oscillating solution does not always lead to a limit cycle.*

In the previous chapter, we studied several distinct systems, each of which manifested a Hopf bifurcation, and each from different domains: electronics (RLC circuit with a tunnel diode, see section 5.1 on the van der Pol oscillator); population dynamics (prey-predator models, see section 5.2 for the Lotka–Volterra model, the classic logistic model, and the refined logistic model); chemistry (chemical kinetics, see section 5.3 for the Brusselator model). These all follow the same nonlinear evolution when close to the bifurcation point. We would like to elaborate on the universal nature of the corresponding generic amplitude equation in this chapter.

© Springer Science+Business Media B.V. 2017
C. Misbah, *Complex Dynamics and Morphogenesis*,
DOI 10.1007/978-94-024-1020-4_6

6.1. Derivation of the Complex Amplitude Equation

To find the generic amplitude equation for all systems generating a Hopf bifurcation, we will begin with a specific model[1], the van der Pol oscillator. Our starting point is the evolution equation for the system, $\ddot{I}-(\epsilon-I^2)\dot{I}+I = 0$ (see equation (5.4)), where I is the current and ϵ the control parameter.

The first nonlinear term in the universal amplitude equation we are looking for is cubic. We might be tempted to think that this cubic term comes out of the third order Taylor expansion we originally used to describe the dynamics of the current (see section 5.1). To see how this is not the case, we can add a quadratic term to the reference equation (5.4), leaving us a modified van der Pol model:

$$\ddot{I} - (\epsilon - I^2)\dot{I} + I + \gamma I^2 = 0, \tag{6.1}$$

where γ is a new control parameter; when zero, it gives us the classic van der Pol oscillator (equation (5.4)) again. This new modified equation has a trivial fixed point, $I_0 = 0$. If we then investigate the linear stability, searching for a solution to the linearized equation of the form $I = \delta I \sim e^{i\omega t}$ (δI assumed to be small), we come to the following equation:

$$\omega^2 - \epsilon\omega + 1 = 0. \tag{6.2}$$

The solution for this dispersion relation is given by:

$$\omega_\pm = \frac{1}{2}\left(\epsilon \pm \sqrt{\epsilon^2 - 4}\right), \quad \text{for } \epsilon > 2,$$

$$\omega_\pm = \frac{1}{2}\left(\epsilon \pm i\sqrt{4 - \epsilon^2}\right), \quad \text{for } \epsilon < 2. \tag{6.3}$$

The value $\epsilon = 0$ corresponds to a critical point: if the parameter is positive, fixed point $I = I_0 = 0$ is unstable (because the real component of ω is positive) while for negative values of parameter ϵ, the fixed point is stable (corresponding to a negative real component of ω). For $\epsilon \simeq 0$ (with $\epsilon > 0$), ω is a purely imaginary number, equal to $\pm i$, which is the signature of a Hopf bifurcation.

6.1.1. Multiscale Analysis

Now, focusing on the neighborhood of the Hopf bifurcation threshold (i.e. for small ϵ), let us expand the eigenvalues found for the modified van der Pol

---

1. At the end of this chapter (subsection 6.4.2), we will also tackle the solution for the Brusselator model.

system (equation (6.3)) to first order in ϵ:

$$\omega_\pm = \frac{1}{2}\epsilon \pm i. \tag{6.4}$$

In the linear regime, we can modify the expression for current $I \sim e^{\omega t}$ to get:

$$I \simeq ae^{it}e^{(\epsilon/2)t}, \tag{6.5}$$

where we have used the above approximation of ω (equation (6.4)) and a is an arbitrary amplitude[2]. This linear expression for current (equation (6.5)) is an adequate approximation so long as we remain in the weakly nonlinear regime, in which parameter ϵ is very small ($\epsilon \ll 1$).

Equation (6.5), which describes the current $I(t)$ in the weakly nonlinear regime, is the product of two functions: an oscillating function (of unit angular velocity), and a term that grows exponentially over time (for $\epsilon > 0$). Parameter ϵ in front of t tells us that the exponential function's growth is slow. This motivates us to use multiscale analysis[3] to consider current $I(t)$ as a function of two independent variables, time t (the fast variable) and variable $T = \epsilon t$ (the slow variable), as follows:

$$I = I(t, T). \tag{6.6}$$

Since, in this notation, current I is a function of two independent variables t and T, we must use partial derivatives to express its first and second derivatives:

$$dI = \frac{\partial I}{\partial t}dt + \frac{\partial I}{\partial T}dT,$$

$$d^2I = \frac{\partial^2 I}{\partial t^2}dt^2 + 2\frac{\partial^2 I}{\partial t \partial T}dtdT + \frac{\partial^2 I}{\partial T^2}dT^2,$$

and we can write the derivative of current I with respect to fast variable t as:

$$\frac{dI}{dt} = \frac{\partial I}{\partial t} + \epsilon\frac{\partial I}{\partial T},$$

$$\frac{d^2I}{dt^2} = \frac{\partial^2 I}{\partial t^2} + 2\epsilon\frac{\partial^2 I}{\partial t \partial T} + \epsilon^2\frac{\partial^2 I}{\partial T^2}. \tag{6.7}$$

By way of example, let us use the following function f to describe current I:

$$I(t) = f(t) = t\sin(\epsilon t) + (\epsilon t)^2 + \epsilon t e^t. \tag{6.8}$$

---

2. Linear equations multiplied by any arbitrary constant remain invariant. As such, it is not possible to determine the amplitude for current when working in the linear regime. Only a nonlinear analysis can place a constraint on the amplitude's value.

3. See subsection 2.3.3, where we introduced multiscale analysis to deal with variables evolving at very different timescales.

The time derivative of this function is written:

$$\frac{df}{dt} = \sin(\epsilon t) + \epsilon t \cos(\epsilon t) + 2\epsilon^2 t + (\epsilon + \epsilon t)e^t. \tag{6.9}$$

Identifying the two variables t and $T = \epsilon t$ so that we may use multiscale analysis, we rewrite function $f(t)$ as $f(t, T)$:

$$f(t, T) = t \sin(T) + T^2 + Te^t. \tag{6.10}$$

Calculating the partial derivatives, we have: $\partial f / \partial t = \sin(T) + Te^t$ and $\partial f / \partial T = t \cos(T) + 2T + e^t$, which we then substitute into the expression of derivative f as a function of its partial derivatives (equation (6.7)) to give:

$$\frac{df}{dt} = \frac{\partial f}{\partial t} + \epsilon \frac{\partial f}{\partial T} = \sin(T) + T \cos(T) + 2\epsilon T + (\epsilon + T)e^t. \tag{6.11}$$

This expression is identical to the one obtained from a direct calculation of f's derivative (equation (6.9)).

The advantage of using the multiscale perspective to separate the slow and fast variables lies in the emergence of parameter ϵ as an explicit variable in the evolution equation, which then allows for an easy classification of terms based on their power of ϵ.

We will now look again at the expression for the van der Pol oscillator in the linear regime (equation (6.5)). For a negative value[4] of parameter ϵ ($\epsilon < 0$), current I evolves toward zero at large time t; the fixed point $I = 0$ thus represents a stable solution. If we vary parameter ϵ to become zero, and then positive, the fixed point $I = 0$ becomes neutral, and then unstable as $\epsilon > 0$, the current evolving into an oscillation with an amplitude that grows exponentially with time (see equation (6.5)) in the linear regime and is expected to saturate due to nonlinear terms. In the vicinity of the bifurcation, we expect the amplitude of the current to be proportional to ϵ, with a certain power. We also know that the value of this power cannot be less than $1/2$ ($I \sim \epsilon^\alpha$, with $\alpha < 1/2$), otherwise the nonlinear effects of the modified van der Pol's evolution equation (6.1) would be drowned out by the too-powerful term in ϵ, canceling any contribution from I^2 (the evolution would reduce to a classic RLC circuit, see equation (5.1)). Following this reasoning, we choose $\epsilon^{1/2}$ as the smallest[5] possible power on which to base our power series expansion of the current, and as such the first dominant contribution to the current is of order square root of ϵ, that is $I(t, T) \sim \epsilon^{1/2}$. We write the general

---

4. Recall that when parameter ϵ is negative, the effective resistance of the circuit is positive (see equations (5.5) and (5.9)).

5. The choice of which power to use to expand a series requires a certain art, and one can only be completely certain of the choice after an in-depth inspection of the final results.

expression for current as an expansion in an analytic series of power $\sqrt{\epsilon}$:

$$I(t,T) = \epsilon^{1/2}I_1(t,T) + \epsilon I_2(t,T) + \epsilon^{3/2}I_3(t,T) + \cdots . \qquad (6.12)$$

To determine the contributions of each successive power of ϵ, we substitute the above equation for current (equation (6.12)) and those of the current's derivatives as obtained through the multiscale analysis (equation (6.7)) into the modified van der Pol equation (6.1).

6.1.2. Expansion to Order $\epsilon^{1/2}$

At this order, the evolution equation of the modified van der Pol equation (6.1) is written as:

$$\frac{\partial^2 I_1}{\partial t^2} + I_1 = 0, \qquad (6.13)$$

with the solution given by:

$$I_1(t,T) = A(T)e^{it} + A^*(T)e^{-it}, \qquad (6.14)$$

where A and its complex conjugate A^* are a priori functions of slow variable T, given that the function is a result of integrating the partial differential equation (6.13).

6.1.3. Expansion to Order ϵ

At this order, the evolution equation of the modified van der Pol equation (6.1) becomes:

$$\frac{\partial^2 I_2}{\partial t^2} + I_2 = -\gamma I_1^2 = -\gamma \left(A^2 e^{2it} + c.c. + AA^* \right), \qquad (6.15)$$

where $c.c.$ is an abbreviation of "complex conjugate". The solution to this equation is the sum of: (i) the solution of the homogeneous equation (that is, the same equation but with a zero right-hand side, i.e. $\partial^2 I_2/\partial t^2 + I_2 = 0$) given by $B(T)e^{it} + c.c.$ (B being an integration factor akin to A); and (ii) the particular solution to the complete equation. Given that the form of the homogeneous problem is identical to the one obtained above for the modified van der Pol equation developed to order $\epsilon^{1/2}$ (see equation (6.14)), we can absorb B into A through the simple change of variables: $A \to A + \sqrt{\epsilon}B$. And so, the solution of interest is the particular solution with the form (obtained by identification with the term on the right in equation (6.15)):

$$I_2 = \frac{\gamma}{3}A^2 e^{2it} + c.c. - \gamma AA^*. \qquad (6.16)$$

6.1.4. Expansion to Order $\epsilon^{3/2}$

At this order, the evolution equation of the modified van der Pol equation (6.1) is written as:

$$\frac{\partial^2 I_3}{\partial t^2} + I_3 = -2\frac{\partial^2 I_1}{\partial t \partial T} + \frac{\partial I_1}{\partial t} + I_1^2 \frac{\partial I_1}{\partial t} + 2\gamma I_1 I_2. \qquad (6.17)$$

Using the solutions from the previous two expansions, we can rewrite the right-hand side of the equation as a sum of the two contributions:

$$\frac{\partial^2 I_3}{\partial t^2} + I_3 = \left[2i\frac{\partial A}{\partial T} - iA + iA^2 A^* \left(1 + i\frac{10\gamma^2}{3} \right) \right] e^{it} + G(T)e^{3it}$$

$$\equiv F(T)e^{it} + G(T)e^{3it}, \qquad (6.18)$$

with

$$F(T) = 2i\frac{\partial A}{\partial T} - iA + iA^2 A^* \left(1 + i\frac{10\gamma^2}{3} \right)$$

and $G(T)$ a function proportional to A^3. At this order, $G(T)$ is of no special interest, as we shall see. The solution to the $\epsilon^{3/2}$ order equation (6.18) is, as before, the sum of two solutions: the solution to the associated homogeneous differential equation, of type $C(T)e^{it} + c.c.$, and the particular solution for the complete equation. In this case, the right-hand side of this equation ($F(T)e^{it} + G(T)e^{3it}$) contains the term e^{it}, whose frequency is equal to one, which is also the eigenvalue of the operator for the homogeneous equation. For this reason, it is called a *resonant* solution. This is similar to the phenomenon of resonance in a linear oscillator with a driving force equal to its natural frequency; the phenomenon of resonance can (in the absence of friction, as would be formally analogous to this problem) lead to a divergence of the oscillator's amplitude. To avoid a similar divergence into values which would no longer describe a physical solution (i.e. to avoid a non-physical response of the system), we must annul function $F(T)$, the prefactor of e^{it}. From this condition we find:

$$\frac{\partial A}{\partial T} = \frac{1}{2}A - \alpha |A|^2 A, \qquad (6.19)$$

with $\alpha \equiv 1/2 + i10\gamma^2/3$ being a complex number. Note that we do not investigate the particular solution corresponding to the behavior of the term e^{3it} (third harmonic) because it is not a resonant term[6]. This explains why G is of no special interest, as anticipated.

---

6. The calculation of a particular solution in response to the behavior of this harmonic would be necessary if we wished to pursue the expansion of the current two orders higher in ϵ.

We may also use an alternative (but equivalent) reasoning to obtain the same amplitude equation (6.19). The particular solution for the $\epsilon^{3/2}$ order expansion of the van der Pol equation and with $G = 0$ (that is, $\partial^2 I_3/\partial t^2 + I_3 = Fe^{it}$) is $(a + bt)e^{it}$, with $b = -iF/2$, and parameter a undetermined because any term of form ae^{it} is a solution to $(\partial^2/\partial t^2 + 1)ae^{it} = 0$. So, we can write the solution to our initial equation as:

$$I_3 = \left(a - \frac{iF}{2}t\right)e^{it} + c.c. \tag{6.20}$$

The resonant term e^{it} introduces a linear contribution in time, described as *secular*, because it grows unbounded. The only way to eliminate the secular term is to set F equal to zero, making our expression equivalent to the amplitude equation (6.19) obtained earlier.

6.1.5. Solvability Condition

The amplitude equation (6.19) obtained in the last section is a particular example of a more general condition, called the solvability condition, generally used to resolve equations of the form:

$$\mathcal{L}u(t) = f(t), \tag{6.21}$$

where $\mathcal{L}$ is any linear operator, and $u(t)$ and $f(t)$ are two functions of variable[7] t. In our present example we have:

$$\mathcal{L} \equiv \frac{\partial^2}{\partial t^2} + 1. \tag{6.22}$$

To determine the conditions which function $f(t)$ must satisfy to ensure a solution for equation (6.21), we first multiply each side of the equation by an arbitrary function $v(t)$, and integrate over a time of period T:

$$\int_0^T \mathcal{L}u(t)\, v(t)dt = \int_0^T f(t)v(t)dt. \tag{6.23}$$

In this expression we identify the scalar product of two functions $\mathcal{L}u(t)$ and $v(t)$, which we can write as[8]:

$$\int_0^T \mathcal{L}u\, vdt = \langle \mathcal{L}u, v\rangle. \tag{6.24}$$

---

7. To relate this to our analysis of the van der Pol oscillator (carried out by expansion into a power series, see the last section), draw the following equivalences: $u(t) \equiv I_3(t)$ and $f(t) \equiv F(T)e^{it}$ (see equation (6.18)), and allow $G = 0$ since we have showed that this term, proportional to e^{3it}, is not important, in contrast to resonant term $f(t) \propto e^{it}$.

8. This notation for the scalar product is typical in discussions of linear algebra; it is also found in quantum mechanics, where linear algebra plays a central role.

We can further denote $\mathcal{L}^+$, the adjoint operator of $\mathcal{L}$, which must by definition satisfy:

$$\langle \mathcal{L}u, v \rangle = \langle u, \mathcal{L}^+v \rangle. \tag{6.25}$$

We are seeking solutions $u(t)$ that are periodic[9] with period T, and so we choose a periodic function of the same form for v.

We then use integration by parts to twice integrate equation (6.23) (vis-à-vis term $v(\partial^2/\partial t^2)u(t)$) to give:

$$\begin{aligned} \langle \mathcal{L}u, v \rangle &= \int_0^T dt\, v(t) \left(\frac{\partial^2}{\partial t^2} + 1 \right) u(t) \\ &= \int_0^T u(t) \left(\frac{\partial^2}{\partial t^2} + 1 \right) v(t) = \langle u, \mathcal{L}v \rangle, \end{aligned} \tag{6.26}$$

where we have used the periodic nature of u and v to cancel out the integrated terms to zero. For example, in the first integration by parts we obtain a term of the type $(v\partial u/\partial t)_0^T = v(T)(\partial u/\partial t)_{t=T} - v(0)(\partial u/\partial t)_{t=0}$ which is equal to zero because functions u and v (and their derivatives) are periodic.

Referring to the definition of an adjoint operator (equation (6.25)), note that operator $\mathcal{L}$ is equal to the conjugate operator $\mathcal{L}^+$ ($\mathcal{L} = \mathcal{L}^+$); operator $\mathcal{L}$ can thus be said to be self-adjoint (or Hermitian).

However, since not all operators are necessarily self-adjoint, let us keep our notations for the two operators $\mathcal{L}$ and $\mathcal{L}^+$ distinct, in order to ensure that we write the fully general solvability conditions for an equation of type $\mathcal{L}u(t) = f(t)$ (equation (6.21)). Using equation (6.23) and the definition of an adjoint operator (equation (6.25)), we can summarize:

$$\langle \mathcal{L}u, v \rangle = \langle f, v \rangle = \langle u, \mathcal{L}^+v \rangle. \tag{6.27}$$

Note that this result is valid for any linear operator $\mathcal{L}$. Remember that v is any function with period T; we can decide to choose a function that belongs to the kernel of adjoint operator $\mathcal{L}^+$. All functions v of the kernel of $\mathcal{L}^+$ must therefore satisfy, by definition, the following condition:

$$\mathcal{L}^+v = 0. \tag{6.28}$$

Combined with equation (6.27), we can fully define the solvability conditions for function f:

$$\langle f, v \rangle = 0, \tag{6.29}$$

for all function v that belongs to the kernel of adjoint operator $\mathcal{L}^+$. In other words, all equations with the form $\mathcal{L}u(t) = f(t)$ have a solution if and only if the function $f(t)$ on the right-hand side of the equation is orthogonal to

---

9. Along the lines of function I_3 from the example given in the preceding section.

the kernel of the adjoint operator (that is, if it is orthogonal to all functions $v(t) \neq 0$ such that $\mathcal{L}^+ v(t) = 0$). This theorem, called the *Fredholm alternative theorem*, is a general result which can be applied to any linear equation of the form $\mathcal{L}u(t) = f(t)$.

Let us demonstrate this result using the example presented in the last section, defining operator $\mathcal{L}$ as $\mathcal{L} \equiv (\partial^2/\partial t^2) + 1$ and the right-hand side of the equation, $f(t)$, such that $f(t) \equiv F(T)e^{it}$. Given $\mathcal{L} = \mathcal{L}^+$, function v satisfies $\mathcal{L}v(t) = 0$. In the last section we found the general solution of the equation, given by $v = ae^{it} + c.c.$, where a is an integration constant. Using the Fredholm condition $\langle f, v \rangle = 0$ we obtain the following:

$$\int_0^{2\pi} dt F e^{it} \left(ae^{it} + a^* e^{-it} \right) = 0, \tag{6.30}$$

where we have used the period of function $v(t)$, $T = 2\pi$. The integral over one time period T for function e^{2it} is equal to zero, reducing the condition that must be satisfied by function F to:

$$\int_0^{2\pi} F A^* dt = 2\pi a^* F. \tag{6.31}$$

Since the integration constant a is non-zero ($a \neq 0$), the Fredholm condition $\langle f, v \rangle = 0$ requires that function F be equal to zero ($F = 0$). From this result we establish the same amplitude equation (6.19) which in turn determines amplitude $A(t)$.

6.1.6. A Few Useful Facts Concerning the Amplitude Equation

If we keep only the linear term of the amplitude equation (6.19), we find that the solution is $A \sim e^{T/2} = e^{t\epsilon/2} = e^{t\omega_r}$, where ω_r is the real component of ω (see equation (6.4)). More generally, all problems which have a Hopf bifurcation are described in the linear regime by the equation: $\partial A/\partial t = \omega_r A$, and the first nonlinear term of the amplitude equation (as we have seen above for the modified van der Pol system) we would add is of the form $(a + ib)|A|^2 A$ (where a and b are real constants). Altogether, the generic form for the amplitude equation of a Hopf bifurcation is the following:

$$\frac{\partial A}{\partial T} = \bar{\omega}_r A - (a + ib)|A|^2 A. \tag{6.32}$$

Note that we have redefined parameter ω_r from the linear equation through parameter $\bar{\omega}_r$, such that $\omega_r = \epsilon \bar{\omega}_r$ in order to make explicit that for $\epsilon = 0$, the real part of ω becomes zero ($\bar{\omega}_r$ being a quantity of order one). Thus the sign of the linear term is determined by that of ϵ. The sign of the cubic term's coefficient cannot be determined by any such simple rule; it might as easily be positive or negative and, with the exception of some rare cases where the sign can be intuited, the sign must be calculated for each system. If the constant a turns out to be positive ($a > 0$), the nonlinear term evolves in opposition to the linear growth, leading to a saturation of the amplitude. If instead constant a is negative ($a < 0$), the nonlinear term contributes to the linear growth and the system ends up with an amplitude diverging toward infinity. To study the dynamics of the system in this situation, the amplitude equation must be expanded to higher orders of ϵ, as we have already seen in our discussion of subcritical bifurcations in chapter 3.

We can reduce the number of coefficients in the amplitude equation by making the variables adimensional through these two transformations: $A \rightarrow A/\sqrt{\bar{\omega}_r/a}$ and $T \rightarrow T\bar{\omega}_r$. This allows us to write the amplitude equation using the single parameter α:

$$\frac{\partial A}{\partial T} = A - (1 + i\alpha)|A|^2 A, \tag{6.33}$$

with $\alpha = b/a$.

It is sometimes useful to keep parameter ϵ explicitly in front of the linear term in the final equation. To do that, we use the following transformations for the amplitude $A \rightarrow \sqrt{\epsilon}A/\sqrt{\bar{\omega}_r/a}$ and for temporal variable[10] $T \rightarrow t\bar{\omega}_r\epsilon$. The evolution equation (6.32) acquires the form:

$$\frac{\partial A}{\partial t} = \epsilon A - (1 + i\alpha)|A|^2 A. \tag{6.34}$$

We can use either version of the evolution equation ((6.33) or (6.34)) depending on our goal. Under the canonical form (equation (6.33)), the amplitude equation shows that the coefficients are of order one, while the second form (equation (6.34)) explicitly shows the dependence on ϵ of both the amplitude and the growth of the instability (the nonlinear term): the amplitude behaves as $\epsilon^{1/2}$, and the growth rate as ϵ.

6.1.7. Deriving the Amplitude Equation from Symmetry

In this section we provide a simple explanation justifying the form of the amplitude equation (6.33) by symmetry. To begin, note that the van der Pol equation (6.1) is invariant under translation $t \rightarrow t + t_0$, where t_0 is a real constant; this invariance must be conserved in the amplitude equation. This kind of linear transformation in time is equivalent to using the rotation operation

$$A \rightarrow A e^{it} \tag{6.35}$$

---

10. Recall that T is the "slow" temporal variable and t the fast one, see subsection 6.1.1.

on the solution for the amplitude equation (6.14). Under this rotation, all the linear terms of the amplitude equation are multiplied by a factor e^{it_0}. As such, in order to conserve the invariance in the amplitude equation, we can only admit nonlinear terms that also undergo the same transformation. Let us make a list of all possible nonlinear terms up to cubic order and see how each term transforms under the substitution:

$$A^2 \longrightarrow A^2 e^{2it_0}$$

$$AA^* \longrightarrow AA^*$$

$$A^{*2} \longrightarrow A^{*2} e^{-2it_0}$$

$$A^3 \longrightarrow A^3 e^{3it_0}$$

$$(a) \quad A^2 A^* \longrightarrow A^2 A^* e^{it_0}$$

$$A^{*2} A \longrightarrow A^{*2} A e^{-it_0}$$

$$A^{*3} \longrightarrow A^{*3} e^{-3it_0}.$$

Only the term labeled (a) transforms in the same manner as a linear term, so we may keep it; the other terms fail to satisfy this condition and must be discarded. From this argument we can contract and understand the origin of the form and the universal character of the nonlinear amplitude equation.

6.1.8. Properties of the Complex Amplitude Equation

The denomination "complex" refers to the fact that the coefficients in the amplitude equation are complex numbers, owing to the Hopf bifurcation. The first important characteristic to note about the canonical amplitude equation (6.33) is that it is non-variational; the equation cannot be written as the derivative of a potential (i.e. $\partial A / \partial t = -\partial V / \partial A^*$, with $dV/dt \leqslant 0$ and V a "potential" – that is, a real function – or a Lyapunov function, all with the exception of the particular case $\alpha = 0$). This stands in contrast to the studies we have taken up in the preceding chapters (in particular, see equation (2.29) where we defined an associated Lyapunov function). The presence of a Hopf bifurcation thus signals a rupture with the variational character of the system: our first manifestation of non-equilibrium behavior.

Writing the complex amplitude A as $A = \rho e^{i\theta}$, where ρ is the magnitude of the amplitude and θ the phase, we can rewrite the amplitude equation (6.33) as:

$$\dot{\rho} = \rho - \rho^3 \tag{6.36}$$

and

$$\dot{\theta} = \alpha \rho^2. \tag{6.37}$$

We can find three fixed points from the first equation, $\rho_0 = 0$ and $\rho_\pm = \pm 1$. The first fixed point is unstable[11] and the other two are stable. After an initial transient period, the evolution of ρ will approach one of the two stable fixed points (the choice of which depends on initial conditions), $\rho_\pm = \pm 1$. Once ρ settles on a fixed point, the second amplitude equation (6.37) will tend toward the following equation: $\dot{\theta} = \alpha$, whose solution is given by:

$$\theta = \theta_0 + \alpha t, \tag{6.38}$$

with θ_0 the initial phase. Thus, after a long enough time, the complex amplitude, when represented graphically in the plane of real and imaginary parts, corresponds to a circle of radius one, centered at the origin, which we will call:

$$A(t \to \infty) = \pm e^{i(\theta_0 + \alpha t)}. \tag{6.39}$$

The two possible solutions differ by a phase of π ($-1 = e^{i\pi}$). In phase space (X, Y) (with $A = X + iY$), starting from some initial condition, the trajectory spirals around the unit circle (see figure 6.1). This circle is called the limit cycle, or an attractor, because whatever the initial conditions, the trajectory moves toward this circle: it is a stable limit cycle.

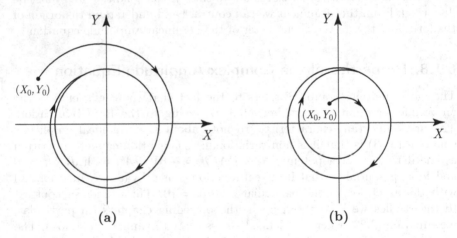

(a)   (b)

Figure 6.1 – Schematic view of the trajectory in the $(X–Y)$-plane. (a) We start from an initial condition (X_0, Y_0) far away from the origin. As time progresses, the trajectory approaches the circle, and settles there indefinitely. (b) The initial condition is close to the origin, and the trajectory spirals outward to join the circle, the stable limit cycle.

---

11. By analogy with mechanics, we would write the amplitude equation for the real component of the amplitude (equation (6.36)) as $\dot{\rho} = -\partial_\rho V$ with $V = -\rho^2/2 + \rho^4/4$. The function V possesses a maximum at $\rho = 0$ (unstable point) and two minima at $\rho = \pm 1$ (stable points).

Now let us use the second form of the amplitude equation (6.34), with its explicit statement of a term's dependence on parameter ϵ. In this case, the amplitude equations divide into functions of ρ and θ (equations (6.36) and (6.37)) as follows:

$$\dot{\rho} = \epsilon\rho - \rho^3,$$

$$\dot{\theta} = \alpha\rho^2. \tag{6.40}$$

For $\epsilon > 0$, the fixed points are $\rho_0 = 0$ and $\rho_\pm = \pm\sqrt{\epsilon}$. For $\epsilon < 0$, $\rho = 0$ is the only fixed point, and is stable. For $\epsilon > 0$, $\rho_0 = 0$ is unstable and $\rho_\pm = \pm\sqrt{\epsilon}$ are stable fixed points. The behavior at infinite time is the limit cycle with radius $\sqrt{\epsilon}$.

6.2. Unstable Limit Cycle

The initial evolution equation (6.19), as obtained from the modified van der Pol equation, has a cubic coefficient, the real part of which is negative, but this is not guaranteed to be generally true for other systems. There are many systems for which the cubic coefficient is positive, and the system above made up of two evolution equations of variables ρ and θ (equations (6.36) and (6.37)) can be accordingly modified to give:

$$\dot{\rho} = \epsilon\rho + \rho^3,$$

$$\dot{\theta} = \alpha\rho^2. \tag{6.41}$$

When parameter ϵ is positive ($c > 0$), there is a single, unstable, fixed point $\rho_0 = 0$ while for a negative value, the system has three fixed points $\rho_0 = 0$ and $\rho_\pm = \pm\sqrt{-\epsilon}$. However, unlike before, the associated limit cycle is unstable; the fixed points $\rho_\pm$ are unstable and the amplitude grows unbounded when $\epsilon > 0$: we must include contributions from higher order terms (in A) into our amplitude equation. Expanding to a higher order (beginning, for example, from equation (6.1), or from any other model containing a Hopf bifurcation) gives us:

$$\frac{\partial A}{\partial T} = \epsilon A + (1 - i\alpha)|A|^2 A - (1 + i\beta)|A|^4 A. \tag{6.42}$$

The form of this quintic equation, notably the absence of terms of order four, and of a term in the form $A^3 A^{*2}$, results from properties of symmetry (see subsection 6.1.7). Note that we have assumed that the real component of the complex coefficient to fifth order term (i.e. $-(1 + i\beta)$) is negative in order to ensure a saturation of the amplitude when parameter ϵ is negative ($\epsilon > 0$). Taking up the decomposition of A into $A = \rho e^{i\theta}$, the evolution equations for modulus ρ and phase θ of the complex amplitude are written as:

$$\dot{\rho} = \epsilon\rho + \rho^3 - \rho^5,$$

$$\dot{\theta} = \alpha\rho^2 + \beta\rho^4. \tag{6.43}$$

A detailed discussion of fixed points for the equation of ρ (6.43) is given in section 3.3. For $-1/4 < \epsilon < 0$, there exist five fixed points ($\rho = 0$ and two pairs of solutions with $\rho \neq 0$). In this interval of ϵ, the fixed point $\rho = 0$ coexists with the limit cycle (see figure 6.2). For $\epsilon < \epsilon' = -3/16$, the fixed point is stable but the limit cycle has become metastable. For $\epsilon > \epsilon'$, the inverse is true. Finally, for $\epsilon > 0$, there exists a pair of solutions with $\rho \neq 0$ corresponding to a stable limit cycle, while $\rho = 0$ has become unstable. In each case where $\rho \neq 0$, the phase behaves linearly in time.

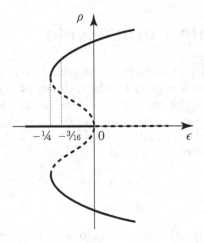

Figure 6.2 – Bifurcation diagram for a subcritical Hopf bifurcation with amplitude ρ. We notice that the behavior is identical to that of a stationary bifurcation (figure 3.20). The only difference lies in the fact that phase grows with time.

6.3. More on the Concept of Limit Cycles

As pointed out in our study of the Lotka–Volterra cycle in chapter 5, it is important to distinguish limit cycles from closed trajectories in phase space. To fully understand this, let us consider a simple example described by the following nonlinear dynamical system:

$$\ddot{x} + x + 2x^3 = 0. \tag{6.44}$$

This equation has a simple periodic solution. If we multiply each term of the equation by the term $2\dot{x}$, we can rewrite it as:

$$\partial_t \left(\dot{x}^2 + x^2 + x^4 \right) = 0. \tag{6.45}$$

Integration yields

$$\dot{x}^2 + x^2 + x^4 = H = \dot{x}_0^2 + x_0^2 + x_0^4, \tag{6.46}$$

where H is an integration constant, analogous to the total energy in mechanics, and $x_0 = x(t = 0)$. This equation (6.46) can be represented by a closed trajectory in the phase space $(\dot{x}, x)$; for each initial condition, associated with a specific value of H, we can draw a closed trajectory in the $(\dot{x}-x)$-plane (see figure 6.3). Note that the orbits are not isolated; they form a continuum. In other words, the choice of a continuous ensemble of H corresponds to a continuous ensemble of closed trajectories. A limit cycle, on the other hand, is a closed orbit without any other closed orbit in its immediate neighborhood. A stable limit cycle is an isolated orbit all trajectories beginning from an initial condition within a certain distance from the cycle[12] merge toward. Up until now, the example we have chosen has only one limit cycle and, as such, the basin of attraction is made up of the ensemble of points on the plane.

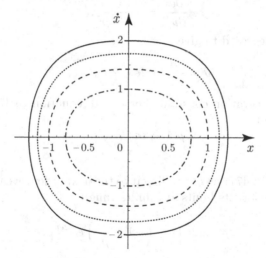

Figure 6.3 – The portrait of orbits obtained from equation (6.46) for different initial conditions, or for different values of H. From the outside in, $H = 4$, 3, 2 and 1. It is not a limit cycle because the orbit is not isolated, but rather has a continuum of orbits in its neighborhood.

---

12. There are situations in which, even if a stable limit cycle exists, certain initial conditions will not, over time, end in a limit cycle. This is the case only for points found in the basin of attraction, which is simply the ensemble of initial conditions which converge to a limit cycle. Other initial conditions, not converging to the cycle, do not belong to the basin of attraction.

6.4. Solved Problems

6.4.1. Time-Dependent Solution
for the Complex Amplitude Equation

Solve the system of equations stemming from the cubic complex amplitude equation (6.40).

Solution. The first evolution equation for magnitude ρ of the complex amplitude in system (6.40) can be solved by following the same procedure used in subsection 2.3.5. We can thus write $\rho(t)$ (equation (6.41)):

$$\rho = \left[\left(-\frac{1}{\epsilon} + \frac{1}{\rho_i^2} \right) e^{-2\epsilon t} + \frac{1}{\epsilon} \right]^{-1/2}. \tag{6.47}$$

We have only kept the positive solution (because ρ is the magnitude of complex number A). $\rho_i = \rho(t=0)$ is the initial value of the amplitude. Using the two equations of system (6.40), and using the identity $(\dot{\rho^2}) = 2\rho\dot{\rho}$, we obtain the following equation:

$$\frac{\partial \theta}{\partial \rho^2} = \frac{1}{2\epsilon - 2\rho^2}, \tag{6.48}$$

which can be integrated to give:

$$\theta = -\frac{1}{2} \ln(2\epsilon - 2\rho^2) + c, \tag{6.49}$$

where c is an integration constant. For $\epsilon > 0$, equation (6.47) gives us the following behavior:

$$\rho(t \to \infty) = \sqrt{\epsilon} \tag{6.50}$$

for all $\rho_i \neq 0$.

Taking equation (6.47) and expanding it to the next order reveals the system's asymptotic behavior (its limits at infinite time):

$$2\rho^2 \simeq 2\epsilon \left[1 - \left(-1 + \frac{\epsilon}{\rho_i} \right) e^{-2\epsilon t} \right]. \tag{6.51}$$

Substituting this result into equation (6.49) we find:

$$\theta(t \to \infty) = \epsilon t + \theta_0. \tag{6.52}$$

The complete solution at infinite time thus takes the form:

$$A(t \to \infty) = \sqrt{\epsilon} e^{i(\epsilon t + \theta_0)}, \tag{6.53}$$

where θ_0 is the initial phase. This equation describes a circle of radius $\sqrt{\epsilon}$ centered at the origin. The functions X and Y (the real and imaginary parts of A) oscillate sinusoidally over time; the frequency of the oscillations is equal to ϵ.

6.4.2. Derivation of the Complex Amplitude Equation for the Modified "Brusselator" Model

Derive the amplitude equation in the neighborhood of the Hopf bifurcation threshold in the modified Brusselator model (system of equations (5.50)).

Solution. In chapter 5 we saw that model for the Brusselator system gives rise to a Hopf bifurcation whose critical point is described by $b = a^2 + 1$ (see subsection 5.3.4).

For this problem, let us introduce a small parameter ϵ as the distance to the instability threshold:

$$\epsilon \equiv \frac{b - (1 + a^2)}{2}. \tag{6.54}$$

The homogeneous solution is stable for $\epsilon < 0$ and becomes unstable vis-à-vis an oscillating solution for $\epsilon > 0$. The growth rate obeys the characteristic equation (5.53), taking the following form for small ϵ:

$$\omega = \epsilon \pm ia. \tag{6.55}$$

Besides introducing slow variable T and adopting the partial derivative formulations described in equation (6.7) for our differential operator, we further rewrite the system (equation (5.50)) using variables denoted by a bar to represent the deviation of the solution from the homogeneous one, defined as follows:

$$X = X_0 + \bar{X}, \quad Y = Y_0 + \bar{Y}, \tag{6.56}$$

with (X_0, Y_0) representing the homogeneous solution (equation (5.52)). We obtain:

$$\frac{\partial \bar{X}}{\partial t} + \epsilon \frac{\partial \bar{X}}{\partial T} = (a^2 + 2\epsilon)\bar{X} + a^2\bar{Y} + 2a\bar{X}\bar{Y}$$

$$+ \frac{1 + a^2 + 2\epsilon}{a}\bar{X}^2 + \bar{X}^2\bar{Y},$$

$$\frac{\partial \bar{Y}}{\partial t} + \epsilon \frac{\partial \bar{Y}}{\partial T} = -(a^2 + 2\epsilon + 1)\bar{X} - a^2\bar{Y} - 2a\bar{X}\bar{Y}$$

$$- \frac{1 + a^2 + 2\epsilon}{a}\bar{X}^2 - \bar{X}^2\bar{Y}. \tag{6.57}$$

We expand $\bar{X}$ and $\bar{Y}$ in power series of $\sqrt{\epsilon}$:

$$\bar{X}(t, T) = \epsilon^{1/2}\bar{X}_1 + \epsilon\bar{X}_2 + \cdots, \tag{6.58}$$

$$\bar{Y}(t, T) = \epsilon^{1/2}\bar{Y}_1 + \epsilon\bar{Y}_2 + \cdots. \tag{6.59}$$

Now we systematically substitute these results into equation (6.57) to derive the set of equations for each order in ϵ.

Order $\epsilon^{1/2}$

For this order, we begin with:

$$\frac{\partial \bar{X}_1}{\partial t} = a^2 \bar{X}_1 + a^2 \bar{Y}_1,$$

$$\frac{\partial \bar{Y}_1}{\partial t} = -(a^2 + 1)\bar{X}_1 - a^2 \bar{Y}_1. \tag{6.60}$$

For both variables, we are looking for solutions of the form $e^{\sigma t}$. Calling upon the constraint of a zero value determinant gives us $\sigma = \pm ia$. This in fact is the expression for ω (equation (5.53) at $\epsilon = 0$, the bifurcation point). The general solution for the system of equations is:

$$\bar{X}_1(t, T) = A(T)e^{iat} + c.c.$$

$$\bar{Y}_1(t, T) = B(T)e^{iat} + c.c. \tag{6.61}$$

where A and B are integration factors that depend on the slow variable T. Given that the determinant of the system (equation (6.60)) is zero, factors A and B are related by:

$$B = \frac{ia - a^2}{a^2}A \equiv \gamma A. \tag{6.62}$$

Order ϵ

At this order, we have:

$$\frac{\partial \bar{X}_2}{\partial t} = a^2 \bar{X}_2 + a^2 \bar{Y}_2 + 2a\bar{X}_1\bar{Y}_1 + \frac{1+a^2}{a}\bar{X}_1^2, \tag{6.63}$$

$$\frac{\partial \bar{Y}_2}{\partial t} = -(a^2 + 1)\bar{X}_2 - a^2 \bar{Y}_2 - 2a\bar{X}_1\bar{Y}_1 - \frac{1+a^2}{a}\bar{X}_1^2. \tag{6.64}$$

Substituting the expressions we just found for variables X_1 and Y_1 (equation (6.61)), we obtain the following inhomogeneous system for X_2 and Y_2:

$$\frac{\partial \bar{X}_2}{\partial t} - a^2 \bar{X}_2 - a^2 \bar{Y}_2 = 2a \left(\gamma A^2 e^{2it} + AA^* \gamma^* + c.c. \right)$$

$$+ \frac{1+a^2}{a} \left(A^2 e^{2it} + AA^* \gamma^* + c.c. \right),$$

$$\frac{\partial \bar{Y}_2}{\partial t} + \left(a^2 + 1 \right) \bar{X}_2 + a^2 \bar{Y}_2 = -2a \left(\gamma A^2 e^{2it} + AA^* \gamma^* + c.c. \right)$$

$$- \frac{1+a^2}{a} \left(A^2 e^{2it} + AA^* \gamma^* + c.c. \right). \tag{6.65}$$

As before, the solution is a sum of the homogeneous version of this equation's solution, plus a particular solution. Given that the homogeneous equations associated to this system of equations is identical to the system described by equations (6.60), they will also have the same solutions (equations (6.61)). We need only redefine the integration constants to absorb the homogeneous solution of the system (6.65) into equations (6.61). Now, looking at the terms on the right-hand side of equations (6.65) in the system, we look for a particular solution of the form:

$$X_2 = \alpha|A|^2 + \beta A^2 e^{2iat}, \tag{6.66}$$

$$Y_2 = \alpha'|A|^2 + \beta' A^2 e^{2iat}. \tag{6.67}$$

Bringing these expressions into equations (6.65), we can identify four algebraic equations with four unknowns, $(\alpha, \alpha', \beta, \beta')$. Their solutions are easily found to be:

$$\alpha = 0, \tag{6.68}$$

$$\alpha' = -\frac{1 + a^2 + a^2(\gamma + \gamma^*)}{a^3}, \tag{6.69}$$

$$\beta = -\frac{2}{3} \frac{i\left(1 + a^2 + 2\,a^2\gamma\right)}{a^2}, \tag{6.70}$$

$$\beta' = \frac{1}{3} \frac{1 + a^2 + 2\,a^2\gamma + 2\,ia + 2\,ia^3 + 4\,ia^3\gamma}{a^3}. \tag{6.71}$$

Order $\epsilon^{3/2}$

At this order, we have

$$\frac{\partial \bar{X}_3}{\partial t} - a^2 \bar{X}_3 - a^2 \bar{Y}_3 = -\frac{\partial \bar{X}_1}{\partial T} + 2\bar{X}_1 + 2a(\bar{X}_1\bar{Y}_2 + \bar{X}_2\bar{Y}_1)$$
$$+ 2\frac{1 + a^2}{a}\bar{X}_1\bar{X}_2 + \bar{X}_1^2\bar{Y}_1,$$

$$\frac{\partial \bar{Y}_3}{\partial t} + (a^2 + 1)\bar{X}_3 + a^2 \bar{Y}_3 = -\frac{\partial \bar{Y}_1}{\partial T} - 2\bar{X}_1 - 2a(\bar{X}_1\bar{Y}_2 + \bar{X}_2\bar{Y}_1)$$
$$- 2\frac{1 + a^2}{a}\bar{X}_1\bar{X}_2 - \bar{X}_1^2\bar{Y}_1. \tag{6.72}$$

Our results for the previous two orders allows us to write:

$$\frac{\partial \bar{X}_3}{\partial t} - a^2 \bar{X}_3 - a^2 \bar{Y}_3 = f_1(T)e^{iat} + g_1(T)e^{3iat} + c.c.,$$

$$\frac{\partial \bar{Y}_3}{\partial t} + (a^2 + 1)\bar{X}_3 + a^2 \bar{Y}_3 = f_2(T)e^{iat} + g_2(T)e^{3iat} + c.c., \tag{6.73}$$

where

$$f_1 = -\frac{\partial A}{\partial T} + 2A + \left[2a(\beta' + 2\alpha' + \gamma^*\beta + 2\alpha\gamma) \right.$$
$$\left. + \frac{2(1+a^2)}{a}(\beta + 2\alpha) + \gamma^* + 2\gamma \right] |A|^2 A \qquad (6.74)$$

and

$$f_2 = -\frac{\partial A}{\partial T}\gamma - 2A - \left[2a(\beta' + 2\alpha' + \gamma^*\beta + 2\alpha\gamma) \right.$$
$$\left. + \frac{2(1+a^2)}{a}(\beta + 2\alpha) + \gamma^* + 2\gamma \right] |A|^2 A. \qquad (6.75)$$

We do not need to explicitly write the forms of functions g_1 and g_2, because they do not include resonant terms and thus do not contribute to any secular form of evolution. As before, the particular solution of the system is of interest (equation (6.73)). We look for solutions of the form $H_1 e^{iat} + H_2 e^{3iat}$ and $G_1 e^{iat} + G_2 e^{3iat}$. Bringing those solutions into equations (6.73) and identifying the terms we want to keep, the ones with resonant term e^{iat}, we find:

$$(ia - a^2)H_1 - a^2 G_1 = f_1,$$
$$(a^2 + 1)H_1 + (ia + a^2)G_1 = f_2. \qquad (6.76)$$

The determinant of this system, associated with unknowns (H_1, G_1), is zero. This is clear from the fact that the operator (equations (6.73)) has eigenvector e^{iat} with eigenvalue of zero (a situation identical to that found at order $\epsilon^{1/2}$). As a result, the above system (6.76) must satisfy the following solvability condition:

$$(a^2 + 1)f_1 - (ia - a^2)f_2 = 0. \qquad (6.77)$$

Using the expressions for f_1 and f_2, we obtain as our final answer:

$$\frac{\partial A}{\partial T} = A + (a_1 + ib_1)|A|^2 A \qquad (6.78)$$

with

$$a_1 = -\frac{19a^3 + 8a - a^5}{3a^3(4 + a^2)},$$

$$b_1 = -\frac{4(a^6 + a^4 - a^2 + 2)}{3a^3(4 + a^2)}. \qquad (6.79)$$

We can verify that $a_1 < 0$ for all $a > 0$. This is a supercritical pitchfork bifurcation.

6.5. Exercises

6.1

Consider the following system of equations:

$$\dot{x} = \mu x - y + xy^2, \tag{6.80}$$

$$\dot{y} = x + \mu y + y^3. \tag{6.81}$$

1. Show that this system presents a Hopf bifurcation. Determine the condition satisfied by μ for the Hopf bifurcation to take place.

2. Using the method developed in this chapter, derive the complex amplitude equation.

3. Is the bifurcation subcritical or supercritical?

6.2

Same questions for the following systems:

$$
\begin{aligned}
\dot{x} &= y + \mu x, & \dot{y} &= -x + \mu y - x^2 y, \\
\dot{x} &= y + \mu x - x^3, & \dot{y} &= -x + \mu y + 2y^3, \\
\dot{x} &= y + \mu x - x^2, & \dot{y} &= -x + \mu y + 2x^2.
\end{aligned} \tag{6.82}
$$

6.3

Consider the following amplitude equation:

$$\dot{A} = (\epsilon - |A|^2)A + a, \tag{6.83}$$

where a and ϵ are two real parameters.

1. Set $A = re^{i\theta}$, and determine the equations satisfied by r and θ.

2. Determine the fixed points and their stability.

3. Plot the bifurcation diagram in the $(r$–$\epsilon)$-plane, for $a > 0$ and $a < 0$.

Chapter 7

Parametric Instabilities and Other Nonlinear Behaviors

Abstract *In this chapter we present parametric instability in which a parameter, such as the stiffness of a spring, is itself a function of time. We will use the pendulum as our case model, and show how the nonlinear equation describing this instability differs from that of a Hopf bifurcation, giving instead what is called a transcritical bifurcation. We will then discuss "phase locking" (or synchronization) in nonlinear oscillators driven by an external sinusoidal force.*

The parametric oscillator is a harmonic oscillator whose parameters (such as the damping ratio, and apparent gravity in the case of a pendulum) vary with time. This type of system, boasting of time-dependent parameters, introduces parametric instabilities – one of the families of generic instabilities. Usually manifesting as resonant phenomena, the parametric instability often signals a transition to chaos, and though we are speaking of resonance it is of a fundamentally different sort to the classical phenomenon of resonance found in resonant oscillators. A linear oscillator of natural frequency ω_0 subjected to an external driving force (which may also be called an excitation or perturbation) of frequency ω_e has an optimal response for $\omega_e = \omega_0$. The simple system we will study in this chapter is characterized by an optimal response for $\omega_0 = \omega_e/2$, where ω_e is the frequency of the *parametric excitation* to which the system is subject. In other words, the sensitivity of the oscillator is maximized when its natural frequency is half that of the driving frequency; this is called *subharmonic resonance*.

© Springer Science+Business Media B.V. 2017
C. Misbah, *Complex Dynamics and Morphogenesis*,
DOI 10.1007/978-94-024-1020-4_7

7.1. Simple Example of a Parametric Excitation

Consider a pendulum with a vertical oscillation of frequency ω_e at its suspension point. For the analysis which follows, we will neglect the effects of friction, introducing frictional forces only later in the chapter.

The instantaneous position of the suspension point is given by $a(t) = -a_0 \cos(\omega_e t)$ (the minus sign is of no essential importance), and its acceleration is given by $\ddot{a} = a_0 \omega_e^2 \cos(\omega_e t)$. The suspended mass m is thus subject, in addition to the usual gravitational acceleration g_0, to an acceleration $\ddot{a}$. For our calculations we can define the total effective gravity as felt by the mass as follows:

$$g(t) = g_0 + g_1 \cos(\omega_e t), \tag{7.1}$$

where we have defined $g_1 = a_0 \omega_e^2$. In the regime of small amplitudes, the pendulum's equation is written as:

$$\ddot{\theta} + \omega_0^2 \left[1 + h \cos(\omega_e t)\right] \theta = 0, \tag{7.2}$$

where h is a parameter such that $h = g_1/g_0$, and $\omega_0 = \sqrt{g_0/\ell}$ is the natural frequency of the pendulum, with ℓ the length of the pendulum, or more precisely, the length between the center of mass and the suspension point of the pendulum. Remember that for a classic forced oscillator, the equation of motion is written as $\ddot{\theta} + \omega_0^2 \theta = f_e \cos(\omega_e t)$, where f_e is the magnitude of the driving force. For the system studied in this chapter, the driving force intervenes as a product with variable θ in the evolution equation (7.2). Everything happens as if the physical parameter, gravity, which characterizes the oscillator, depended on time; hence the origin of the term *parametric* for talking about this kind of excitation.

7.2. Subharmonic Instability

The evolution equation for a parametric oscillator such as above (equation (7.2)) is called non-autonomous with respect to time, because one of the coefficients is time-dependent. For a differential evolution equation which *is* autonomous (its coefficients being independent of time), and linear, the solution is of the form $e^{\omega t}$. If instead one of the coefficients in the equation is a periodic function of time, the solution will rather look like:

$$\theta(t) = f(t)e^{\omega t}, \tag{7.3}$$

where ω is a complex number, and $f(t)$ is a periodic function with respect to time, whose period is the same as that of the coefficients in the differential equation (i.e. $f(t + 2\pi/\omega_e) = f(t)$). This last fact is a result of the Floquet–Bloch theorem (see [77]).

To make a fully general analysis, we would need to decompose the function $f(t)$ into a Fourier series containing each harmonic. However, since we will be using perturbation theory for our calculations, we only need to take the first harmonic into account. In the absence of a parametric perturbation ($h = 0$) the solution to the evolution equation (7.2) is given by:

$$\theta(t) = \theta_0 \cos(\omega_0 t + \phi), \tag{7.4}$$

where θ_0 is an amplitude and ϕ a phase which can be determined from the initial conditions. Using the general solution (equation (7.3)) as our starting point, we can further write a solution for small h of the form:

$$\theta(t) = \theta_0 \cos\left(\frac{\omega_e}{2} t + \phi\right) e^{\omega t}. \tag{7.5}$$

We have anticipated that for small values of h the frequency of the solution will be close to ω_0, that is, $\omega_e/2 \simeq \omega_0$. In other words, we have assumed that the instability will be subharmonic, with the responsibility of later determining the conditions under which our hypothesis is valid[1]. For now, we substitute the general solution (equation (7.5)) into the parametric evolution equation (7.2) to give:

$$\left\{ \frac{\partial^2}{\partial t^2} + \omega_0^2 \left[1 + h \cos(\omega_e t) \right] \right\} \cos\left(\frac{\omega_e}{2} t + \phi\right) e^{\omega t}$$
$$= \left[\left(\omega_0^2 - \frac{\omega_e^2}{4} + \omega^2 \right) \times \cos\left(\frac{\omega_e}{2} t + \phi\right) - \omega_e \omega \sin\left(\frac{\omega_e}{2} t + \phi\right) \right.$$
$$\left. + \frac{h}{2} \omega_0^2 \cos\left(\frac{\omega_e}{2} t - \phi\right) + \frac{h}{2} \omega_0^2 \cos\left(\frac{3\omega_e}{2} t + \phi\right) \right] e^{\omega t}. \tag{7.6}$$

Note that the parametric character of the excitation induces higher harmonics (the third harmonic, of frequency $3\omega_e/2$). Later, in subsection 7.7.1, we will show that the effect of such a harmonic is negligible with respect to the harmonic $\omega_e/2$, but we can already leave it out of our calculations.

Expanding the second half of equation (7.6), we can separate the terms into $\cos((\omega_e/2)t)e^{\omega t}$ and $\sin((\omega_e/2)t)e^{\omega t}$. Since $\cos((\omega_e/2)t)$ and $\sin((\omega_e/2)t)$ are two independent functions, equation (7.6) is satisfied only if the coefficients multiplying each function are zero simultaneously. From this we can write

---

1. Nothing forces us to make this choice and it is possible to proceed in a more general fashion, but this assumption allows us to simplify our presentation of the problem.

down the following system:

$$\left(\omega_0^2 - \frac{\omega_e^2}{4} + \omega^2 + \frac{h}{2}\omega_0^2\right)\cos(\phi) - \omega_e\omega\sin(\phi) = 0,$$

$$\omega_e\omega\cos(\phi) + \left(\omega_0^2 - \frac{\omega_e^2}{4} + \omega^2 - \frac{h}{2}\omega_0^2\right)\sin(\phi) = 0. \qquad (7.7)$$

These two equations, linear in $\cos(\phi)$ and $\sin(\phi)$, have a non-zero solution only if their determinant is zero, that is:

$$\omega^4 + 2\left(\omega_0^2 + \frac{\omega_e^2}{4}\right)\omega^2 + \left(\omega_0^2 - \frac{\omega_e^2}{4}\right)^2 - \frac{h^2}{4}\omega_0^4 = 0. \qquad (7.8)$$

This is the dispersion equation for the parametric system. A second order equation in $x \equiv \omega^2$, the sum of its two solutions (equal to $-2(\omega_0^2 + \omega_e^2/4) < 0$) is always negative. This means the equation has two real roots, at least one of which is negative. A negative value of ω^2 corresponds to a purely imaginary number (which means $e^{\omega t}$ is not increasing with time, and the solution is stable). Accordingly, the instability emerges for a certain positive value of ω^2. Given that the sum of both roots is negative, the existence of a positive root is possible only if the product of the two roots, given by

$$\left(\omega_0^2 - \frac{\omega_e^2}{4}\right)^2 - \frac{h^2}{4}\omega_0^4,$$

is negative. We can express the necessary condition for instability as:

$$h > 2\left|1 - \left(\frac{\omega_e}{2\omega_0}\right)^2\right|. \qquad (7.9)$$

This instability condition ($\omega^2 > 0$) corresponds to the first cross-hatched area in figure 7.1, on the left. For a weak or nearly zero amplitude excitation h, the parametric instability occurs if $\omega_0/\omega_e \simeq 1/2$. Any excitation with a frequency outside this domain (in cross-hatching) will die out after a short time. However, the condition $\omega^2 > 0$ holds for an infinite number of domains in addition to the one just described. Indeed, it can be shown (see solved problem 7.7.1) that the instability occurs, in addition to the subharmonic one, in any domain centered around points characterized by $\omega_0/\omega_e = n$, with n any positive integer. For example, $n = 1$ corresponds to an excitation with a frequency equal to the natural frequency, such as in the typical case of resonance for a forced oscillator. Thus, in addition to the first hatched area in figure 7.1 (on the left), there exists an infinite number of hatched domains representing the area of instability. Finally, note that the boundaries of figure 7.1 are no longer linear when friction is taken into account, and that the instability no longer occurs for $h = 0$, requiring a finite amplitude.

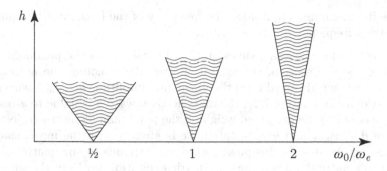

Figure 7.1 – The cross-hatched sections represent the domains of parametric instability. This qualitative figure shows the domains corresponding to the subharmonic and the two harmonics $\omega_0 = \omega_e$ and $\omega_0 = 2\omega_e$.

Why is the case of resonance with a natural frequency equal to exactly one half of the external frequency of special interest? It requires the smallest excitation amplitude h of any realistic parametric oscillator (in other words, whenever friction is taken into account, see solved problem 7.7.1). Also, the instability region associated with subharmonic resonance is wider than that associated with other resonances (see figure 7.1). We will see in solved problem 7.7.1 that the bandwidth of the harmonic resonance (of order $m = 2$, if $m = 1$ refers to the subharmonic resonance) varies proportionally to h^2 while that associated with $m = 1$ varies proportionally to h. Generally, the bandwidth of resonance of order m varies with the excitation amplitude proportionally to h^m.

7.3. An Intuitive Picture of Parametric Resonance

To clearly differentiate classic resonance from parametric resonance, let us investigate the conditions underlying classic resonance.

In a basic description, the emergence of classic resonance in oscillators occurs when the driving force (assumed to be harmonic, i.e. sinusoidal) changes sign after a time equivalent to half the period corresponding to the natural frequency of the oscillator (this means a forcing frequency of $\sin(\omega_0 t)$, which creates a response in the oscillator of the same frequency $\sin(\omega_0 t)$). In other words, the period of the driving force must be equal to the natural period of the oscillator, and simply displaced by half a phase. For example, when pumping a swing: if we apply a harmonic force in order to optimize the movement, the force has to change sign the moment the angle of the swing

reaches its maximum amplitude. The frequency of the force must be equal to
the natural frequency of the oscillator.

This is not so for a parametric excitation. In the case of the pendulum discussed in the last section, the excitation acts like a modulation of gravity.
Suppose a mass is dropped from the maximum amplitude $\theta = \theta_{max}$ with zero
initial velocity (see figure 7.2); the mass heads toward $\theta = 0$ due to a restoring force equal to the projected weight of the pendulum along the trajectory.
Between $\theta = \theta_{max}$ and $\theta = 0$, this force is aligned with the movement, so
the mass accelerates. In this phase, which corresponds to one quarter of the
pendulum's natural period, the parametric excitation works in the same direction as the restoring force. When the mass begins the second quarter of
the trajectory, the phase lies between $\theta = 0$ and $\theta = -\theta_{max}$, the restoring
force of the weight is now in the opposite direction of movement (see figure 7.2), causing a deceleration of the mass. The parametric excitation will
have a tendency to lessen this deceleration. In fact, the acceleration created
from the suspension point intervenes in the effective gravity by changing sign
at the moment the mass begins this second phase. When the mass attains
$\theta = -\theta_{max}$ and begins its third phase (between $\theta = -\theta_{max}$ and $\theta = 0$), the
driving excitation contributes once again to the restoring force (which corre-
sponds again to a change in sign of the excitation). Thus, over the first half
of one period, from θ_{max} to $-\theta_{max}$, the parametric excitation has changed
its sign twice, meaning it has a period two times smaller than that of the
pendulum, or that its frequency is two times higher than the pendulum's
natural frequency: $\omega_e = 2\omega_0$; as you can see, subharmonic resonance.

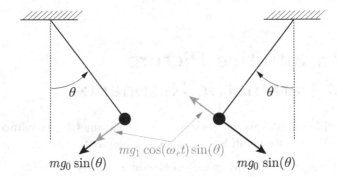

Figure 7.2 – Left: for $0 < \theta < \theta_{max}$, the force associated with the perturbation (proportional to g_1) works in the same direction as gravity. Right: for
$-\theta_{max} < \theta < 0$, the force works in the opposite sense to attenuate the
deceleration cause by weight. Thus, the force has a frequency twice as high
as the system's natural frequency (the direction of action of the force has
changed twice more often than that of the projected weight).

7.4. Universal Amplitude Equation in the Neighborhood of Subharmonic Resonance

Now we want to find the universal amplitude equation in the neighborhood of the threshold value for the parametric instability (this and the questions which follow are discussed with similar terminology in the article by S. Fauve, which I strongly recommend [32]). In a nonlinear study, we have to begin again from the complete pendulum equation (i.e. without making the small angle approximation $\sin\theta \simeq \theta$). Including friction, that nonlinear evolution equation for the simple pendulum looks like:

$$\ddot{\theta} + 2\lambda\dot{\theta} + \omega_0^2 \left[1 + h\cos(\omega_e t)\right]\sin\theta = 0. \tag{7.10}$$

Previously, we have seen how this equation gives rise to a *linear* subharmonic instability. To analytically solve for the *nonlinear effects*, we must place ourselves in the neighborhood of the parametric resonance by setting $\delta = (\omega_0^2/\omega_e^2) - 1/4$. δ measures our distance from the resonance threshold; in analytic expansions it is useful to reformulate this as $\delta = \epsilon\Delta$, with Δ of order one and ϵ the small parameter associated with the proximity to the resonance. The three parameters at our disposition are thus λ, characterizing friction; h, defining the amplitude of the driving force; and δ, the distance from the instability threshold. The situation is interesting when all three effects are equally present (without one dominating the others). To this end, we want to study situations where the friction and forcing are of order ϵ, so we set $\lambda = \epsilon\Lambda$ and $h = \epsilon H$, where Λ and H are of order one. Now we want to find a solution θ of form[2]:

$$\theta(t) = \sqrt{\epsilon}\left[\theta_0(t, T) + \epsilon\theta_1(t, T) + \cdots\right], \tag{7.11}$$

where T is a slow temporal variable. The dissociation of variables into slow and fast can be explained by analysis of the dispersion equation in the presence of friction (equation (7.56)). Since parameters λ and h depend on ϵ, the growth rate of the instability, denoted ω, has a real part which equally

---

2. We have decided to look for a solution of this form because the nonlinear (in θ^3) and linear (in θ) terms from the expansion of $\sin\theta$ make a comparable contribution in ϵ. The linear contribution is preceded by ϵ to reflect that the growth rate of the instability is of order ϵ. The instability corresponds to $\omega = 0$; in the neighborhood of the instability, ω is small and we can neglect the ω^4 in front of ω^2 in the dispersion relation (equation (7.8)). Then comparing the term ω^2 to $(\omega_0^2 - \omega_e^2/4)^2 \sim \epsilon^2$, we can deduce $\omega \sim \epsilon$. In the neighborhood of the instability and in the linear regime, we have $\dot{\theta} = \omega\theta \sim \epsilon\theta$. Comparing this term against the cubic one θ^3, we can anticipate the expansion of θ to a power of ϵ (see equation (7.11)). We can verify after the fact whether our choice is appropriate or not.

depends upon ϵ. In other words, the instability evolves on a timescale of order ϵ^{-1}, and we can thus introduce a slow variable T such that $T = \epsilon t$. At the bifurcation, though, there may be a finite imaginary part of ω (as in the case of the Hopf bifurcation, see chapter 5). This is why, in principle, we should have both slow scale T and fast scale t.

Following the approach presented in chapter 6, we can write down the differentiation operator as follows:

$$\frac{d}{dt} \longrightarrow \frac{\partial}{\partial t} + \epsilon \frac{\partial}{\partial T}. \tag{7.12}$$

We then substitute this expression into the pendulum's nonlinear evolution equation (with friction included, equation (7.10)), as well as using the expressions of the parameters as a function of ϵ and the form of solution θ (equation (7.11)), to find the successive contributions in powers of ϵ. To zeroth order (i.e. in ϵ^0), we obtain the following:

$$\mathcal{L}\theta_0 \equiv \left[\frac{\partial^2}{\partial t^2} + \omega_0^2 \right] \theta_0 = 0, \tag{7.13}$$

with the solution

$$\theta_0 = A(T)e^{i\omega_0 t} + A^*(T)e^{-i\omega_0 t}, \tag{7.14}$$

where $A(T)$ is the amplitude and $A^*(T)$ its conjugate. To order one in ϵ, we have:

$$\mathcal{L}\theta_1 = -2\frac{\partial^2 \theta_0}{\partial t \partial T} - 2\Lambda \frac{\partial \theta_0}{\partial t} - \omega_0^2 H \theta_0 \cos(\omega_e t) + \frac{\omega_0^2}{6}\theta_0^3. \tag{7.15}$$

Now we use the following result from the dispersion relation (7.8): $\omega_e = 2\omega_0 - 4\omega_0 \epsilon \Delta + O(\epsilon^2)$, and keep only the resonant terms ($e^{i\omega_0 t}$) from the second half of the first order (in ϵ) evolution equation (7.15). We have already seen the importance of resonant terms in determining the solvability condition in chapter 6. The homogeneous equation $\mathcal{L}\theta_1 = 0$ has eigenfunction $e^{i\omega_0 t}$. The presence of this term, also called the secular or resonant term, in the second half of the evolution equation (7.15), is what imposes the solvability condition (see chapter 6 for more details), which comes down to eliminating the secular term's coefficients. From this, we derive the evolution equation for amplitude $A(T)$:

$$\frac{\partial A}{\partial T} = -\Lambda A - \frac{i\omega_0}{4}|A|^2 A + \frac{i\omega_0 H}{4} A^* e^{-4i\omega_0 \Delta T}. \tag{7.16}$$

If we introduce variable B by the transformation $A = Be^{-2i\omega_0 \Delta T}$, the amplitude equation for variable B takes an autonomous form (i.e. the coefficients are not time-dependent) as follows[3]:

$$\frac{\partial B}{\partial T} = (-\Lambda + i\nu)B + i\beta|B|^2 B + i\mu B^*, \tag{7.17}$$

where we have defined $\nu = 2\omega_0\Delta$, $\mu = \omega_0 H/4$ and $\beta = -\omega_0/4$.

7.4.1. Determining the Nonlinear Equation Using Symmetry

We could also have exploited symmetries of the problem in order to find the amplitude equation for a subharmonic instability (equation (7.17)). The first point to notice would be the following invariance of the pendulum with friction's nonlinear evolution equation (7.10): $t \to t + 2\pi/w_e$. If we then consider the solution θ_0 obtained for the zeroth order equation (in ϵ, equation (7.14)): that symmetry operation is equivalent to using the transformation $B \to -B$ on the solution θ_0 expressed as a function of variable B (remember $A = Be^{-2i\omega_0\Delta T}$). This invariance of $B \to -B$ is the only constraint we need to determine the amplitude equation we are searching for. It follows that, in addition to the usual terms with B, the conjugate terms B^* are also allowed.

Note that when we were studying the amplitude equation for the stationary as well as the Hopf bifurcation (chapter 5), the equation had to be invariant under the transformation $A \to Ae^{i\theta}$, where θ is an arbitrary phase constant. This constraint did not allow for any terms of the form A^* in the final equation, in contrast to the equation we are studying now.

One solution for the amplitude equation of the subharmonic instability (equation (7.17)) is the trivial solution of $B = 0$. A linear stability analysis (in searching for a solution of form $B = X + iY$), with $X, Y \sim e^{\omega t}$ leads us to the following dispersion equation:

$$\omega^2 + 2\Lambda\omega + \Lambda^2 - \mu^2 + \nu^2 = 0. \tag{7.18}$$

The trivial solution is unstable if $\omega > 0$, that is, if

$$\mu > \mu_c = \sqrt{\Lambda^2 + \nu^2}. \tag{7.19}$$

The stability diagram in plane (μ–ν) (see figure 7.3) shows similar results to the ones found by studying the instability zones for subharmonic resonance (see figures 7.1 and 7.7).

---

3. This means that we are in the rotating frame of the frequency $w_e/2$, in contrast to the situation described by the expression of θ_0 (see equation (7.14)), associated with a frequency equal to w_0.

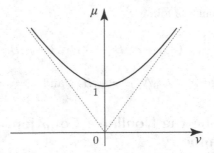

Figure 7.3 – Diagram of the parametric instability found from an amplitude equation. In the area below the curve, the solution $B = 0$ is unstable. The dotted lines show the limiting frictionless case, while the continuous curve traces the solution including friction.

7.5. Nonlinear Instability

The amplitude equation associated with the subharmonic instability (equation (7.17)) has another, non-trivial solution (i.e. $B \neq 0$). Defining $B = Re^{i\theta}$, we can separate the equation into its real and imaginary parts:

$$\partial_t R = -\Lambda R + \mu \sin(2\theta) R, \tag{7.20}$$

$$\partial_t \theta = \nu + \beta R^2 + \mu \cos(2\theta). \tag{7.21}$$

In addition to the trivial solution $R_0 = 0$, this system of equations has two fixed points θ_0 and R_0 satisfying

$$\sin(2\theta_0) = \frac{\Lambda}{\mu}, \quad |\beta| R_0^2 = -\nu \pm \sqrt{\mu^2 - \Lambda^2}, \tag{7.22}$$

with $\mu > \Lambda$.

We consider the case $\beta > 0$ (the case $\beta < 0$ leads to the same conclusion and this physically corresponds to work with the complex conjugate amplitude equation). The solution branches for $\nu < 0$ and $\nu > 0$, traced in the plane $(\mu\text{-}\nu)$ (see figure 7.4; the thin gray lines are not physical solutions since they correspond to $R_0^2 < 0$), comprise a stable part for $R_0^2 > 0$ (solid line), and an unstable part (dotted line). It can be shown that the solution $R_0 = 0$ becomes unstable at $\mu = \mu_c = \sqrt{\nu^2 + \Lambda^2}$.

The trivial solution bifurcates continuously for $\nu > 0$ toward the upper branch (the branch found above the μ-axis, see figure 7.4(b)): this represents a supercritical bifurcation.

For $\nu < 0$, we have instead a subcritical bifurcation (as introduced in chapter 3). The solution $R_0 = 0$ can transition (see figure 7.4(a)) toward the upper stable branch even before the instability threshold $\mu = \mu_c$ is reached.

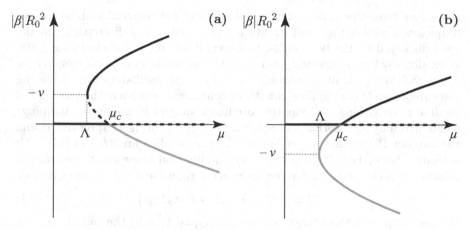

Figure 7.4 – Solution branches for (a) $\nu < 0$; and (b) $\nu > 0$. The solid lines correspond to stable solutions. The dotted lines represent unstable branches. The thin gray lines are non-physical branches since they correspond to $R_0^2 < 0$.

The point $\nu = 0$ is called the tricritical point in virtue of the fact that the bifurcation is subcritical for $\nu \to 0^-$ (approaching zero from the negative side) and supercritical for $\nu \to 0^+$. The tricritical point has a particular property; if we set $\mu = \mu_c + \epsilon$ and $\nu = 0$, we will see that $R_0 \sim \epsilon^{1/4}$ (as opposed to the pitchfork bifurcation where $R_0 \sim \epsilon^{1/2}$).

We could also use the traditional approach and analyze the stability by adding a perturbation to each term, $\theta = \theta_0 + \theta_1(t)$ and $R = R_0 + R_1$, expanding the two amplitude equations corresponding to the real and imaginary parts of the subharmonic instability (equation (7.21)) to order one in R_1 and in θ_1, and finally searching for a solution of form $e^{\omega t}$. We would easily come to the following equation:

$$\omega^2 + 2R_0 \Lambda \omega + 4\beta R_0^2 (\nu + \beta R_0^2) = 0. \tag{7.23}$$

The solution is unstable when $\nu - |\beta| R_0^2 > 0$, with $\beta < 0$; this corresponds to part of the solution branch just below the turning point (see figure 7.4(a)).

7.6. Phase Locking
or Phase Synchronization

Synchronization dynamics often play a role in nonlinear science. Imagine we set two clocks at time t to indicate the exact same time, and then let them run independently of each other (i.e. not coupled). In a real experiment, the clocks will never have the same exact frequency, and this inevitable difference in frequency, however tiny, will eventually be evident in a discrepancy in the time displayed by the two clocks. However, if we couple the clocks (or more generally, any two oscillators), even weakly, a phase-locking phenomenon may occur and bring about synchronization (i.e. both oscillators may have the same frequency). We will illustrate this phenomenon as seen in the van der Pol oscillator (describing a dissipative oscillator subject to nonlinear damping) when driven by an external voltage of frequency ω_e. In the chapter describing the van der Pol oscillator (see section 5.1) we studied an RLC circuit with a tunnel diode. For this situation, we modify and complete the evolution equation (5.4) by adding a forcing term to the right-hand side of the equation:

$$\ddot{I} - (2\mu\epsilon - I^2)\dot{I} + \omega_0^2 I = F\cos(\omega_e t). \tag{7.24}$$

We have replaced the friction coefficient (equal to ϵ in the initial van der Pol evolution equation (5.4)) by $2\mu\epsilon$. We use this different notation in order to later facilitate the comparison between this problem and the parametric oscillator.

In the neighborhood of $\epsilon \simeq 0$, the dynamics are slow (see the discussion of critical slowing down introduced in chapter 2). The amplitude follows an exponential behavior $e^{\epsilon t}$, multiplied by a temporally oscillating function (i.e. $\sin t$). To find the amplitude equation, we first separate slow variable $T = \epsilon t$ from fast variable t and follow the same approach as earlier in this chapter for the parametric oscillator. We then set[4] $I = I_0 + \epsilon I_1 + \cdots$.

7.6.1. Non-resonant Forcing

First let us consider the case of non-resonance, where no phase locking occurs ($\omega_e \neq \omega_0$). The evolution equation for the forced van der Pol oscillator (equation (7.24)) expanded to zeroth order (ϵ^0) is as follows:

$$\mathcal{L}I_0 \equiv \left[\frac{\partial}{\partial t^2} + \omega_0^2\right]I_0 = F\cos(\omega_e t), \tag{7.25}$$

with solution:

$$I_0 = A(T)e^{i\omega_0 t} + c.c. + \frac{F}{\omega_0^2 - \omega_e^2}\cos(\omega_e t). \tag{7.26}$$

To order ϵ we obtain:

$$\mathcal{L}I_1 = \left(2\mu - I_0^2\right)\frac{\partial I_0}{\partial t} - 2\frac{\partial^2 I_0}{\partial t \partial T}. \tag{7.27}$$

---

4. Note the difference between this and the van der Pol oscillator we worked with in chapter 6, where the expansion about I begins with I_0 and not I_1. This is due to the fact that in this case, there is a forced zero order solution, $F\cos(\omega_e t)$, while in the case of the van der Pol oscillator, the solution to the lowest order is $I = 0$; leaving the leading term to be in ϵ.

Using the zero order solution, we obtain the solvability equation by setting the factor in front of the secular term to zero ($e^{i\omega_0 t}$):

$$\frac{\partial A}{\partial T} = \xi A - \frac{1}{2}|A|^2 A, \tag{7.28}$$

where we have defined

$$\xi = \mu - \frac{F^2}{4[\omega_e^2 - \omega_0^2]^2}. \tag{7.29}$$

This equation is formally identical to that obtained from a van der Pol equation in the absence of forcing (see equation (6.18)), with $\gamma = 0$ and $\mu = 1$. At this order, the only effect of the forcing is a delay in the onset of instability (see figure 7.5). In other words, the forcing has a tendency to stabilize the system; eventually, a perturbation could completely stabilize an unstable solution. This is the case for the free inverted pendulum; the position of the mass in the vertical position pointing upwards is unstable. However if the pendulum is subject to a perturbation, this position can become stable. You can perform a simple experiment yourself by placing a rod vertically at the center of your hand; at rest, this position is unstable. You will discover that the rod is stabilized if you vibrate your hand in the right way (for a recent study on the stability of the inverted pendulum see [94]).

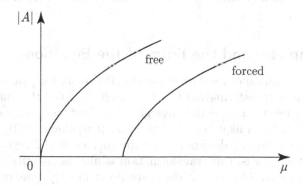

Figure 7.5 – Solution branches in the absence and presence of forcing.

7.6.2. Resonant Forcing

We will now consider the case of resonant forcing. To simplify the analysis, we will assume that the forcing is of order ϵ and define $f = \epsilon F$, with F of order one. This choice implies that solution I_0 (equation (7.26)) does not have a term in F, which makes the analytic expression simpler without modifying any general conclusion. To show explicitly that the forcing is resonant, we expand the external frequency to first order in ϵ, as $\omega_e = \omega_0 + \epsilon\omega_1$ (we suppose we are close to resonance), with ω_1 of order one. From the evolution equation of the forced van der Pol oscillator (equation (7.24)), we obtain, to leading order:

$$I_0 = A(T)e^{i\omega_0 t} + c.c. \tag{7.30}$$

To the next order, the condition for canceling the factor of $e^{i\omega_0 t}$ lets us set up the amplitude equation:

$$\frac{\partial A}{\partial T} = \mu A - \frac{1}{2}|A|^2 A - \frac{iF}{4\omega_0}e^{i\omega_1 T}. \tag{7.31}$$

Using the change of variable $A = Be^{i\omega_1 T}$, we obtain an amplitude equation for the forced van der Pol oscillator described in the rotating frame:

$$\frac{\partial B}{\partial T} = (\mu - i\omega_1) B - \frac{1}{2}|B|^2 B - \frac{iF}{4\omega_0}. \tag{7.32}$$

If we also rewrite the leading order solution for the forced van der Pol oscillator (equation (7.30)) as a function of variable B, we have

$$I_0 = B(T)e^{i(\omega_0 + \epsilon\omega_1)t} + c.c. = B(T)e^{i(\omega_e)t} + c.c. \tag{7.33}$$

This form expresses the solution in the reference frame turning at the same frequency as the excitation ω_e, whereas the previous form as a function of variable A (equation (7.30)) expresses the solution in the reference frame of the natural frequency.

7.6.3. Symmetry and the Form of the Equation

The evolution equation of the forced van der Pol oscillator (equation (7.24)) is invariant under transformation $t \to t + 2\pi n/\omega_e$, and so the final equation derived from it must conserve this invariance. To leading order, we have seen that the solution looks like: $I_0 = Be^{i\omega_e t} + c.c.$ (equation (7.33)). The transformation $t \to t + 2\pi n/\omega_e$ does not affect the coefficient of B, which depends on $e^{i\omega_e t}$. In other words, this transformation is like the identity operator on variable B: $B \to B$. This means there are no symmetry constraints placed on the evolution equation when it is expressed in terms of B, allowing, for example, the introduction of a term B^2. This last term can be eliminated through another change of variables $B \to B + c$, where c is a constant chosen so that B^2 disappears from the equation. However, this operation comes at a price: the amplitude equation then contains a term which is constant (of the same type as $-iF/4\omega_0$ in equation (7.32)). We will further consider these developments in the following section.

7.6.4. High Order Phase Locking

In this section we will use symmetries in order to derive the amplitude equation in the case of a higher order phase locking. Suppose the van der Pol oscillator is excited by a force $F\cos(n\omega_0 t/p)$, with integers n and p. Going step by step through the same calculations as before, we obtain a solution:

$$I_0 = B(T)e^{ip\omega_e t/n} + c.c. \tag{7.34}$$

The invariance of the original equation under transformation $t \to t + 2\pi/\omega_e$ imposes an invariance under transformation $B \to Be^{ip\omega_e t/n}$ on the amplitude equation expressed as a function of B. We can verify that the following terms $dB/dT, B, |B|^2 B$ all transform in the same fashion:

$$(dB/dT, B, |B|^2 B) \to (dB/dT, B, |B|^2 B)e^{2ip\pi/n}.$$

The transformation also allows terms of the form B^{*n-1}. In fact, operation $B \to Be^{ip\omega_e t/n}$ results in the following transformation: B^{*n-1} to $B^{*n-1}e^{2ip\pi}e^{2ip\pi/n} = B^{*n-1}e^{2ip\pi/n}$, which is identical to the transformation of terms $(dB/dT, B, |B|^2 B)$. Thus, the general form of the equation that B must satisfy is:

$$\frac{dB}{dT} = (\mu + i\nu)B - \beta|B|^2 B - \alpha B^{*n-1}, \tag{7.35}$$

where we have chosen a real[5] number $\beta > 0$ (as in the case of the van der Pol oscillator). α, which qualifies the forcing, is also chosen to be real[6]. The scenario discussed in the last section corresponds to $n = 1$. The parameter ν measures the displacement from the resonance frequency (the same way that ω_1 did for $n = 1$). For the sake of simplicity, we will not go beyond the case $n = 2$, and will set $\beta = 1$ (without loss of generality).

The real and imaginary parts of the general amplitude equation expressed in B for a van der Pol oscillator forced by $F\cos(n\omega_0 t/p)$ (equation (7.35)) are:

$$\partial_t R = \mu R - R^3 - \alpha R\cos(2\theta),$$
$$\partial_t \theta = \nu + \alpha\sin(2\theta). \tag{7.36}$$

We have eliminated the trivial solution $R = 0$ from the equation associated with the phase (because for that case, the equation is automatically satisfied, no matter the phase!). There exists another set of fixed points which we denote θ_0, described by: $\sin(2\theta_0) = -\nu/\alpha$. This solution exists only for $|\nu/\alpha| < 1$. The two fixed points can also be written as

$$2\theta_0 = -\arcsin\left(\frac{\nu}{\alpha}\right), \quad 2\theta_0 = \pi + \arcsin\left(\frac{\nu}{\alpha}\right). \tag{7.37}$$

---

5. The fact that β is real is not a general result, but a constraint adopted here for the sake of simplicity.

6. In the evolution equation (7.32), the forcing term is purely imaginary, and one must simply change the time origin to obtain a forcing term that looks like sin instead of cos, thereby passing from one case to the other.

A linear stability study of these points does not present any difficulties (the phase equation does not contain R) and we find that the first branch is unstable while the second is stable.

We are left with the task of examining the stability for solution $R = 0$. The linearized equation for $R(t)$ of the system (equation (7.36)) around fixed point $R = 0$ gives:

$$\partial_t R_1 = [\mu - \alpha \cos(2\theta_0)] R_1, \qquad (7.38)$$

whose solution is $R_1 \sim e^{\omega t}$ with $\omega = \mu - \alpha \cos(2\theta_0) = \mu \pm \sqrt{\alpha^2 - \nu^2}$.

The solution is stable only if both of the eigenvalues are negative: for eigenvalue $\mu + \sqrt{\alpha^2 - \nu^2}$ this means that $\mu < 0$. To assure that both solutions are less than zero, or $\omega < 0$, we must impose $\mu < -\sqrt{\alpha^2 - \nu^2}$. These constraints outline the regions of stability: for $\omega = 0$ (with a null imaginary component), the bifurcation is stationary, and above the parabolic curve traced in the plane $(\nu-\mu)$ (see figure 7.6), the solution is stable. The state $R = 0$ below the parabola is called the *forced state*. Indeed, for $R = 0$ (and thus $A = B = 0$), the solution to the forced van der Pol oscillator's zeroth order evolution equation (7.26) shows that the oscillator is simply forced at the frequency of the external force, ω_e.

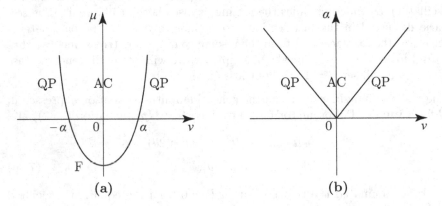

Figure 7.6 – Diagram of possible solutions: QP (quasi-periodic); AC (frequency locked); and F (forced state). (a) $\mu \neq 0$; (b) $\mu = 0$.

When $R \neq 0$, the state is said to be "locked": the oscillator (whose frequency is slightly shifted with respect to its angular frequency ω_0, as for the non-forced van der Pol oscillator) synchronizes with the angular frequency $\omega_e p/n$ of the excitation force (here, we restrain ourselves to the case $n = 2$). The phase difference between the excitation and that of the oscillator is "locked" at value θ_0.

The search for the fixed point when $R \neq 0$ gives

$$R_0^2 = \mu + \sqrt{\alpha^2 - \nu^2}, \tag{7.39}$$

with $\mu + \sqrt{\alpha^2 - \nu^2} > 0$. In other words, this solution exists above the parabola traced in plane $(\nu$–$\mu)$ (see figure 7.6). A linearization lets us establish that this solution (associated with the phase $\theta_0 = \pi + \arcsin(\nu/\alpha)$) is stable.

Once again, looking at the real and imaginary parts of the general amplitude equation (written with variable B for a van der Pol oscillator forced by $F \cos(n\omega_0 t/p)$, equation (7.35)), we have:

$$\partial_t R = \mu R - R^3 - \alpha R \cos(2\theta),$$
$$\partial_t \theta = \nu + \alpha \sin(2\theta). \tag{7.40}$$

In the case where $|\nu/\alpha| > 1$, the solution of the general amplitude equation for the forced van der Pol oscillator (equation (7.36)), θ, becomes an oscillating function (in a situation analogous to the tumbling movement of a pendulum, see subsection 3.1.7). Since R is coupled to θ, R also oscillates in time. When $|\nu/\alpha| < 1$, a transition[7] occurs marking the evolution of the "frequency locked" state toward a quasi-periodic state; the term "quasi" referring to the fact that the frequency of the oscillator intrinsically has no simple (i.e. rational) relationship to the frequency of the excitation (see figure 7.6 for a representation of the solutions and boundaries of the bifurcations).

Finally, let us note that quasi-periodic dynamics, along with subharmonic bifurcations, constitute two of three possible scenarios in the transition toward chaos (see chapter 8, dedicated to the study of chaos). For a great number of interesting examples, see Drazin [29], Strogatz [99], and Demazure [27].

7.7. Solved Problem

7.7.1. Instability of Higher Harmonics and the Effect of Friction

Make a study of the instabilities associated with harmonics $\omega_0 = \omega_e$ and $\omega_0 = 2\omega_e$. Compare the minimum amplitude required to induce the instability with the amplitude necessary to trigger the subharmonic instability.

Solution. In order to simplify the calculation, we will adopt a different method for analyzing the instability than before, and first apply it to the subharmonic instability to check that it gives correct results. The two methods

---

7. A saddle-point bifurcation.

are equivalent as long as the analysis is confined to be in close proximity to the instability point.

We first place ourselves in the neighborhood of the subharmonic instability by imposing

$$\omega_e = 2\omega_0 + \epsilon, \tag{7.41}$$

where ϵ is a small parameter. The parametric evolution equation (7.2) is then written as:

$$\ddot{\theta} + \omega_0^2 \left[1 + h\cos(2\omega_0 + \epsilon)t\right]\theta = 0. \tag{7.42}$$

We now search for a solution of form:

$$\theta(t) = a_0 \cos\left(\omega_0 + \frac{\epsilon}{2}\right)t + b_0 \sin\left(\omega_0 + \frac{\epsilon}{2}\right)t, \tag{7.43}$$

where a_0 and b_0 are two general functions of time, such that $a_0, b_0 \sim e^{\omega t}$ (equation (7.5)). The bifurcation is defined by $\omega = 0$ (ω is a real number at the bifurcation). If we wish to determine the neutral curve separating the stable and unstable regions (as in figure 7.1), we can simply start with $\omega = 0$ (to simplify the algebra). This is equivalent to considering a_0 and b_0 as constants. Plugging the solution for $\theta(t)$ (equation (7.43)) into the evolution equation in the neighborhood of the bifurcation (equation (7.42)) we obtain:

$$\left[-a_0\left(\epsilon\omega_0 + \frac{\epsilon^2}{4}\right) + \frac{h\omega_0^2 a_0}{2}\right]\cos\left(\omega_0 + \frac{\epsilon}{2}\right)t$$

$$+ \left[-b_0\left(\epsilon\omega_0 + \frac{\epsilon^2}{4}\right) - \frac{h\omega_0^2 b_0}{2}\right]\sin\left(\omega_0 + \frac{\epsilon}{2}\right)t = 0. \tag{7.44}$$

Eliminating the coefficients of the sinusoidal functions, we obtain the following relationship:

$$\epsilon = \pm\frac{1}{2}h\omega_0. \tag{7.45}$$

The region of instability is given by:

$$|\epsilon| < \frac{1}{2}h\omega_0, \tag{7.46}$$

where we have kept the leading term in ϵ. Note also that, since h is small, the terms in $h\epsilon$ are of a higher order. This result (equation (7.46)) is analogous to that obtained in the first analysis we made of the subharmonic oscillator (equation (7.9)) in which, by keeping the dominant term in ϵ, we had $\omega_e = 2\omega_0 + \epsilon$. The result we have just written (equation (7.46)) is only valid in the neighborhood of ω_0, and as such it is wise to restrict ourselves to the lowest order expansion in $\omega_e - 2\omega_0 \equiv \epsilon$.

Effect of Higher Harmonics in Subharmonic Resonance

The study of higher order harmonics requires that we expand the evolution equation in the neighborhood of the bifurcation to the next order. The product of $h \cos(2\omega_0 + \epsilon)t$ and $a_0 \cos(\omega_0 + \epsilon/2)t$ gives rise to terms $\cos 3(\omega_0 + \epsilon/2)t$ which were ignored. For coherence, we must now also search for a solution of the form:

$$\theta(t) = a_0 \cos\left(\omega_0 + \frac{\epsilon}{2}\right)t + b_0 \sin\left(\omega_0 + \frac{\epsilon}{2}\right)t + a_1 \cos 3\left(\omega_0 + \frac{\epsilon}{2}\right)t$$
$$+ b_1 \sin 3\left(\omega_0 + \frac{\epsilon}{2}\right)t. \tag{7.47}$$

The evolution equation becomes, in the presence of the third harmonic:

$$\left[-a_0\left(\epsilon\omega_0 + \frac{\epsilon^2}{4}\right) + \frac{h\omega_0^2 a_0}{2} + \frac{h\omega_0^2 a_1}{2}\right] \cos\left(\omega_0 + \frac{\epsilon}{2}\right)t$$

$$+ \left[-b_0\left(\epsilon\omega_0 + \frac{\epsilon^2}{4}\right) - \frac{h\omega_0^2 b_0}{2} + \frac{h\omega_0^2 b_1}{2}\right] \sin\left(\omega_0 + \frac{\epsilon}{2}\right)t$$

$$+ \left[\frac{h\omega_0^2 a_0}{2} - 8\omega_0^2 a_1\right] \cos 3\left(\omega_0 + \frac{\epsilon}{2}\right)t$$

$$+ \left[\frac{h\omega_0^2 b_0}{2} - 8\omega_0^2 b_1\right] \sin 3\left(\omega_0 + \frac{\epsilon}{2}\right)t = 0. \tag{7.48}$$

Setting all four coefficients to zero gives us our results. The two last terms define constants a_1 and b_1 as $a_1 = ha_0/16$, $b_1 = hb_0/16$, and the first two terms give us the following relation: $\omega_0\epsilon \pm h\omega_0^2/2 + \epsilon^2/4 - h^2\omega_0^2/32 = 0$. To the order h^2, the new thresholds of the instability are given by:

$$\epsilon = \pm\frac{1}{2}h\omega_0 - \frac{h^2\omega_0}{32}. \tag{7.49}$$

Comparing this to the instability condition for the subharmonic oscillator (equation (7.46)), we see that taking into account the third harmonic in our study of the subharmonic resonance only changes the threshold of the bifurcation by an order of h^2, creating a subdominant effect with respect to the terms of order h.

Harmonic Resonance

Now let us look at the principal resonance, which emerges when the external frequency is equal to the natural frequency of the oscillator: $\omega_e = \omega_0$. We place ourselves in the neighborhood of this resonance by positing:

$$\omega_e = \omega_0 + \epsilon, \tag{7.50}$$

where ϵ is a small parameter.

The parametric evolution equation (7.2) takes the form:

$$\ddot{\theta} + \omega_0^2 \left[1 + h\cos(\omega_0 + \epsilon)t\right]\theta = 0. \tag{7.51}$$

We seek solutions in the form:

$$a_0 \cos(\omega_0 + \epsilon)t + b_0 \sin(\omega_0 + \epsilon)t$$
$$+ a_1 \cos 2(\omega_0 + \epsilon)t + b_1 \sin 2(\omega_0 + \epsilon)t + c_1 \tag{7.52}$$

where a_0, a_1, b_0, b_1, c_1 are constants that will be determined later. In contrast to subharmonic resonance where ϵ is of order h (equation (7.46)), one may notice that, at the instability threshold, ϵ is of the order h^2 ($\epsilon \simeq h^2$). Thus we need to take the next harmonic (i.e. $2\omega_e = \omega_0$) into consideration, in order to keep the expansion coherent.

Also, we have introduced a "zero"th harmonic term, because the product of the function $h\cos(\omega_0 + \epsilon)t$ and $a_0 \cos(\omega_0 + \epsilon)t$ gives us, besides a second harmonic, a constant term which we call c_1.

If we substitute the form of the solution we are searching for (equation (7.52)) into the expanded parametric evolution equation just elaborated (equation (7.51)), we obtain:

$$\left(-2a_0\epsilon\omega_0 + \frac{h\omega_0^2 a_1}{2} + h\omega_0^2 c_1\right)\cos(\omega_0 + \epsilon)t$$

$$+ \left(-2b_0\epsilon\omega_0 + \frac{h\omega_0^2 b_1}{2}\right)\sin(\omega_0 + \epsilon)t + \left(-3a_1\omega_0^2 + \frac{\omega_0^2 a_0 h}{2}\right)\cos 2(\omega_0 + \epsilon)t$$

$$+ \left(-3b_1\omega_0^2 + \frac{h\omega_0^2 b_0}{2}\right)\sin 2(\omega_0 + \epsilon)t + \left(c_1\omega_0^2 + \frac{h}{2}\omega_0^2 a_0\right) = 0, \tag{7.53}$$

where the solution determines (by zeroing the terms between brackets in equation (7.53)) the relation between the different constants: $a_1 = ha_0/6$, $b_1 = hb_0/6$, $c_1 = -ha_0/2$, and the two limits – inferior and superior – of the instability zone are given by:

$$\epsilon_- = -\frac{5h^2\omega_0}{24}, \quad \epsilon_+ = \frac{h^2\omega_0}{24}. \tag{7.54}$$

Comparing these domains of instability with those of the subharmonic instability (equation (7.46)) shows clearly that the harmonic resonance has a narrower instability zone (due to its dependence on h^2) than that of the subharmonic resonance (of dependence h).

The Effect of Friction

Friction is taken into account by adding $2\lambda\dot{\theta}$ to the evolution equation (7.2), where λ is the friction coefficient. This gives us the following:

$$\ddot{\theta} + 2\lambda\dot{\theta} + \omega_0^2 \left[1 + h\cos(\omega_e t)\right]\theta = 0. \tag{7.55}$$

The Floquet–Bloch theorem remains applicable, and following an analogous calculation to the one in the last section, we obtain the condition of a null determinant which allows us to establish the new dispersion equation, naturally more complicated than the one for subharmonic instability (equation (7.8)):

$$
\omega^4 + 4\lambda\omega^3 + \left(\frac{\omega_e^2}{2} + 4\lambda^2 + 2\omega_0^2\right)\omega^2 + \left(\omega_e^2\lambda + 4\omega_0^2\lambda\right)\omega
$$
$$
+ \left(\omega_0^2 - \frac{\omega_e^2}{4}\right)^2 - \frac{h^2}{4}\omega_0^4 + \lambda^2\omega_e^2 = 0. \tag{7.56}
$$

To verify the result, we need just remark that in the case of subharmonic resonance, and if we limit ourselves only to the dominant order, the first equation of the evolution equation of the system (7.44) acquires the supplementary term $2\lambda\omega_0 b_0$, while the second equation has an additional term $-2\lambda\omega_0 b_0$. Annulling the determinant of the system, to avoid trivial solution $a_0 = b_0 = 0$, leads to the dispersion equation (7.56). The instability threshold is found by positing $\omega = 0$ in the dispersion equation (7.56). We find:

$$
\epsilon = \pm\sqrt{\left(\frac{h\omega_0}{2}\right)^2 - 4\lambda^2}. \tag{7.57}
$$

The minimum amplitude of the subharmonic instability of a system with friction is thus given by:

$$
h_{\min} = \frac{4\lambda}{\omega_0}. \tag{7.58}
$$

We are now in a position to show that the subharmonic instability has a lower threshold than the principal harmonic. First we determine the threshold resonance of the principal harmonic including friction by using, for variable θ, the expression we used for the study of harmonic resonance without friction (equation (7.52)). The effect of friction modifies the evolution of the system (equation (7.53)) by including new contributions $2\lambda\omega_0 b_0$, $-2\lambda\omega_0 a_0$, $4\lambda\omega_0 b_0$, and $2\lambda\omega_0 a_0$ which modulate the coefficients in front of the sine and cosine functions. Canceling out the coefficients determines the expression for constants a_0, b_0, a_1, and b_1; the value for constant c_1 is identical to its previous value (i.e. $c_1 = -ha_0/2$).

The compatibility condition determines the zones of the instability. With the goal of simplifying the presentation of results, we make an expansion limited to the lowest order in h and λ (neglecting also $h^2\lambda^2$ terms in front of $h^4\omega_0^2$). The two limits, upper and lower, in $\epsilon(h)$ of the harmonic instability zone, including friction, are given by:

$$
\epsilon_+ = \frac{h^4\omega_0^2 - 96\lambda^2}{24h^2\omega_0}, \qquad \epsilon_- = \frac{-5h^4\omega_0^2 + 96\lambda^2}{24h^2\omega_0}. \tag{7.59}
$$

The two branches cross for the minimum value $h'_{\min}$ of the instability threshold h given by:

$$h'_{\min} = 2(2)^{1/4} \left(\frac{\lambda}{\omega_0}\right)^{\frac{1}{2}}. \tag{7.60}$$

The minimum amplitude required for harmonic resonance, $h'_{\min}$, is of order $\lambda^{1/2}$ (λ is assumed to be small); an amplitude larger than the minimal amplitude, $h_{\min}$, is needed for subharmonic instability (equation (7.58)).

Thus, compared to the principal harmonic, subharmonic resonance requires a smaller perturbation amplitude. Generally, the minimal amplitude required to attain a resonance of order m (i.e. $2\omega_0 = m\omega_e$, with m an integer) is of order $\lambda^{1/m}$. Plotted in the plane $((\omega_0/\omega_e)–h)$, the zones of instability correspond to different harmonics which are called *Arnold tongues* (see the cross-hatched domains in figure 7.7).

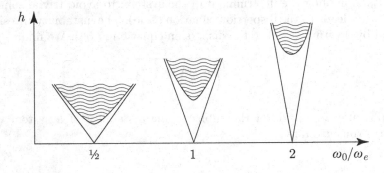

Figure 7.7 – The cross-hatched sections represent the domains of parametric instability with friction. We also show the domain found for a system without friction (the conical domains figure 7.1). In this qualitative figure we show the domain which corresponds to the subharmonic and the two harmonics $\omega_0 = \omega_e$ and $\omega_0 = 2\omega_e$.

7.8. Exercises

7.1

Consider the following nonlinear equation, known as the Duffing equation:

$$\ddot{x} + k\dot{x} + ax + bx^3 = \Gamma \cos(\omega t). \tag{7.61}$$

We wish to analyze the effect of an external forcing (right-hand side term) close to and far from the resonance condition.

1. Consider first the linear problem in the absence of friction and show that the solution is given by:

$$x(t) = A\cos(\sqrt{a}t) + B\sin(\sqrt{a}t) + \frac{\Gamma\cos(\sqrt{a}t)}{a - \omega^2}, \qquad (7.62)$$

provided that $\omega^2 \neq a$.

2. As can be seen above, the solution diverges (resonance) if $\omega^2 = a$. In that case, show that the solution has the form:

$$x(t) = A\cos(\omega t) + B\sin(\omega t) + \frac{\Gamma t\sin(\omega t)}{2\omega}. \qquad (7.63)$$

3. Introduce now the friction term (by still focusing on the linear regime). We assume that $k > 0$ and that $k^2 < 4a$ (weak friction). Show that the general solution takes the form:

$$x(t) = e^{-kt/2}\left[A\cos\left((a - k^2/4)^{1/2}t\right) + B\sin\left((a - k^2/4)^{1/2}t\right)\right]$$
$$+ \frac{\Gamma[(a - \omega^2)\cos(\omega t) + k\omega\sin(\omega t)]}{(a - \omega^2)^2 + k^2\omega^2}. \qquad (7.64)$$

Under which condition do we have resonance? Suppose now that t is large enough so that the transient term vanishes. Plot the amplitude of the oscillation as a function of ω.

4. Let us now introduce the nonlinear terms. We focus on the weakly non-linear regime by setting $b = -\epsilon$ (where ϵ is a small parameter). For simplicity, let us set $\omega = 1$ (which simply means that we measure time in units of ω^{-1}). We use the following notation: $a \equiv \Omega^2$, and we disregard, in a first step, the friction term, so that the Duffing equation becomes:

$$\ddot{x} + \Omega^2 x - \epsilon x^3 = \Gamma\cos t. \qquad (7.65)$$

We seek solutions in the following power series of the small parameter ϵ:

$$x(t) = x_0(t) + \epsilon x_1(t) + \epsilon^2 x_2(t) + \cdots. \qquad (7.66)$$

Write the equations obeyed by x_0, x_1 and x_2. Show then that the solution for x_0 is given by:

$$x_0(t) = A\cos(\Omega t) + B\sin(\Omega t) + \frac{\Gamma\cos t}{(\Omega^2 - 1)}, \qquad (7.67)$$

if Ω is non-integer ($\Omega \neq \pm 1$). Explain this condition. We restrict ourselves in what follows to 2π-periodic solutions.

5. Show that x_1 is given by:

$$x_1(t) = \frac{3\Gamma^3 \cos t}{4(\Omega^2 - 1)^4} + \frac{\Gamma^3 \cos(3t)}{4(\Omega^2 - 1)^3(\Omega^2 - 9)}, \qquad (7.68)$$

Explain why this method is not valid near resonance.

In order to continue our analysis, we could adopt the same method used for the van der Pol oscillator in this chapter. However, it is instructive to use an alternative.

6. Recall that when we dealt with the forced van der Pol equation, the zeroth order equations did not contain a right-hand side term. That was the consequence of our hypothesis that the forcing is of order ϵ, so that the forcing did not enter the leading order equation. In the analysis above, there was no possibility to kill the so-called secular terms. Consider again equation (7.61) without friction and set $\tau = \omega t$. Show that the equation becomes:

$$\ddot{x} + x = \epsilon \left[\gamma \cos \tau - \beta x + x^3 \right], \qquad (7.69)$$

where the derivative acts on τ now. Express the relations between the new coefficients (β, γ) and the old ones. We choose β and γ of order unity. The novel point now is that in the limit $\epsilon \to 0$ the oscillation occurs with pulsation unity. How should the scaling of the coefficients with ϵ be chosen in ordre to comply with expression 7.69 Can you comment on the relations between the coefficients?

7. Show by adopting an expansion in the power series of ϵ that:

$$x_0(t) = a_0 \cos \tau + b_0 \sin \tau. \qquad (7.70)$$

8. Determine the equation satisfied by $x_1(t)$ and regroup the terms as common factors of a linear combination in $\sin \tau$, $\cos \tau$, $\sin(3\tau)$, and $\cos(3\tau)$.

9. By eliminating the secular terms, show that we have the relations:

$$a_0 \left[\beta - \frac{3}{4} (a_0^2 + b_0^2) \right] = \gamma, \quad b_0 \left[\beta - \frac{3}{4} (a_0^2 + b_0^2) \right] = 0. \qquad (7.71)$$

Find the solutions for a_0 and b_0 in terms of the forcing coefficient and β. Plot $|a_0|$ as a function of β for $\gamma > 0$, $\gamma = 0$ and $\gamma < 0$. What can you conclude?

10. Let us come back now to our original variables of equation (7.61). By using the above results, show that:

$$\omega^2 = a + \frac{3}{4} b a_0^2 - \frac{\Gamma}{a_0}. \qquad (7.72)$$

Comment on this result. Plot a_0 as a function of ω. Consider the case $\Gamma a_0 > 0$ and $\Gamma a_0 \gg 0$, as well as $\Gamma = 0$.

11. Let us now introduce the friction and show that:

$$\ddot{x} + x = \epsilon \left[\gamma \cos \tau - \kappa \dot{x} - \beta x + x^3 \right]. \tag{7.73}$$

By following the same strategy as above (in particular, by eliminating secular terms), show that:

$$\kappa b_0 + a_0 \left[\beta - \frac{3}{4} \left(a_0^2 + b_0^2 \right) \right] = \gamma, \quad \kappa a_0 - b_0 \left[\beta - \frac{3}{4} \left(a_0^2 + b_0^2 \right) \right] = 0. \tag{7.74}$$

Deduce the following relation:

$$r_0^2 \left[\kappa^2 + \left(\beta - \frac{3}{4} r_0^2 \right)^2 \right] = \gamma^2, \tag{7.75}$$

where $r_0 = (a_0^2 + b_0^2)^{1/2}$. Using the relations between the coefficients of (7.61) and (7.69), express r_0 as a function of ω and Γ. Plot r_0 as a function of ω. Consider the cases where Γ is either small or large. What can you conclude?

7.2

By following the method in this chapter, determine the complex amplitude equation from the following model equation:

$$\ddot{x} + x = \epsilon \left[\gamma \cos \tau + (1 - x^2) \dot{x} - \beta x + x^3 \right]. \tag{7.76}$$

ϵ is a small parameter. Analyze the fixed points and their stability. Represent the solutions in the plane of the amplitude of the fixed point and β.

Chapter 8

Introduction to Chaos

Abstract *In this chapter, we give a general overview of chaos in "dissipative" systems, in which energy (or its analog) is not constant. We will explain the three possible scenarios for transition toward chaos, and introduce concepts useful for the study of the said chaos, namely the strange attractor, the Poincaré section, the fractal dimension, and self-similarity. We will discuss the difference between randomness and chaos and show that chaos, synonymous with a high sensitivity to initial conditions in deterministic systems (systems devoid of random effects), can be viewed from both probabilistic and statistical perspectives – approaches typically used for non-deterministic systems. We will discuss the concept of a "crisis". We will finish with a brief discussion on controlling chaos.*

8.1. A Typical Example

Consider an animal population that lives on an isolated territory; for example, a population of wild rabbits on a desert island, and with certain natural resources available. What is the evolution of the population size over time?

To respond to this question and understand the formalization of the problem into a model, let us take a step-by-step approach.

Say y_n is the population size reported for year n. A priori, it is natural to think that the greater the initial size of the population, the greater the size of the population the year after. The simplest way of expressing this evolutionary rule is with a proportionality law, that is, set the population size for year $(n + 1)$ to be proportional to that of year n, as follows: $y_{n+1} = ky_n$, where k is a coefficient with proportionality greater than one and whose exact value depends on many factors, especially that of abundance of resources.

© Springer Science+Business Media B.V. 2017
C. Misbah, *Complex Dynamics and Morphogenesis*,
DOI 10.1007/978-94-024-1020-4_8

This law of proportionality implies that the size of the population should grow indefinitely, even exponentially, over the years. However, the resources and land available to the population are limited, so at a certain point the population reaches a maximum size which optimizes the set of conditions to which it is subjected. The critical size y_c corresponds to this sort of natural equilibrium of the population, above which the population will suffer a decline. This decline is, of course, not indefinite: when reduced to a number below the aforementioned critical value, the population size will once again begin to grow. The simplest mathematical expression of this regulatory phenomenon is given by the model:

$$y_{n+1} = ky_n(2y_c - y_n). \tag{8.1}$$

This is called the *logistic* equation. It behaves in two general ways, depending on whether the population is above or below the critical population size. When it is below this critical value, population size y_{n+1} is an increasing function of y_n, and above the critical value, the population will decrease. The resulting curve for the logistic equation (8.1) above is thus a parabola, reaching a maximum when $y_n = y_c$, and then decreasing.

However, in spite of the disconcerting simplicity of the logistic equation (8.1), we have not yet begun to cover the surprisingly rich behaviors it exhibits, intriguing for any first time explorers of such equations. To explore them ourselves, let us first make the following change of variables:

$$k = 2a/y_c, \quad x_n = y_n/(2y_c), \tag{8.2}$$

where a is a positive real number. With the new variables, the logistic equation (8.1) takes the following form:

$$x_{n+1} = 4ax_n(1 - x_n). \tag{8.3}$$

We want to avoid negative numbers because there is no physical meaning to a negative population. To do this, we restrict variable x_n to the interval $[0, 1]$, which signifies that y_n is in turn restricted to the interval $[0, 2y_c]$. In other words, the number of individuals making up the population varies between 0 (concurrent with the disappearance of the population) and a maximum size y_c, determined by a confluence of environmental constraints.

8.2. A Further Example in Which the Future of the Population Depends on a Single Individual!

We dedicate this section to an example study of the evolution of a population over time. To begin, we must analyze which factors may be significant for the evolution of the population size. One quick response would have to do with the reproductive cycle: for example, the gestation time for a rabbit is about 30 days, so the characteristic time for a study of the rabbit population would be equal to about one month. So if we call x_N the number of individual rabbits composing the population at the end of N months, we can use the logistic equation (8.3) to calculate the population size at the end of each month via the following iterations:

$$x_1 = 4ax_0(1 - x_0)$$
$$x_2 = 4ax_1(1 - x_1)$$
$$\vdots$$
$$x_{n+1} = 4ax_n(1 - x_n)$$
$$\vdots$$
$$x_N = 4ax_{N-1}(1 - x_{N-1}). \tag{8.4}$$

Keep in mind that the actual size of the population is given by multiplying variable x_n by the number $2y_c$ (see equation (8.2)).

Now we will get a feel for possible values. Fixing the critical value at one million (i.e. $y_c = 10^6$) and taking an initial population size $x_0 = 0.3$ (equal to a real size of 600 000 individuals, after multiplying by $2y_c$), we endeavor to calculate the size of the population after $N = 400$ months and investigate how the variance control parameter a affects results. We can program the iterative operations on a small calculator for two example values to compare the population sizes after a period of 400 months as a function of the initial size.

1. Population size study for a control parameter value of $a = 0.4$

For $a = 0.4$, iterating over $N = 400$ months leaves us with a population size equal to about 749 998 individuals.

What happens to the $N = 400$ population size for a slightly different initial population size, say $x_0' = 0.300\,001$? This corresponds to a number of individual rabbits equal to $y_0 = 600\,002$, which is two rabbits more than in the previous calculation. Given the tiny modification of the initial population (a difference of $1/100\,000$), we would hypothesize an equally insignificant difference in the final number. Indeed, a quick calculation yields a final population of identical size, 749 998 individuals.

This quantitative conclusion aligns with what we would call common sense for this particular value of a, so we will now choose a different value to reveal some complexity.

2. Population size study for a control parameter value of $a = 0.91$

This time, at the end of 400 months, the population size obtained from an initial population of $x_0 = 0.3$ (or 600 000 individuals) is equal to 635 175.

For a slightly different initial population, in which we have added only two rabbits ($x_0' = 0.300\,001$), we find a radically different population size of more than twice the first outcome: $1\,318\,984$ rabbits. If we go in the other direction, beginning with two less rabbits ($x_0'' = 0.299\,99$, equivalent to a real population of $599\,998$ rabbits), our final count becomes $783\,757$ rabbits.

The stunning disparity of results in final population sizes, despite minimal initial variations (of order one in a hundred thousand), goes against our common sense. The straightforward mathematical expression $4ax(1-x)$ is well *controlled*, in the sense of being free of any random influences. It is a completely deterministic law, housing no intrinsic probabilities, and yet the final result is extremely sensitive to the initial size of the population.

This iterative law corresponds to an evolutionary law, of variable n, which we know to be true of many physical systems (it has been tried and proven, unlike many other evolutionary laws which remain theoretic). In such systems we find that an uncertainty, however tiny, in the initial variables make it impossible to predict the final state of the system. An uncertainty is impossible to avoid in any real life measurements (any experiment – physical, mechanical, biological or even computational, cannot be exactly precise!). In other words, high precision or even absolute knowledge of the evolution equation is useless in helping us predict the final state. This brings us to touch upon the domain in modern science: in traditional mechanics, an imprecise knowledge of the initial conditions does not change the calculation of the system's evolution. But here, we have chaos, whose landscape is defined by *dynamics which are highly sensitive to initial conditions*.

8.2.1. From a Fixed Point to Chaos

To better discern the behaviors arising from the logistic equation $4ax(1-x)$, we present a qualitative survey of the phenomena observed with the variation of parameter a.

Starting from an initial condition x_0, we find each value x_n after n iterations. If we collect the series of pairs $(0, x_1), (1, x_2), (2, x_3), (n, x_{n+1}), \ldots, (N, x_{N+1})$ we can then represent variable x_{n+1} as a function of n for different values of a (see figure 8.1). For small enough values of parameter a (i.e. $a < 3/4$, see below), we find that whatever the initial value of x_0, the solution always converges to the same final state after N iterations (and only a few iterations are necessary for the convergence to occur with a precision of 1% difference; after about 100 iterations, the precision is of order 10^{-4}). As the value of parameter a is increased, we first observe an oscillation of unit period, then an oscillation of period 2, then 4, etc. Beyond a certain threshold value of a (see below), we obtain an irregular signal where not one of the solutions found over the previous iterations seems to be repeated; we can say that the periodicity has become infinite (the signal will not return to the same value again before an infinite number, arbitrarily large, of iterations). If we repeat the same operation with an initial condition x_0 which is slightly different from the first, the series of numbers x_n, equally irregular, quickly diverges from the series found just before. These irregular signals coming from such great sensitivity to initial conditions are called *chaotic*.

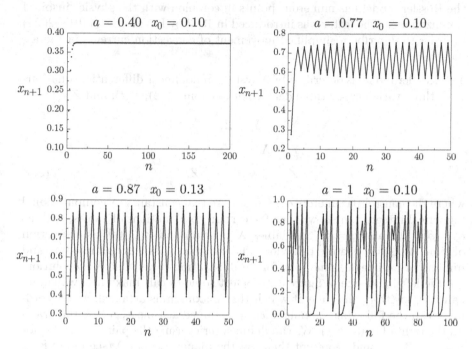

Figure 8.1 – Different solutions x_{n+1} (starting from initial condition x_0) as a function of n. As a grows, we pass a regime of fixed points ($a = 0.4$) to an oscillation of unit period, then double the length, etc. (here we do not show the periods 4, 8, etc.). For $a = 1$, the signal is chaotic. For more details about the choice of a, see subsection 8.9.1.

8.3. Origin and Meaning of the Function $f(x) = 4ax(1 - x)$

Already fascinating, even in purely mathematical terms, the function (which is sometimes called a map, for a reason that will become clear) $f(x) = 4ax(1 - x)$ is not just a thought experiment but actually represents the dynamics of certain real systems.

In previous chapters we have seen that much of the phenomena in nature can be described by nonlinear models; that is, nonlinear systems of differential equations. For example, the Rössler model, named after Otto E. Rössler (1940–) who suggested the model in 1976 (see [9]), is now a prototypical model for continuous chaotic phenomena. Originally conceived in a formal manner, the Rössler model has numerous points in common with the physics-inspired Lorenz model. The latter was introduced in 1963 by E.N. Lorenz (1917–2008) in order to describe a simplified movement of convection currents in the atmosphere[1].

The Rössler model is described by a system of nonlinear differential equations with three variables, or three degrees of freedom: $X(t)$, $Y(t)$, and $Z(t)$:

$$\dot{X} = -Y - Z,$$

$$\dot{Y} = X + \alpha Y,$$

$$\dot{Z} = \beta + XZ - \gamma Z, \tag{8.5}$$

where the control parameters α, β and γ are real constants. A computational numerical integration shows that for certain combinations of values, such as $\alpha = \beta = 0.2$ and $\gamma = 5.7$, variables $X(t)$, $Y(t)$ and $Z(t)$ (all as a function of time t) display chaotic behavior. The variables pass through maxima and minima in an irregular motion (for an illustration of variable X's evolution, see the left-hand side of figure 8.2). Select some (local) maximum value and call it x_n, with t_n the time at which that maximum is attained, and accordingly call the following maximum x_{n+1}, attained at time t_{n+1} (see the graph on the right of figure 8.2). We can define a series using the pairs $[x_n, x_{n+1}]$ for $n = 1, 2, 3, \ldots$, and represent them on the plane $(x_n\text{–}x_{n+1})$ (see figure 8.3). The pairs of points lie on a curve which looks like a parabola, like the logistic curve $x_{n+1} = 4ax_n(1 - x_n)$. Actually, the precise expression of the iterative function itself is not important, but its shape has to be bell-like. For example, we could also have used recurrence law $x_{n+1} = a\sin(2\pi x_n)$, with $x_n \in [0, 1]$. Given that function $a\sin(2\pi x_n)$ has a parabolic form in the interval $[0, 1]$, repeating the above selection of pairs would result in the same qualitative shape[2].

---

1. The system of equations which makes up the Lorenz attractor is the following: $\dot{X} = \alpha(Y - X)$, $\dot{Y} = -XZ + \beta X - Y$, and $\dot{Z} = XY - \gamma Z$, where X describes the velocity of the atmosphere and Y and Z are related to the temperature (see [9]), with α, β, and γ as the control parameters.

2. We advise the reader to try the following exercise: take the iterative law $x_{n+1} = a\sin(2\pi x_n)$ and repeat all the analyses of this chapter in order to verify that the results obtained with the new expression are similar to the results found with $x_{n+1} = 4ax_n(1 - x_n)$.

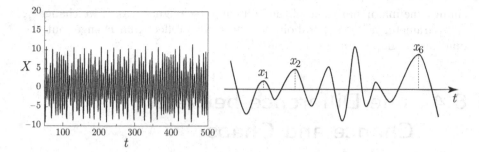

Figure 8.2 – Left: $X(t)$ as a function of t, solution for equation (8.5). Right: we demonstrate the picking out of maxima n and $n+1$ from the chaotic signal to form the graph shown in figure 8.3.

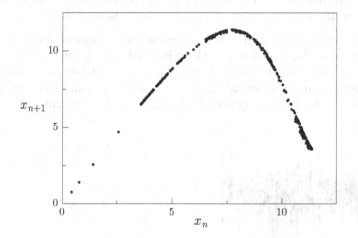

Figure 8.3 – We show the maximum x_{n+1} as a function of x_n, as arising from $X(t)$, the solution to equation (8.5). This bell curve can be approximated by a parabolic function.

The graphical representation of the pairs $[x_n, x_{n+1}]$ in plane (x_n–x_{n+1}) (see figure 8.3) allows us to make two observations: (i) although $X(t)$ is erratic, the curve $x_{n+1} = f(x_n)$ is not erratic at all: it is surprisingly regular! This is the first of a few fascinating results: behind the apparent chaos, there is a hidden regularity which comes from the *deterministic* nature of the evolution equation. But this determinism is fairly fragile: a tiny uncertainty in the initial conditions will give a completely different result; (ii) without being a rigorous parabola, the curve has the qualitative form of a bell curve; to describe it, the parabola is the simplest function that we can adopt. A natural candidate is $x_{n+1} = f(x_n) = 4ax_n(1-x_n)$, which has a maximum at $x = 1/2$, with $x \in [0,1]$ (this choice is always possible if we use an appropriate change of variables). Modifying the Rössler system's parameters (equation (8.5)) will change the maximum value of the bell curve (see figure 8.3). If we change γ, the parameter a in the parabolic law must also reflect that change, but the qualitative shape remains unchanged.

8.4. The Difference between Chance and Chaos

Looking for similarities between random phenomena exhibiting randomness, it may seem legitimate to compare a game of chance, such as roulette[3], with chaotic phenomena, since the two have comparably irregular outcomes. However, the two phenomena reflect two very different situations. To see this, let us compare the results of a game of chance and those of the Rössler system.

Take a selection of computer generated random numbers deriving from a probability distribution such as a Gaussian (other laws, such as the uniform distribution, are equally useful). Every number x_n, picked at random in the nth drawing, can be "arranged" to form a pair $[n, x_n]$. If, as before, we plot x_n as a function of n (see figure 8.4), the curve will be irregular.

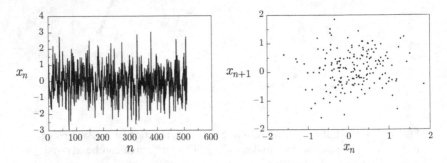

Figure 8.4 – Left: a random signal x_n, determined by a Gaussian probability distribution, as a function of n. Right: a handful of maxima x_{n+1} as a function of x_n. Note the important difference with figure 8.3 concerning the behavior of x_{n+1} as a function of x_n.

---

3. In daily life, roulette is considered a game of chance. We could think, with good reason, that roulette is a complex dynamical system, as is any system which escapes relatively simple modeling. In this chapter, we are only interested in relatively simple systems (those modeled easily by a set of equations) because this reflects a small number of degrees of freedom. For systems not so easily modeled, we can make use of a statistical description, and from this consider them as part of a probabilistic/statistical categorization, which we will do with the case of roulette.

Unlike the Rössler system's signal, the points are not distributed in a regular fashion (compare to figure 8.3), but are instead spread about the plane in a random distribution. This demonstrates the fundamental difference between so-called *deterministic chaotic dynamics* and *stochastic dynamics*. Does it make any sense to compare chance and chaos? Though the analysis we just made seems to tell us no, there *are* important similarities between the two phenomena.

The regular bell curve revealed from the chaotic dynamics of the Rössler system displays the underlying determinism in its evolution. If we knew the exact value for iteration n, the next iteration could be perfectly determined. Since we can only imperfectly master and describe any real system's initial conditions, we cannot a priori know, to perfect precision, what value the signal will attain after some specified time, lending a statistical character to the deterministic dynamics. Accordingly, we will use tools developed in statistical physics to dissect the problem later in this chapter.

Before ending this section, we would like to remind the reader that the precise form of the iteration law (for example, $f(x) = 4ax(1 - x)$) is not essential; only the qualitative form is important as it reveals a *universality* of certain processes.

8.5. Commentaries on Population Dynamics and the Logistic Curve: Warning!

The iterative function $x_{n+1} = 4ax_n(1 - x_n)$ is often used to describe population dynamics. For an isolated population (no competitors or predators) a first growth phase which is quasi-exponential is seen before the evolution slows down as problems associated with overpopulation kick in (lack of sustenance, fighting due to promiscuity, etc.), until finally the population reaches a maximum threshold number. Mathematician Pierre Verhulst (1804–1849) first introduced a simple logistic model in 1838 to explain this phenomenon in the following form: if $N(t)$ refers to the actual population size at instant t, function N obeys:

$$\dot{N} = \rho N \left(1 - \frac{N}{k} \right), \tag{8.6}$$

where k is the upper limit attained by the population size (equivalent to the maximum value of N, it is also called the *biocapacity*), and ρ a parameter. This equation is reminiscent of the prey-predator Lotka–Volterra model we have previously studied (see subsection 5.2.2). The discretized version of logistic equation (8.6) looks as follows:

$$\frac{N\left[(n+1)\Delta\right] - N\left[n\Delta\right]}{\Delta} = \rho N(n\Delta)\left(1 - \frac{N(n\Delta)}{K}\right), \tag{8.7}$$

where Δ is the time step, and the time derivative has been replaced by a finite difference (left-hand term). If we define $4a = 1 + \rho\Delta$, $K = k(\rho\Delta + 1)/\rho\Delta$, and $x_n = N(n\Delta)/K$, we recover the logistic model with which we earlier introduced the notion of chaos:

$$x_{n+1} = 4ax_n(1 - x_n). \tag{8.8}$$

Now, the solution for the continuous equation (8.6) is *not* chaotic: it has a simple solution of the form $N(t) = k/\left[1 + (k/N_0 - 1)e^{-\rho t}\right]$, with N_0 the initial value at $t = 0$. The discrete version of this equation (8.8), which corresponds to a logistic curve, *is* chaotic. This brings up a delicate problem: how, and under what circumstances, is it valid to discretize a continuous equation? In this case, we are saved because the Rössler system (equation (8.5)) itself is continuous in time and the logistic curve which we derive from it is not a pure discretization of the system, but rather the result of relating two successive maxima found at discrete times. The problem is completely avoided in this context. This highlights the fact that when we talk about discrete time, it has not to be confused with a discretization of the continuum evolution equation.

8.5.1. Continuous and Discrete Time, Iterative Laws, and Poincaré Section

In the Rössler system (equation (8.5)), variables $X(t)$, $Y(t)$, and $Z(t)$ are temporal real signals, and t is the familiar continuous time. Index n of variable x_n, also coming from the temporal signal $X(t)$, corresponds to a cutting up of time, and represents the position of the maximum x_n. The index n thus refers to a discrete time. We can define a function f of x_n to create an iterative expression $x_{n+1} = f(x_n)$. Variable $f(x_n)$ gives information on the future of the signal at instant $(n + 1)$. Applying this iteration successively N times to the initial value x_0, $f(f(\ldots f(x_0)))$, we obtain the value of the signal at instant N.

Henri Poincaré (1854–1912) proposed to take on the study of chaotic signals by revealing their values at particular instants rather than for every instant (i.e. continuously). This approach has several advantages: (i) the complexity of the chaotic signal $X(t)$ is reduced, even allowing for a description of the phenomenon by a simple and regular function such as the bell curve; (ii) the information derived from iterative function $f(x)$ is complete, in spite of the change from continuous to discrete time steps. The iterative function $f(x)$ which is defined in this process is called a *recurrence* or *Poincaré* map, or a *Poincaré section*.

8.6. A Geometric Approach to the Poincaré Section

In chapter 5 we introduced the notion of phase space and the representation of trajectories of dynamical systems in phase space. For example, the dynamics of the van der Pol oscillator is represented in a 2-D phase space $(X = I, Y = \dot{I})$, where I is the current in the van der Pol circuit. The representation of the circuit's trajectory in this plane is a closed curve, called a limit cycle (see figure 5.5 (left)). For a two dimensional space, the Poincaré sections correspond to an intersection of this phase space trajectory with a line. For a three dimensional phase space, as in the case of the Rössler system (equation (8.5)), of three variables (X, Y, Z), the Poincaré sections correspond to an intersection of the phase space trajectory with a surface.

The evolution of these intersection points is particularly interesting for its dimensional reduction of the problem; we are left with a simple representation of the nature of the system's dynamics. For a limit cycle, a Poincaré section shows us one or two isolated points (see figure 8.5(b)); a more complex movement shows several isolated points in the intersection with the plane (see figure 8.5(a)). For a bi-periodic regime (for example, $2T$-periodic, where T is the period), the section will show two points (if the dotted line in figure 8.5(b) is chosen so that it intersects the entire structure of the attractor), or four points in the $4T$-periodic case, and so forth.

As a general rule, in the case of a trajectory with multiple periods (which we can call nT-periodic, with integer n), the Poincaré section is a discrete ensemble of isolated points. When the movement becomes more complex, with a quasi-periodic movement[4] (see figure 8.6), the Poincaré section becomes a quasi-continuous curve.

8.7. Strange Attractors

We have already come across a few attractors in recent chapters. They were simple attractors, each described by a fixed point (also known as a source point) and a limit cycle. For a trajectory which is $2T$-periodic, the cycle is doubled; it is quadrupled in the case of a $4T$-periodic regime, and so forth (see figure 8.5). Each of these is an attractor of a relatively simple structure. The question naturally arises: in phase space, what is the shape of an attractor in a chaotic regime (such as the Rössler system)?

---

4. A movement is quasi-periodic when there are two temporal frequencies and when the relationship between the frequencies is irrational; see section 8.8.2.

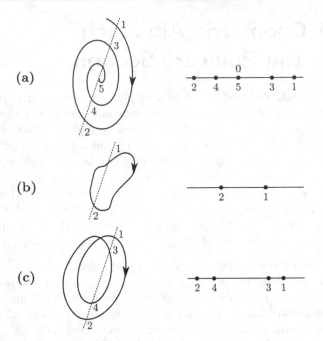

Figure 8.5 – Some illustrations of a Poincaré section for different trajectories of phase space. (a) A stable fixed point at (0,0); the section (in dots) is composed of isolated points converging toward the center with time. (b) A limit cycle with two points of intersection with the dotted line segment. (c) Bi-periodic movement, with a section composed of four points (we can also imagine a section which would only reveal two points).

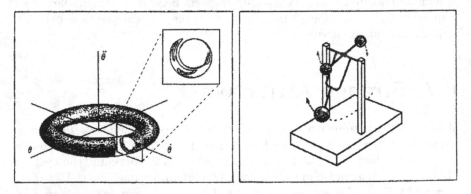

Figure 8.6 – The diagram on the left shows a strange attractor and its Poincaré section. The section is complex; it is also self-similar. The figure on the right is a mechanical schematization of a pendulum which can show chaotic behavior. [© Jose Louis Subias]

It turns out to be something of a strange shape (see figure 8.7)! In the chaotic regime, the trajectories densely populate the attractor. As such, the strange attractor outlines a structure composed of points and empty spaces in the phase space, created from the ensemble of states adopted by the dynamical system throughout the course of its evolution. A different ensemble of initial conditions will give a completely different sequence of points which, nonetheless, will afterward show the same figure in the phase space. Whatever the initial conditions, the system always evolves around the same attractor, but it is impossible to predict a precise position on the attractor in the course of time. Note, furthermore, the fractal and self-similar nature of the strange attractor. If you look at the attractor at different scales (zooming in successively), an identical structure is revealed at each scale (see figure 8.8). We shall come back to this later in the chapter.

8.8. Three Possible Scenarios for Transition into Chaos

Apart from the stationary fixed point, the simplest description for the evolution of a dynamical system is given by a periodic orbit. We can find an increasing complexity in the system by variation of the control parameter, until reaching chaos. Whatever the nature of the system (biological, physical, economical, etc., and regardless of its governing equations), there are three distinct paths, called scenarios or routes, that a system can take as it transitions into chaos: (i) by subharmonic cascades; (ii) by going through a quasi-periodic mode; and (iii) via intermittency.

8.8.1. Transition into Chaos via Subharmonic Cascades

Proposed independently in 1978 by both Feigenbaum [33], and Tresser and Coullet [103], the first scenario for transitioning into chaos is by subharmonic cascades.

Imagine there exists a signal X with a period T (i.e. $X(t + T) = X(t)$). For such a signal, all maxima are identical; there exists a single unique pair of maxima $[x_n, x_{n+1}]$. Thus, the Poincaré section is a single point on the plane, $[x_n, x_{n+1}]$. One function which would give us this signal is the sinusoidal $X(t) = \sin(t)$. All of its maxima are equal to one, and the ensemble of the pairs of maxima is $(x_n, x_{n+1}) = (1, 1)$, $\forall n$. Now, a bi-periodic signal $X(t)$ (for example $X(t) = \sin(t) + 1.2\sin(2t)$) would correspond to a system with two types of maxima, called x_{max1} and x_{max2}. In this case, the representation of pairs $[x_n, x_{n+1}]$ in the plane is associated with two distinct points, with coordinates (x_{max1}, x_{max2}) and (x_{max2}, x_{max1}). A $4T$-periodic movement would give rise to four distinct points, etc.

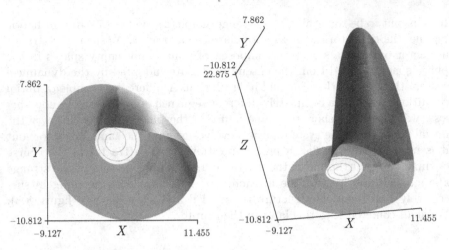

Figure 8.7 – Phase space trajectory for the Rössler system. 2-D on the left, with a projection on the plane (X–Y); 3-D on the right.

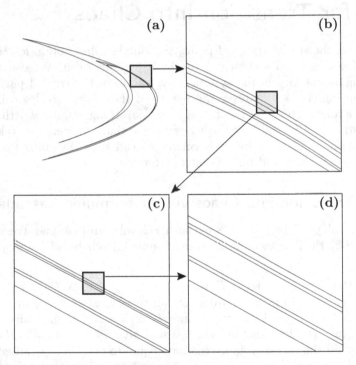

Figure 8.8 – Phase space trajectory of a chaotic system. Increasingly zooming in shows the same structure in the trajectories, the signature of self-similarity. [© Christian Ricordeau, from an image by James P. Crutchfield]

A chaotic movement corresponds, by definition, to a situation where no maximum value is repeated afterward. This is the case for a Rössler system in the chaotic regime, for which over time we obtain a large number of maxima, each with a distinct value. Plotted in plane $[x_n, x_{n+1}]$, the ensemble of pairs (x_n, x_{n+1}) trace out a bell curve (see figure 8.3). We will see below that as a control parameter is varied, the initially periodic motion (with period T) may lose its stability in favor of a bi-periodic motion (with period $2T$). A further variation of the control parameter leads to the loss of stability of the bi-periodic regime in favor of a quadru-periodic regime (with period $4T$), and so on. Each time the system loses its stability, the period is doubled (or frequency halved, hence the denomination subharmonic cascade). This period doubling process continues without end as the the control parameter is varied, until the period becomes infinite, which corresponds to chaos. We shall later treat in detail this transition to chaos by using the logistic function $f(x) = 4ax(1 - x)$ (see section 8.9).

8.8.2. Transition toward Chaos
via a Quasi-periodic Regime

The second scenario of transition into chaos entails going through a regime of quasi-periodicity, as identified in 1971 by Ruelle and Takens [90].

Take a system which is periodic, with angular frequency ω_1. If we vary one of the control parameters, the system acquires a new frequency ω_2. For a subharmonic cascade, the ratio ω_2/ω_1 is a rational number (for example $\omega_2 = \omega_1/2$); by contrast, in a quasi-periodic regime, the ratio ω_2/ω_1 is an *irrational* number (for example, $\sqrt{5}$).

If the ratio between the two frequencies is rational (for example $\omega_2/\omega_1 = 3/4$), the phase space trajectory will, after a certain amount of time (a time equal to four times the smallest common period, in this case 12), close in on itself. Consequently, the trajectory cannot cover phase space in a dense way, since it will continue to double over itself.

If instead the ratio is irrational, the trajectory will never close and thus it will densely cover the attractor in phase space. For a quasi-periodic regime, the attractor has the shape of a torus (see figure 8.9). Qualitatively, this torus represents two types of movement in the trajectory: one movement of frequency ω_1 around a smaller circle and another of frequency ω_2 around a bigger circle (see figure 8.9). A physical system's attractor in the quasi-periodic regime is generally a torus of a variable area. For example, the torus presented on the right of figure 8.9 shows the attractor of a quasi-periodic regime as found from the numerical results in a problem of crystal growth (see [53, 54]).

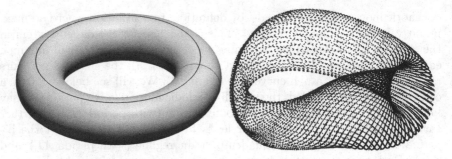

Figure 8.9 – Left: an ideal torus. Right: a real torus obtained from a numerical simulation [53,54] exploring crystalline growth. [From [53] K. Kassner, C. Misbah & H. Müller-Krumbhaar. Transition to chaos in directional solidification, *Phys. Rev. Lett.*, **67**(12): 1551–1554, Sep 1991, © American Physical Society]

Given the dense character of the trajectory, the Poincaré section (see figure 8.10) describes a continuous (or quasi-continuous) curve, which could be a closed trajectory if we picked the right cross-section. By varying a control parameter, the torus in phase space can be destroyed to create a more complex attractor. Then the Poincaré section is no longer a simple continuous loop, but becomes a fine and complex structure of fractal dimension (see figure 8.10 and section 8.12 for this notion).

8.8.3. Transition to Chaos via Intermittency

Intermittence is the last of the three possible scenarios by which one can transition to chaos, and is again some sort of destabilization of a limit cycle, or more generally, destabilization of an attractor. Studied since 1979 by Pomeau and Manneville (see [84]), the scenario describes the evolution from periodic dynamics to episodic apparitions of chaotic bursts upon the variation of a control parameter. These chaotic episodes last a short amount of time before fading to give way to the periodic regime again. The time intervals between each chaotic burst are not regular, but they can become more and more frequent upon adequate variation of a control parameter. When a critical value of the control parameter is reached, the signal becomes completely chaotic (see figure 8.11 for an illustration of a typical path through intermittency toward chaos).

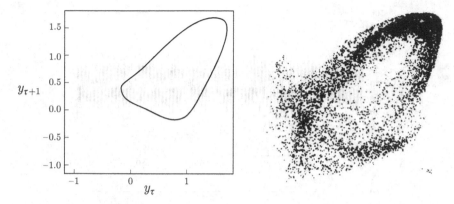

Figure 8.10 – A Poincaré section (numerical simulation [53] from the problem of crystalline growth). Left: we follow a dynamic variable (here the position of the edge of the growth) obtained at time $\tau + 1$ as a function of the position of that same front at time τ. The system is at this point in the quasi-periodic regime, the scenario before chaos. The section is a dense curve (it even seems continuous at the scale presented here), an expression of the irrational ratio between the two frequencies (the attractor of three dimensions is a torus). Right: after changing a control parameter (in this case, the growth rate) the torus is destroyed. The system leaves the quasi-periodic regime, entering instead the chaotic regime which is characterized by a strange attractor (no longer a torus), shown in the Poincaré section (right). [From [53] K. Kassner, C. Misbah & H. Müller-Krumbhaar. Transition to chaos in directional solidification, *Phys. Rev. Lett.*, **67**(12): 1551–1554, Sep 1991, © American Physical Society]

8.9. A Detailed Study of Chaos Attained by a Subharmonic Cascade

Of the three scenarios leading to chaos, it is easiest to make an in-depth study of the transition via subharmonic cascades. In this section, we present, along with the main properties of the logistic model, the specific conditions which allow for the emergence of chaos via this scenario.

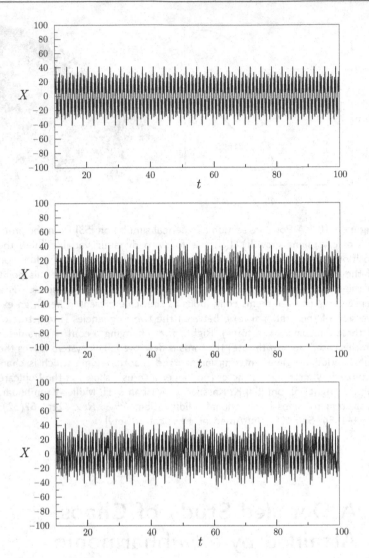

Figure 8.11 – Chaotic dynamics can be arrived at through intermittence as in this example, in which we see a numerical solution of the Lorentz system (see section 8.3). We set $\alpha = 10$, $\gamma = 8/3$, and vary β. Top: we fixed the value at $\beta = 165.8$, with the result of a periodic regime. Middle: $\beta = 166.2$, chaotic bursts begin to interrupt the chaotic regime. Bottom: for $\beta = 166.3$, the regime has become entirely chaotic.

8.9.1. Fixed Points of the Logistic Map
$$x_{n+1} = f(x_n) = 4ax_n(1 - x_n)$$

First, we determine the simplest solutions for the logistic equation as described by fixed points. Any point x_n^* satisfying $f(x_n^*) = x_n^*$ is a fixed point of the model. To simplify notation, we will omit index n in the following expressions; the equation for the fixed points is thus written:

$$f(x^*) = x^*. \qquad (8.9)$$

This equation has a simple geometric interpretation: all fixed points belong to the intersection of curve $f(x)$ with the first bisector $y = x$. The fixed points of model $f(x) = 4ax(1 - x)$ are thus defined by the equation: $4ax^*(1 - x^*) = x^*$, with the solutions:

$$x^* = 0,$$

and

$$x^* = 1 - \frac{1}{4a} \quad \text{for} \quad a > \frac{1}{4}. \qquad (8.10)$$

The condition $a > 1/4$ guarantees that the fixed point is positive ($x^* > 0$). If we re-establish the index notation, we note that the fixed points satisfy $x_{n+1} = x_n$ (the maximum "n" of signal $X(t)$ is equal to maximum "$n + 1$" of the signal, for any n). In other words, a fixed point in the logistic map f corresponds to a periodic regime in terms of continuous variable t; we designate the associated period by T.

8.9.2. Stability of Fixed Points

A fixed point (recall that it corresponds to a periodic movement in continuous time) may be stable or unstable. To verify whether it is stable we can add a small perturbation δx to x^*. The fixed point x^* is stable if the value of δx decays over time, and unstable if it does not. Adding the perturbation we get:

$$x = x^* + \delta x. \qquad (8.11)$$

We then substitute this expression into our model given by the application $f(x)$. Since, by definition, perturbation δx is small, we can make a Taylor expansion:

$$f(x^* + \delta x) \simeq f(x^*) + \delta x f'(x^*), \qquad (8.12)$$

where f' designates the derivative. To determine the evolution of fixed point x^* after a long enough time, f is applied to x^* N times, with $N \gg 1$. Let us carry out this operation in successive steps, first applying f only twice to fixed point x^*. We find:

$$f[f(x)] \simeq f[f(x^*)] + \delta x f'(x^*) f'[f(x^*)] \qquad (8.13)$$

$$= f(x^*) + \delta x f'(x^*) f'(x^*), \qquad (8.14)$$

where we have used the properties of the fixed point (i.e. $f(x^*) = x^*$). For N times, we obtain:

$$f^{(N)} = f(x^*) + \delta x [f'(x^*)]^N. \tag{8.15}$$

Keep careful track of the notations: the symbol (N) on the left-hand side of the equation $f^{(N)}$ is an abridged notation used to indicate that it is an iterated function (i.e. a composite of itself; $f^{(N)}$ signifies that the application "f is carried out N times"). On the right-hand side, the notation $[f'(x^*)]^N$ represents the power of the function (exponent N). Perturbation δx is multiplied at each step by the derivative of f with respect to x^* (i.e. $f'(x^*)$); $|f'(x^*)|$ represents the absolute value of the slope at the point of intersection of curve f with the first bisector.

If $|f'(x^*)| < 1$, perturbation δx is multiplied at each time step by a number with a value less than one, which means the perturbation is always diminished: the fixed point is stable. If on the other hand $|f'(x^*)| > 1$, the perturbation grows with time and the fixed point is unstable. This is a general result and does not depend in any explicit way on the model. In our model $f(x) = 4ax(1 - x)$, we need simply to calculate the slope at x^* to deduce information about its stability. The slopes of the two fixed points (equation (8.10)) are given by:

$$f'(0) = 4a, \quad f'\left(1 - \frac{1}{4a}\right) = 2(1 - 2a). \tag{8.16}$$

When $a > 1/4$, the derivative of f at the fixed point $x^* = 0$ is greater than one ($f'(0) > 1$), meaning that the fixed point $x^* = 0$ for this case is always unstable. Now let us look at the other fixed point. Suppose that $a < 1/2$; we then have $|2(1 - 2a)| = 2(1 - 2a)$. The condition for instability of the fixed point, $2(1 - 2a) > 1$, would require $a > 1/2$, in direct contradiction with our hypothesis. It follows that the fixed point is always stable for $a < 1/2$. For $a > 1/2$, we have $|2(1 - 2a)| = 2(2a - 1)$; the fixed point is unstable (i.e. $2(2a - 1) > 1$) for $a > 3/4$, and otherwise stable. Summing up our results:

$$x^* = 0 \quad \text{is always unstable;}$$
$$x^* = 1 - \frac{1}{4a} \quad \text{is stable for} \quad a < \frac{3}{4};$$
$$\text{and unstable for} \quad a > \frac{3}{4}. \tag{8.17}$$

Thus, the critical value $a_c = 3/4$ of parameter a marks the loss of stability.

What happens after passing this critical value? To determine this empirically, we pick a value of parameter a greater than $3/4$, for example, 0.78 and make a numerical calculation of the values of function f. After a large enough number of iterations, we obtain an oscillation of function f between the two values 0.772 and 0.547, call them x_1 and x_2 respectively, and note that these oscillations continue on indefinitely over time. This also means that x_1 and x_2 satisfy:

$$f(x_1^*) = x_2^*, \quad f(x_2^*) = x_1^*. \tag{8.18}$$

In other words, to get back to the same point, we have to apply function f twice:

$$f[f(x^*)] = x^*. \tag{8.19}$$

The two points x_1 and x_2 are thus the fixed points of the composed application $f(f)$ (and not of f). In terms of continuous time, this result means that the temporal signal $X(t)$ finds the same maximum after a time equal to the duration of two periods, that is, at the end of a time of length $t + 2T$. Thus it is a bi-periodic regime of period $2T$ and the frequency is equal to $\omega/2$, which corresponds to a subharmonic frequency. When we approach the critical value $3/4$ of parameter a, the fixed point $x^* = 1 - 1/4a$ associated with the periodic movement loses its stability in favor of the two new stable fixed points (bi-periodic regime): we say the system has undergone a *subharmonic bifurcation*.

To study the composite application $f(f(x))$, let us write $X = f(x) = 4ax(1-x)$ such that $f(X) = f(f(x)) = 4aX(1-X)$. The use of the rule for derivatives of composite functions means we must write partial derivatives as:

$$\frac{\partial f(X)}{\partial x} = \frac{\partial f}{\partial X}\frac{\partial X}{\partial x} = 16a^2(1-2X)(1-2x). \tag{8.20}$$

This derivative becomes zero in two cases: $x = 1/2$ and $X = 1/2$. If we use the definition of variable X as a function of x, the equation for $X = 1/2$ (i.e. $x^2 - x + (1/8a) = 0$) has two solutions:

$$x_{\pm} = \frac{1 \pm \sqrt{1 - 1/(2a)}}{2}, \tag{8.21}$$

with $x_+ > 1/2$ while $x_- < 1/2$.

The point $x = 1/2$ corresponds to the minimum of the composite function $f(f(x))$, while the points $x_{\pm}$ correspond to the maxima of $f(f(x))$. The fixed points of f^2 correspond geometrically to the intersection of the function $y = f^2(x)$ with the line $y = x$. The function f^2 (see figure 8.12, middle graph, for $a = 0.8$) has four fixed points, one of which is trivial (i.e. $x^* = 0 \equiv x_0^*$). The slope of $f^2(x)$ (a shorthand notation for $f(f(x))$) at the origin is greater than one, confirming the instability of the trivial fixed point $x = 0$. The stability of the other three points (denoted x_1^*, x_2^*, and x_3^*, with $x_0 < x_1^* < x_2^* < x_3^*$) requires the calculation of the slope at the intersection points of $y = f^2$ with $y = x$. A completely analytic calculation can be done but proves to be inconvenient, leading to very fastidious expressions. For our purposes, it is enough to use a calculator to obtain and understand the results.

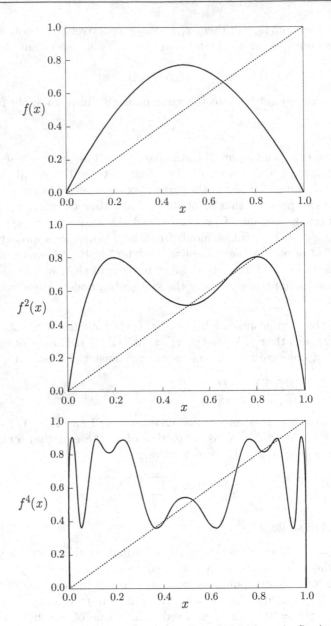

Figure 8.12 – On the graphs, from top to bottom, we show the fixed points of $f(x)$ (2 fixed points), $f^2(x)$ (4 fixed points) and $f^4(x)$ (8 fixed points). We have chosen $a = 0.8$ for the case of f^2 and $a = 0.888$ for the case of f^4. The geometric meaning of the fixed point is given by the intersection with the first bisector, shown in dots.

For example, use $a = 0.8$. The calculation shows (by iterating the application N times, N being large enough) that fixed points x_1^* and x_3^* are stable, while x_0^* and x_2^* are unstable. Thus we have alternation between stable and unstable points. Note that by iterating with a computer, we only see the stable fixed points.

8.9.3. Accumulation Point of the Cascade

If we augment the value of control parameter a to $a_4 = 0.86237$, we find that this demarcates a new threshold where the two fixed points x_1^* and x_3^* become unstable (we can verify that the absolute value of slope for f^2 at these points is greater than one). Iterations will show that now that $a > a_4$ there are four (stable) fixed points which the system moves through. We will call them $\tilde{x}_1^*$, $\tilde{x}_2^*$, $\tilde{x}_3^*$, and $\tilde{x}_4^*$, introducing notation $\tilde{x}$ to avoid confusion with the fixed points of f^2. These points obey the following rule:

$$f(\tilde{x}_1^*) = \tilde{x}_2^*, \quad f(\tilde{x}_2^*) = \tilde{x}_3^*, \quad f(\tilde{x}_3^*) = \tilde{x}_4^*, \quad f(\tilde{x}_4^*) = \tilde{x}_1^*. \qquad (8.22)$$

In other words, $\tilde{x}_i^*$ are the fixed points of f^4:

$$f^4(\tilde{x}_i^*) = \tilde{x}_i^*, \quad i = 1, 2, 3, 4, \qquad (8.23)$$

where f^4 is an abbreviation for function f composited four times with itself: $f(f(f(f(x))))$. This brings us to the quadru-periodic regime (see figure 8.12). The intersection of this function with the first bisector (equation $y = x$) reveals eight fixed points: four of these are unstable (the slope at the intersection is greater than one) and the other four stable (even if, with numerical iterations, we only see the stable points).

This regime lasts until parameter a reaches the threshold value $0.886\,02$, written a_8, where we once again see the emergence of a new bifurcation associated with period doubling. This bifurcation schema associated with period doubling is reproduced for different threshold values of parameter a; we have thus a subharmonic cascade of frequency halving at each step (see table 8.1).

Three important remarks can be made about the results reported in table 8.1. (i) The system loses stability with period doubling; (ii) there is an accumulation point for the cascading threshold value: the distance between two successive critical values a_i and a_{i+1} of parameter a diminishes quickly ($a_{32} - a_{16} \ll a_{16} - a_8 \ll a_4 - a_2$). As a is increased the period becomes very large, until it becomes infinite. This occurs for a certain value of a, denoted by (a_∞). The distance between two successive values of the parameter a approaches zero. This infinite period[5] regime describes chaos;

---

5. This result cannot be found through a direct numerical calculation because the period is infinite, but the theoretical demonstration is not recapitulated here because it lies beyond the mathematical scope of this book.

Table 8.1 – Summary table for the subharmonic cascade.

Interval of the control parameter	Period of the regime
$a < a_2 = 0.75$	$T = 1$
$a_2 < a < a_4 = 0.862\,37$	$T = 2$
$a_4 < a < a_8 = 0.886\,02$	$T = 4$
$a_8 < a < a_{16} = 0.892\,18$	$T = 8$
$a_{16} < a < a_{32} = 0.892\,4728$	$T = 16$
$a_{32} < a < a_{64} = 0.892\,4835$	$T = 32$
$\vdots$	
$a = a_\infty = 0.892\,486\,418\ldots$	$T = \infty$

and (iii) when the period become very large, the difference between values of a for successive bifurcations is reduced each time by a nearly constant factor. An analysis reveals the following:

$$\lim_{n \to \infty} \frac{a_n - a_{n-1}}{a_{n+1} - a_n} = 4.669\,201\,609\ldots \tag{8.24}$$

This scaling factor is actually a universal number, independent of the details of the logistic function f we are considering here. We can already verify, using the results of table 8.1, that for the first iterations $n \leqslant 32$, the scaling factor is almost constant and close to the asymptotic result (equation (8.24)). Since the constant is universal, each chaotic system will enter chaos via a subharmonic cascade with the same rate of variation of the control parameter a. Indeed the constant $4.669\ldots$ designates the proportionality factor between the difference in the control parameter values of two successive bifurcations. That is, if the control parameters for bifurcation n and $n-1$ have a given difference, the difference in parameter a between the two next successive bifurcations n and $n+1$ is equal to the previous difference divided by $4.669\ldots$, regardless of the details of the map $f(x)$.

8.9.4. Bifurcation Diagram and the Notion of Self-Similarity

Let us plot the fixed points of the logistic model $x_{n+1} = f(x_n) = 4ax_n(1-x_n)$ that we found in subsections 8.9.1–8.9.3 against parameter a (bifurcation diagram, figure 8.13). When a is less than the first threshold value a_2 ($a < a_2 = 3/4$), $f(x)$ has a single unique fixed point (not counting the origin, $x = 0$). The value of this fixed point $x^*(a)$ is a function of a. When the value of the control parameter augments ($a > a_2$), the branch associated with the fixed point $x^*(a)$ splits into two distinct branches: this is the first period doubling bifurcation. For a value of parameter a above the second threshold value a_4 ($a > a_4$), a period doubling bifurcation occurs again in each of the two branches, giving us a total of four new branches. For each new threshold value, we see another period doubling.

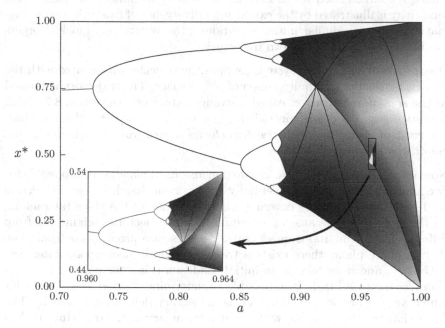

Figure 8.13 – A bifurcation diagram showing the accumulation point of subharmonic cascades and self-similar structure (shown in the zoomed-in section). [Adapted from [33] M.J. Feigenbaum. Quantitative universality for a class of non-linear transformations, *Journal of Statistical Physics*, **19**: 25–52, 1978]

The bifurcation diagram (figure 8.13) also reflects the decreasing difference in value between successive threshold values of parameter a, until it reaches the accumulation point, defined in the last paragraph $a = a_\infty = 0.892\,486\,418\ldots$ At this stage there exists, in principle, an infinite number of fixed points (associated with function f, infinitely composite: f^∞). The period is also infinite, which means each new iteration never returns the same value. If we zoom in on this region, we find an equivalent diagram (see figure 8.13): the bifurcation diagram is said to be self-similar.

8.10. Critical Dimension
for Attaining Chaos

Chaos is characterized by an extreme sensitivity to initial conditions. This sensitivity is illustrated by the exponential divergence of two trajectories originating from very similar initial conditions[6]; the system very quickly forgets the initial proximity of the two trajectories.

Dissipation of energy in a system confines the trajectory associated with the chaotic movement to a finite region of phase space. The trajectory is limited to the area of an attractor, called a strange attractor (see section 8.7). This condition seems a priori contradictory to the first property which is characteristic of chaos, in which *two trajectories issued from neighboring initial conditions will diverge in time.*

Now we will take a closer look. In two dimensions (a plane) it is impossible for two trajectories to grow exponentially apart: on one hand, as seen above, due to dissipation trajectories covering a limited region of the plane (attractor); on the other hand, because of determinism two trajectories originating from different initial conditions never cross. We present a proof by contradiction. If, in a phase plane, there exists a trajectory which at one point intersects with itself, and if we take as an initial condition this same point of intersection (see figure 8.14), it is impossible to determine a priori from this point which was the previous evolution of the trajectory (left or right branch). The future indeterminacy linked with the indeterminacy of the trajectory is thus in contradiction with the principle of determinism. In a chaotic regime, two initially close trajectories will diverge exponentially with time. However, due to the dense character of trajectories in a chaotic regime, a trajectory that has to deviate from that which was initially close is surrounded by many other trajectories that will repel it due to the non-crossing condition. Furthermore, a trajectory that is at the periphery of all other trajectories, cannot move far apart due to dissipation. Thus in a dynamical system with two degrees of freedom, chaos cannot take place.

---

6. We must specify that the trajectories may approach each other for certain periods of time before diverging again.

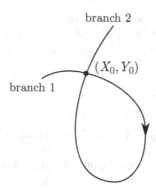

Figure 8.14 – An initial condition such as (X_0, Y_0) could belong to either branch of the trajectory. This is in contradiction with deterministic dynamics.

On the other hand, two trajectories in a 3-D phase space have the necessary room to avoid crossing each other; they can thus diverge infinitely far. The critical dimension of the phase space is in fact three – the minimum number of degrees of freedom a system needs to develop chaotic dynamics. For example, the Rössler system (see equation (8.5)) can become chaotic, while a van der Pol oscillator (see equation (5.4)) cannot; the second degree temporal evolution equation describing the van der Pol system is equivalent to two first order equations, which provides two degrees of freedom to the system and thus it is impossible for the system to go into a chaotic regime. For the van der Pol oscillator, even the scenario of a subharmonic bifurcation is impossible: all regimes which correspond to a period doubling are prohibited. The trajectory in phase space associated with this regime would need to go around twice before closing in on itself, which would be impossible without it crossing its own path, which is prohibited by the logic of determinism.

A simple pendulum (governed by a second order equation) with two degrees of freedom is subject to the same constraints as the van der Pol system. However, if one of the oscillators is driven by an external time-dependent excitation, the system gains an extra degree of freedom in the form of a temporal variable, so the pendulum may enter a chaotic regime.

To illustrate, consider the case of a sinusoidal excitation, where the pendulum equation takes the form $\ddot{\theta} + \dot{\theta} + \sin\theta = A\cos(\omega t)$, where A and ω are respectively the amplitude and the frequency of the external force. Even if the temporal variable is not a degree of freedom in the usual sense of the term, it takes care of an important condition: to determine the unique trajectory of the system. For this we need to know the details of one more variable. In this case, besides θ and $\dot{\theta}$ of a given moment, we must also know the value and sign of the force $A\cos(\omega t)$ at that moment.

To investigate explicitly how the temporal variable acts as a degree of freedom, let us define $X = t$, $Y = \theta$ and $Z = \dot{\theta}$ and take a look at the new system, determined by three equations, that define the evolution equation for the forced pendulum:

$$\dot{X} = 1,$$
$$\dot{Y} = Z,$$
$$\dot{Z} = -Z - \sin(Y) - A\cos(\omega X). \qquad (8.25)$$

Thus, for a non-autonomous equation[7] the time variable acts as a degree of freedom in the same way as dynamic variable X (or its derivative).

8.11. Lyapunov Exponents

8.11.1. Definition

We have seen that chaos in a system is synonymous to a loss of memory of initial conditions over time. If x_0 and $x_0 + dx_0$ are two neighboring initial conditions, the study of chaotic systems leads to the following statement: the initial difference dx_0 is amplified, on average (for large enough n, with n a discrete temporal variable), according to the law:

$$dx_n \sim dx_0 e^{\nu n}, \qquad (8.26)$$

where dx_n represents the distance between the two values (which were initially separated by dx_0) after time n (i.e. after n iterations) and where parameter ν is a positive number which we call the *Lyapunov exponent*, named after A.M. Lyapunov (1857–1918). This exponent expresses how quickly two trajectories which began from infinitely close initial conditions separate (in other words, how dx_n becomes increasingly far from dx_0). The larger the Lyapunov exponent ν, the more quickly any memory of the initial conditions of the system is lost. In reality, there are as many Lyapunov exponents as there are degrees of freedom, because we can always decompose the trajectory in terms of its projection along each axis of the phase space.

8.11.2. Properties of Lyapunov Exponents

First of all, it is useful to specify that we are interested in the chaos of dissipative systems, such as an electronic circuit (ohm dissipative) or a mechanical

---

7. A differential equation is called autonomous if the independent variable (time) does not appear in the equation, and if the formal relationship does not depend on this variable.

system which experiences friction[8]. Dissipative systems will experience a global diminishment of their initial volume in phase space over time. This does not mean that the three sides (imagine a volume in phase space having a cubic shape) of the volume evenly contract. In general, one of the three sides will dilate while the other two contract in a manner such that overall, the initial volume decreases over time. This is an essential point in chaos theory, as it allows trajectories to diverge in the direction of the dilation (associated with a positive Lyapunov exponent) and contract in the other two directions, thereby confining the dynamics to follow the shape of an attractor. This notion is illustrated for a limit cycle in two dimensions in figure 8.15.

The sign of the volume's variation in phase space can be determined by the sum of a system's Lyapunov exponents (see [9]). Since the overall volume is decreasing because of dissipation, the sum must be negative. Accordingly, at least one of the Lyapunov exponents must be negative. Another of the Lyapunov exponents must be zero, to ensure the trajectory "stays" localized for the whole time it is within the strange attractors (creating a bounded region). Finally, for a chaotic regime, at least one of the exponents must be positive[9], to let the system manifest the sensitivity to the initial conditions.

8.12. Self-Similarity and Fractals

We have presented two typical examples in which we observe the self-similar property of chaos (see figures 8.8 and 8.13).

The property of self-similarity means the geometric figure that emerges from the dynamics looks the same at all scales: zooming into any one part reveals the same global shape. This type of structure, popular today, is known as a fractal. The most celebrated of fractals is undoubtedly the Mandelbrot set named after B. Mandelbrot (1924–2010) (see figure 8.16). Self-similarity means these figures have a particular mathematical property: *fractal dimension*.

---

8. There exists a separate branch of physics concerned with non-dissipative chaos, also called Hamiltonian chaos, or conservative chaos. One example of such a case is found in the dynamics of the solar system, where the dissipation is not taken into account due to its very minimal effect.

9. This does not exclude the occasional existence of two positive exponents, in which case the system is said be in a state called *super-chaotic*. This can only happen in a phase space of four or more dimensions, since, as we have already specified, one of the exponents must be negative, and one must be zero (to ensure the diminishing volume of the trajectory in phase space and localization respectively). Though in many situations there is only one positive exponent, there are examples which have shown to have several positive exponents (see, for example, [54]), which lead to *super-chaos*.

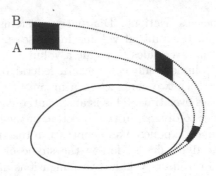

Figure 8.15 – Illustration of area diminution in phase space for a dissipative system. Two trajectories issued from two initial conditions A and B are such that the area between the two trajectories (indicated by the dark color) diminishes with time, and the two trajectories spiral around an attractor, drawn here as a limit cycle for simplicity.

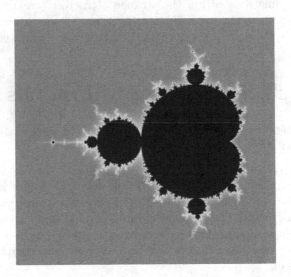

Figure 8.16 – This fractal, called the Mandelbrot fractal, is found in the following way. $u_{n+1} = u_n^2 + C$, where n is a whole number, and C is a complex number (u_n thus complex). This series might be divergent or convergent depending on the value of C. The Mandelbrot set corresponds to a set of points on the plane – real and imaginary components of C – corresponding to values of C, leading to a convergence of the series. These points belong to the domain represented in black in the figure; these domains are self-similar. [© Georg Lehnerer #75342555 Fotolia]

8.12.1. Definition of a Fractal Dimension

We traditionally classify the usual geometric objects by some integer number dimension. All lines (curved or straight) have one dimension, all surfaces have two, all volumes three, etc. The construction of a geometric figure which is self-similar at each scale, however, forces us to reconsider this way of conceptualizing the definition of an object's dimensions.

Consider a shape of two dimensions. One way of defining the dimension of a geometrical object represented in a two dimensional space is by first covering the shape by many squares having a lateral direction ϵ (see figure 8.17), and then in a second step, counting the number of these squares $N(\epsilon)$ needed to completely fill in the curve. Of course, this number is a function of ϵ, since it represents the size of the square (if ϵ is small we need a large number of squares). The fractal dimension D is thus defined by the following expression:

$$D = \lim_{\epsilon \to 0} \frac{\ln(N(\epsilon))}{\ln(1/\epsilon)}. \tag{8.27}$$

Alternative definitions are also possible, but this choice is useful (i) for its simplicity; and (ii) for its compatibility with the usual Euclidean dimension. Thus, for a line of length L, the number of segments of size ϵ needed to cover the line is L/ϵ, and we have a dimension equal to one: $D = \lim_{\epsilon \to 0} \ln(L/\epsilon)/\ln(1/\epsilon) = 1$. The number of squares needed to cover a surface of area S is equal to S/ϵ^2, resulting in a fractal dimension of two ($D = 2$, equation (8.27)).

Now let us apply the definition to self-similar objects, taking as our case study the geometric figure we call a *Koch snowflake*, defined by H. von Koch (1870–1924) well before the invention of the term fractal (see figure 8.18). We start with an equilateral triangle (represented in the far left of figure 8.18). Each side of the triangle is divided into three equal segments, and the middle segment serves as the base for a new, smaller, equilateral triangle, its sides one-third the size of the original. We repeat the process ad infinitum to obtain the lacy geometric figure shown on the far right of figure 8.18.

In the first iteration, we substitute each side of the initial equilateral triangle of length 1 with four segments one-third the size (second image of figure 8.18). From each small segment there emerges a new segment one ninth the size of the first, and so on. In the mth iteration, the figure is composed of 4^m segments, each of size $(1/3)^m$. From the definition of fractal dimension D (equation (8.27), where we noted that $\epsilon \to 0$ is equivalent to $m \to \infty$) we obtain:

$$D = \lim_{m \to \infty} \frac{\ln(4^m)}{\ln(3)^m} = \frac{\ln(4)}{\ln(3)} \simeq 1.26\ldots \tag{8.28}$$

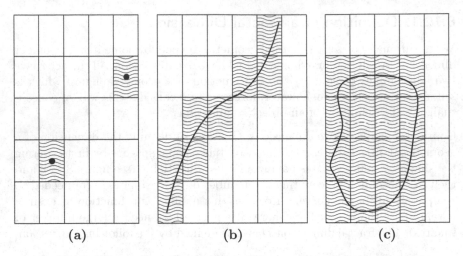

Figure 8.17 – We cover a geometric figure with the help of cubes (or squares, in two dimensions) having sides of size ϵ and we note $N(\epsilon)$ as the number of cubes needed to cover the line. We have chosen three examples: (a) an object composed of two isolated points; (b) a line in the plane; and (c) a surface enclosed by a closed curve. The cross-hatched squares show the number necessary to cover each object.

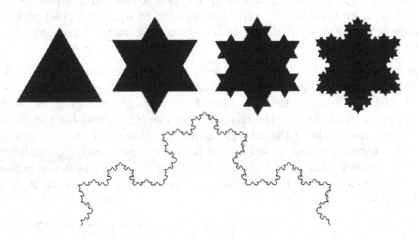

Figure 8.18 – We take each side of the triangle of unit length and divide it into three segments such that the middle segment becomes the base for a new equilateral triangle with length one-third the original, and so forth. This fractal is known as the *snowflake*.

The dimension of this object, even though it seems to be made of linear segments, is equal to a non-integer number (a fraction of an integer). The qualitative adjective *fractal* which Mandelbrot attributed to it comes from the Latin root *fractus* which signifies fractured, irregular. The Sierpinski triangle named after P.F. Sierpinski (1882–1969) (see figure 8.19) is an example of another popular fractal whose dimension is equal to about 1.58.

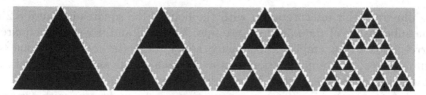

Figure 8.19 – This fractal was discovered by Sierpinski in 1915 and has been found in Italian art from the thirteenth century. It is created by putting a white triangle in the center of the initial triangle, colored black. We repeat the operation with each of the three peripheral black triangles, and so forth. Let N_m be the number of triangles found after m iterations. We have $N_m = 3^m$. We must cover each segment of the shape. The side of the black triangle of the mth iteration has a length equal to $(1/2)^m$. The fractal dimension is thus $\ln(3)/\ln(2) \simeq 1.58\ldots$

8.12.2. Caution: A Self-Similar Pattern Is Not Necessarily a Fractal

Take the case of a square (see figure 8.20) divided into four sections, which we further divide into four, and so on, a large number of times. If the original length is of unit length, the first iteration creates squares with sides of length $(1/2)$. For m iterations, the smallest squares are of length $(1/2)^m$. According to the definition of a fractal (equation (8.27)), $D = \ln(4^m)/\ln(2^m) = \ln(4)/\ln(2) = 2$, exactly equivalent to the Euclidian dimension.

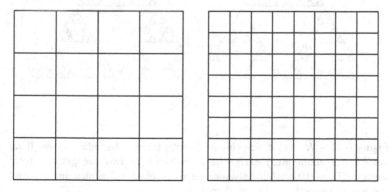

Figure 8.20 – A self-similar but non-fractal shape.

A more intuitive interpretation of the fractal dimension defined by equation (8.27) can be given if we consider it as the logarithm of the ratio of the number of self-similar parts over the amplification factor, as follows:

$$D = \frac{\ln(\text{total number of self-similar parts})}{\ln(\text{amplification factor})}. \qquad (8.29)$$

We illustrate this interpretation with the help of the squares in figure 8.20. The subdivision of the initial square into N^2 small and self-similar squares corresponds to an amplification factor equal to N (in going from one to four squares, the amplification factor of each small square is equal to two). This definition of the fractal dimension (equation (8.29)) gives us $D = \ln(N^2)/\ln(N) = 2$. The same operation can be undergone with a cube, resulting in the fractal dimension $D = 3$. What then, besides self-similarity, is needed to create a fractal with a non-integer dimension? First let us look at what the new expression for the fractal dimension gives (equation (8.29)) for the Sierpinski triangle (see figure 8.21).

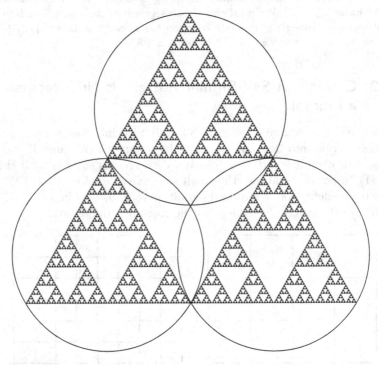

Figure 8.21 – We can divide this shape into three self-similar shapes. If we amplify (or zoom into) each side by a factor of two, we get the total surface. The relationship between the number of self-similar images and the amplification factor is equal to $3/2$.

This geometrical shape is created by subdividing the initial triangle into four triangles, three of which are self-similar (the triangles in black). The length of the sides of each self-similar triangle is two times smaller than that of the initial triangle. Thus, after one iteration, the amplification factor is equal to two for each self-similar triangle. The same operation is reiterated with these three triangles, and so forth. Consequently, the fractal dimension is:

$$D = \frac{\ln(3)}{\ln(2)} = 1.58\ldots \tag{8.30}$$

To some extent, the fractal dimension contains information about self-similarity but also about complexity: the shape of the Sierpinski triangle (figure 8.19) is more complex than that of the self-similar square (figure 8.20).

8.13. Crisis

We have now seen that the chaotic state of a system can be represented by an attractor composed of an ensemble of points in phase space, created by a trajectory characteristic of chaos. We can nonetheless have periodic unstable orbits in the attractor, that is, curves that are more or less complex and entangled with each other. If we vary one control parameter, the attractor can change shape drastically. Such events, called crises, happen for example when there is a collision between a strange attractor and one of these unstable periodic orbits. Three types of crises have been identified: (i) boundary crises, in which a strange attractor is destroyed; (ii) interior crises, in which the size of the strange attractor changes suddenly when the control parameter reaches a certain value; and (iii) an attractor merging crisis, in which two (or more) attractors fuse to form a single attractor. The abrupt change in phase space marking the crisis is also observable in the chaotic signal, as we will see below.

We will use the van der Pol oscillator for our example. Subject to an external periodic excitation or force (see section 5.1), the van der Pol oscillator obeys the following evolution equation:

$$\ddot{x} + \mu(x^2 - 1)\dot{x} + x = a\sin(\omega t), \tag{8.31}$$

where μ is a positive parameter, and a is the amplitude of the perturbation, with frequency ω. The phase space of the oscillator has three dimensions (the three variables are t, x and $\dot{x}$; see section 8.10). Since the dynamical system has three degrees of freedom, a chaotic scenario can emerge (again, see section 8.10), as is easily verified by following the evolution of velocity variable $\dot{x}$ upon variation of parameter a (see figure 8.22). The phenomenon known as crisis is reflected in abrupt changes in the chaotic dynamics (sudden large amplitudes visible in the temporal signal).

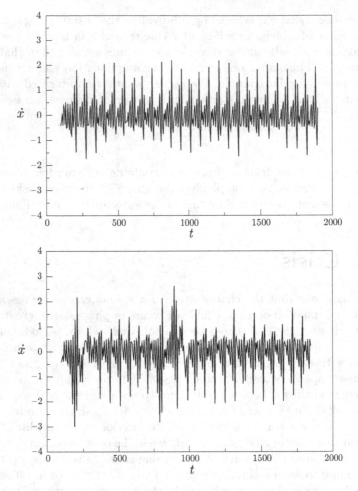

Figure 8.22 – Top: chaotic signal $\dot{x}(t)$ with $a = 0.987\,65$. Bottom: we see the emergence of a crisis for $a \simeq 0.987\,77$, characterized by sharp bursts in the signal. $\omega = 0.45$ and $\mu = 1$.

In conclusion, a chaotic regime requires three dimensions, has as many Lyapunov exponents as degrees of freedom, at least one exponent is positive (expressing sensitivity to initial conditions), one negative (reflecting dissipation), and one zero (expressing the existence of an attractor). Furthermore, chaos gives rise to complex self-similarity characterized by fractal dimensions, and can lead to a crisis for some values of control parameters. Finally, there exists three generic transitions toward chaos: subharmonic cascade, quasiperiodicity and intermittency.

8.14. Randomness and Determinism

The first time one observes a chaotic signal, a natural question may arise: is there any difference between chaos and fluctuations, or randomness? As we have seen, chaos takes place within systems which are completely deterministic, in that there is no source of randomness. Nevertheless, the two concepts can be viewed in the same way, as we shall see below. Let us discuss if there is any link between randomness and determinism. Before going into the details, let us introduce some preliminary ideas through some new examples.

8.14.1. Tent Map

Previously, we made a detailed study of the logistic map (see section 8.9) which allowed us to introduce the first transition toward chaos by subharmonic cascade (see subsection 8.9.1). The tent map, defined by the following equation:

$$x_{n+1} = 1 - 2 \left| x_n - \frac{1}{2} \right| \equiv f(x_n), \qquad (8.32)$$

is more appropriate for analyzing the statistical side of chaos.

As its name indicates, this map has the form of a tent (see figure 8.23). We can show that the interval $[0, 1]$ is inscribed into itself (the application of f on any number in that interval returns values found in the same interval). In what follows, we will also restrict ourselves to this interval.

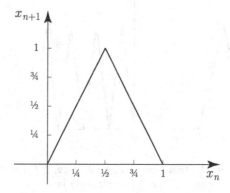

Figure 8.23 – Tent map.

For a value of x_n which is less than $1/2$, the expression between the absolute value brackets is negative and we can formulate the tent map as: $x_{n+1} = 2x_n$. For all values of x_n greater than $1/2$, the tent map gives us: $x_{n+1} = 2(1 - x_n)$.

Thus, if we start with an initial condition $x_0 < 1/2$, the first iteration gives $2x_0$ (in other words, there is a "stretching out" of the interval from x_0 to $2x_0$). After n iterations, if the value of x_n is still below $1/2$, we continue to see a stretching of the interval. After a certain number of iterations, the value of x_n goes above the threshold value $1/2$ (due to stretching), and in this case, the expression of the map changes and becomes $x_{n+1} = 2(1 - x_n)$, which will lead to a contraction instead. Suppose, for example, that $x_0 = 0.52$; we have thus, after one iteration, $x_1 = 2 \times 0.48 = 0.96$, then $x_2 = 2 \times (1 - 0.96) = 0.08 < 1/2$. Thus, if the x value is greater than 0.5, the application systematically brings the value back into the interval $[0, 1/2]$. In short, we can say that the evolution of the tent map is described by two operations: (i) a stretching of the interval $[0, 1/2]$ by a factor of two; and (ii) a folding back (contraction) into the interval of origin, and so forth. This is a general property of chaotic maps: a stretching, corresponding to a divergence of the phase space trajectories, and a folding, which keeps the trajectories bound onto the attractor.

8.14.2. Sensitivity of the Tent Map to Initial Conditions

Consider the iterative map of the tent function f. The shape associated with the function twice iterated: $f(f(x_n)) \equiv f^2(x_n)$, has the form of two "tents" posed against each other in the interval $[0, 1]$ (figure 8.24, left). Similarly, the shape of the tent map iterated m times: $f^m(x_n) = x_{n+m}$, is made up of m "tents" which are bounded on the interval $[0, 1]$ (figure 8.24, right).

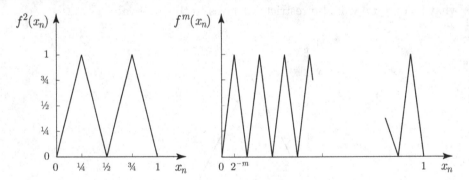

Figure 8.24 – Left: the tent map composed twice, $f^2(x_n)$. Right: the tent map composed m times, $f^m(x_n)$ (m relatively large).

To convince ourselves, it is enough to note that if variable x_n has a value 0, $1/2$, or 1, we have $x_{n+2} = 0$, whereas if variable x_n has a value $1/4$ or $3/4$, we have $x_{n+2} = 1$. Realizing that x_{n+2} is a linear function of x_n between these special points, the graph of the tent map composed once with itself (i.e. $f^2(x_n)$) is easily deduced (figure 8.24, left). The same reasoning applied to the tent function composed $(m-1)$ times with itself (i.e. $f^m(x_n)$), would give us the graph on the right in figure 8.24.

Thus, picking an initial condition whose value is known with uncertainty up to $\pm 2^{-m}$ (this margin of error is very small, when m is taken to be relatively large) the map $f^m(x_n)$ may take any value within the interval $[0,1]$. In other words, the knowledge of initial conditions within interval $[x_0 - 2^{-m}, x_0 + 2^{-m}]$ gives no indication of the final state of the system. This result can also be verified numerically with a calculator, just as with the logistic map[10].

The tent map is actually found to be easily related to the logistic map defined by equation $x_{n+1} = 4ax_n(1 - x_n)$ when we make a small change of variables. Let us look at the case where the control parameter[11] $a = 1$; the equation for the logistic map is then: $x_{n+1} = 4x_n(1 - x_n)$.

If we make the change of variable from x to y as defined by:

$$x = \sin^2\left(\frac{\pi y}{2}\right) = \frac{1}{2}\left[1 - \cos(\pi y)\right], \qquad (8.33)$$

and substitute this expression into the logistic map $(x_{n+1} = 4x_n(1 - x_n))$, we obtain: $\sin^2(\pi y_{n+1}/2) = 1 - \cos^2(\pi y_n) = \sin^2(\pi y_n)$.

We can immediately deduce the following relationship: $\pi y_{n+1}/2 = \pm \pi y_n + p\pi$ where p is a natural integer number. Since variable y_n is such that $y_n \in [0,1]$, when the value of y is such that: $0 \leqslant y_n < 1/2$, and $y_{n+1} = 2(1 - y_n)$, we obtain the map: $y_{n+1} = 2y_n$ (the sign in front of y_n is positive and $p = 0$). When the value of variable y is such that: $1/2 \leqslant y_n \leqslant 1$, we obtain: $y_{n+1} = 2(1 - y_n)$ (the sign in front of y_n is negative and $p = 1$). This concludes the mapping of the logistic map onto the tent map. We are now in a position to discuss the notion of determinism against randomness.

8.14.3. Playing Roulette or Chaos?

At the advent of rational mechanics we began to trust in the triumph of determinism. Roulette fans would have been thrilled to test our deterministic theories at their game tables. However, this deep faith, which too quickly connected the idea of an absolute predictability and determinism with Newtonian mechanics, is foiled by the existence of an unpredictable chaos in those same Newtonian systems, even in systems whose descriptive equations are relatively simple. It is now proven that these chaotic systems constitute the vast majority of mechanical systems, rather than being the exception.

---

10. Remember that we established in subsection 8.9.1 that generally, a fixed point is unstable if the derivative of f at the point of intersection with the first bisector has an absolute value greater than one. Thus, the fixed points of the tent map are unstable.

11. Recall that we have seen, in subsection 8.9.1, that chaos occurs when the control parameter a is such that $a > 0.892\ldots$

As we have already discussed, if we were capable of determining with infinite precision the initial state of the system, we would be able to predict the subsequent evolution of the system. It is our incapacity to attain such a degree of precision concerning the initial conditions which is at the root of the collapse of predictability. We could thus attribute the origin of chaos to a lack of information. However, we must not believe that this missing information is due to a temporary theoretical ignorance, or to a contextual experimental inaptitude that we will one day overcome as science progresses. Rather, it is a fundamental uncertainty, as described below.

Defining an initial state for a mechanical system of one particle is to define, for example, the initial position and velocity of said particle. The values of these initial conditions are given by real numbers which are in turn defined by an infinite sequence of numbers, for example, $1/3 = 0.33333\ldots$, $\sqrt{3} = 1.732\ldots$ If we wanted to designate all the possible values of an initial condition, we would need to be able to specify every number in the infinite sequence, which is beyond human capability.

When the dynamics of the system are not chaotic, this impossibility of being infinitely precise in stating the initial conditions is not important, because the final state of the system is not sensitive to small uncertainties in the initial conditions. In a chaotic regime, however, the incertitude is amplified exponentially. Chance or uncertainty is thus encoded in the lack of information about the initial state.

So, is it still possible to establish a statistic description of chaos, as we can for a game of dice?

8.14.4. Invariant Measure

To obtain the first element of a response to this question, let us study the simple system, constructed as follows. Take the interval $[0,1]$, subdivide this interval into a certain number of equally sized subintervals, for example, one hundred: $[0, 0.01[, [0.01, 0.02[, \ldots, [0.99, 1]$ and call each of these intervals $I_1, I_2, \ldots, I_{100}$. Then pick a real number x_0 from the interval $[0,1]$ at random; necessarily, it must belong to one of the subintervals. This number is taken as the initial condition for a logistic or tent map (or any other type of map that is known to show chaos); then carry out N iterations, with N a large number, say $N = 1000$. The final result x_N also belongs to one of the intervals I_j $(1 < j < 100)$. We repeat the process several times, say 1000, and make a histogram of the events, in which we read intervals I_j in the horizontal axis and the number of times that each interval was visited in the vertical axis. If N is large enough, we will come to a very interesting conclusion: the histogram is a curve (which we will see in what follows) that is robust (in the sense that it is independent of N if N is large enough, greater than

a few hundred), and which has the same status as any other probability law
(a Gaussian probability law, for example).

Now consider the tent application (see equation (8.32)) and divide the interval
$[0, 1]$ into N equal subintervals, calling them $[(m - 1)N, m/N]$ with $m = 1, 2, \ldots, N$. In the continuum description, let us call the variable x along the
axis of subintervals. Since the tent application is linear, none of the intervals
will be privileged over any other. Consequently, as the number of iterations
grows large, the number of times an interval is visited approaches $1/N$; the
probability of landing in an interval becomes $1/N$. This is a law of uniform
distribution. Given that x is a continuous variable, it is useful to introduce
a probability density $\rho(x)$, where $\int_a^b \rho(x)dx$ represents the number of times
that interval $[a, b]$ is visited. As we have said above, this number is equal to
the interval itself, that is, $(b - a)$. This means we have a constant density ρ:

$$\rho(x) = 1. \tag{8.34}$$

Density ρ is thus called the *natural invariant density*, or the *invariant measure*. This density is uniform, in the same way as the probability density of
some state in a game of chance, such as throwing dice (each face of the die
has the same chance of landing face up, $1/6$).

Returning again to the logistic map and invoking density $\rho(x)$, we see that the
number of times that an interval dx is visited is equal to ρdx. We have verified
that the logistic equation is related to the tent equation through a change
of variables (see equation (8.33)); designate the variable for the tent map
by y and for the logistic map by x. We have $\rho(x)dx = \rho(x(y))(dx/dy)dy$.
The uniformity of the distribution in terms of y gives us the following
condition: $\rho(x(y))(dx/dy) = 1$, that is, $\rho(x(y)) = dy/dx$. Now we use a
change of variable (equation (8.33)) to express y as a function of x. We
find $dy/dx = 2/(\pi \sin(\pi y))$. Since we also have at our disposal the relation
$\sin(\pi y) = 2\sqrt{x(1 - x)}$, we can establish the expression of density $\rho(x)$ for the
logistic map:

$$\rho(x) = \frac{1}{\pi\sqrt{x(1 - x)}}. \tag{8.35}$$

This is the natural invariant density for chaos in the logistic map (see figure 8.25); each map has its own invariant density. The term invariant refers to
the fact that if we repeat the experiment or simulation and look at the statis-
tics for the visited intervals, we always find the same density. This contrasts
with the non-statistical description where by fixing the initial conditions and
repeating an experiment never returns the same value for any time or number
of iterations. Also keep in mind that each map giving rise to chaos has its
own invariant measure (or probability density). In particular, we have seen
that the tent and the logistic map have different densities.

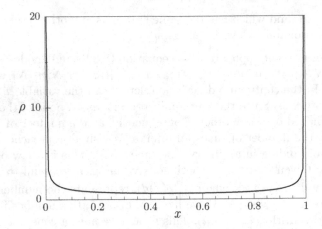

Figure 8.25 – Invariant density $\rho(x)$ of the logistic map.

8.15. Controlling Chaos

The study of the modes of chaos, and of possibilities of control over it, has given rise to many scientific publications. This research is justified, on one hand, by the interest in certain cases in knowing how to avoid chaos, and on the contrary, to benefit from the onset of chaos for specific applications. For example, in a combustion chamber, it is interesting to be able to spark the turbulent regime of reactants in order to increase the efficiency of the mixture. On the other hand, the control of chaotic regimes in certain electronic devices (especially those which can create problems) is essential in order to avoid chaotic situations which would forfeit their functionality.

As we have seen, chaos is characterized by the presence of a strange attractor in phase space. In our study of subharmonic cascades, we have stressed that the fixed point of a function or map $f(x)$ (recall that a fixed point of the map corresponds to a limit cycle for real dynamics) can become unstable and bifurcate into two branches, which corresponds to an oscillation with a doubled period. There exist in total two stable and two unstable fixed points (see figure 8.12). In the chaotic regime, the strange attractor lies at the heart of a considerable number of unstable limit cycles, or unstable orbits, which densely populate the phase space (see [80]).

By definition, the unstable orbits are repulsive. Is it possible for them to become attractive through an external action? We have already seen one case where one may stabilize an unstable state with the help of a perturbation (see section 7.6): a rigid pole maintained vertically on the palm of one's hand is unstable because of gravity but can be stabilized by a certain oscillation of the hand. Similarly, an unstable orbit can become stable when subject to an oscillation.

To control chaos one follows more or less the same reasoning. A relatively weak perturbation can be used in a chaotic system (i) to make chaos disappear, giving way to a periodic regime due to the proximity of periodic orbits – we know they are dense – now stable, which were originally unstable; or (ii) to avoid regimes of crisis, while the chaotic regime persists.

To illustrate the interest in controlling chaos, apart from being a fundamental problem of research, exciting as it may be, we take an example from daily life. A boat floating on the water is bound to encounter chaos if it encounters certain unfavorable atmospheric conditions. Nonetheless, every ship's captain seeks to avoid a crisis – such as a shipwreck. Since the captain cannot control the meteorological conditions, instead she introduces perturbations to help control the navigation of the machine.

The control of chaos, as an object of study and research, has been of keen interest since the end of the twentieth century and many interesting applications have been found, in different domains, such as in mechanical systems [28], lasers [36], heart tissue [35], chemical reactions [82], and too many others to make an exhaustive list.

8.16. Exercises

8.1

Consider the following Poincaré application:

$$f(x) = \frac{2x}{1 + x^2}. \tag{8.36}$$

1. Show that this maps the interval $]-\infty, +\infty[$ onto $]-1, 1[$.

2. Find the fixed points of f and analyze their stability.

3. Show that if $x_{n+1} = f(x_n)$ for $n = 0, 1, 2, \ldots$, and if $-\infty < x_0 < +\infty$, then there exists a real number z_1 such that $x_1 = \tanh(z_1)$ provided that $x_0 \neq \pm 1$.

4. Deduce that $x_n = \tanh(2^{2n-1} z_1)$. Find the attractors of f and their basins of attraction.

8.2

Consider the following Poincaré map ($a > 0$):

$$f(x) = a\pi^{-1} \sin(\pi x). \tag{8.37}$$

1. Show that $x = 0$ is a fixed point for any value of a. Show that $x = \pm u(a)/\pi$ are two other fixed points, where u is a solution of the equation: $u - a\sin(u) = 0$.

2. Plot the graph showing the values of a for which these two fixed points exist. Are there other fixed points?

3. For which values of a is the fixed point $x = 0$ stable?

4. Show that the two other fixed points are stable if $a < a_1$ with $a_1 = u_1/\sin(u_1)$, where u_1 is a solution of the equation $\tan(u) + u = 0$.

8.3

Consider the following Poincaré map:

$$f(x) = ae^x, \tag{8.38}$$

where $-\infty < x < +\infty$ and $a \leqslant 0$.

1. Show that f has a unique fixed point $X(a)$ which is stable if $-e \leqslant a \leqslant 0$ and unstable if $a < -e$.

2. Show that the fixed point $X(a) \sim -a$ when $a \to 0$, $X(a) \sim -1+(a+e)/2e$ when $a \to -e$, and $X(a) \sim -\ln(-a)$ when $a \to -\infty$.

3. We define the second generation map g as $g = f(f(x))$ (composed map). Show that $g_x = fg$, $g_{xx} = f(1+f)g$, $g_{xxx} = f(1+3f+f^2)g$, where the subscript designates differentiation with respect to x (for example, g_x, g_{xx} represent the first derivative and second derivatives respectively). Deduce that $g_x(X) = X^2$, $g_{xx}(X) = X^2(1+X)$, $g_{xxx}(X) = X^2(1+3X+X^2)$. We have used the identities $f(X) = g(X) = X$.

4. With the help of a graphical representation, analyze the intersection between the straight line $y = x$ and the curve $y = f(x)$, and show that a bi-periodic limit cycle emerges from the fixed point X when a decreases to $-e$. Can you use the above results to conclude analytically on this issue?

5. Show that if $x = g(x)$ and $x \neq X$, then we have the relation:

$$(X+1)(X-1) + \frac{1}{2}X^2(X+1)(x-X)$$

$$+ \frac{1}{6}X^2(1+3X+X^2)(x-X)^2 + \cdots = 0 \tag{8.39}$$

when $x \to X$. Deduce that the bi-periodic cycle is characterized by the two values:

$$X_1, \; X_2 = -1 \pm \left(\frac{-6(a+e)}{e}\right)^{1/2} + O(a+e)$$

when $a \to -e$.

8.4

Consider the following Poincaré map:

$$f(x) = ax(1 - x^2). \tag{8.40}$$

1. Show that f maps the interval $[0, 1]$ onto itself if $0 \leqslant a \leqslant 3\sqrt{3}/2 \simeq 2.598$.

2. Show there exists a fixed point $X = 0$ for any a, and two other fixed points $X = \pm[(a-1)/a]^{1/2}$ for any a such that $a > 0$ and $a < 1$.

3. Show that $X = 0$ is stable for $-1 < a < 1$ and $X = [(a-1)/a]^{1/2}$ is stable for $1 < a < 2$.

4. Use a computer (or calculator) to study the map f and find the subharmonic cascade for different values of a_r ($r = 1, 2, \ldots$) (a_r designates the bifurcation value of parameter a corresponding to an r-periodic regime), and check that

$$\frac{a_{r-1} - a_r}{a_r - a_{r+1}} \longrightarrow \delta \tag{8.41}$$

when r is large enough, where δ is the universal constant of Feigenbaum, and $a_\infty \simeq 2.3023$ (the value of the accumulation point of the cascade).

8.5

We have seen in this chapter that a self-similar object is not necessarily fractal. We would like to show that a fractal object is not necessarily self-similar. Such an object can be constructed as follows. Consider a square that is divided into nine identical smaller squares (see figure 8.26). The next step is to randomly select one square from the nine small ones that remain intact, while the other eight are subject to the same nine subdivisions made with the original square, and so on. Then we take a random nine sides which will eliminate the following subdivision operations (shown in the figure). After several operations one obtains the result shown on the right of figure 8.26.

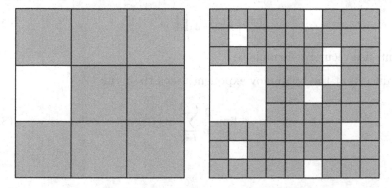

Figure 8.26 – A fractal object which is not self-similar.

1. We now take one side of the initial square (whose length is taken to be equal to one). Show that the surface on the left side of figure 8.26 is covered by $N = 8$ squares having a lateral length equal to $\epsilon = 1/3$.

2. Show that the surface on the right of that figure is covered by $N = 8^2$ squares having $\epsilon = (1/3)^2$ as a lateral length.

3. Deduce the number of total squares and their lateral sizes after N iterations. Using the definition of fractal dimension, evaluate the fractal dimension of the object represented in figure 8.26.

8.6

We would now like to calculate the Lyapunov exponents for some examples giving rise to chaos. We have seen the definition of a Lyapunov exponent in section 8.11. From definition (8.26), we readily conclude that the Lyapunov exponent can be written as:

$$\nu = \frac{1}{n} \ln \left| \frac{dx_n}{dx_0} \right|, \tag{8.42}$$

where we recall that dx_n represents the difference between two values of the dynamic variable obtained after a time interval equal to n (discrete time) from two close initial conditions, separated by a distance dx_0 at the initial time.

1. Let $x_{n+1} = f(x_n)$ be a Poincaré map. Express dx_n in term of f. Let the two neighboring initial values be denoted by x_0 and $x_0 + dx_0$.

2. Show that if dx_0 is small enough (i.e. $dx_0 \to 0$), then we can write:

$$\nu = \frac{1}{n} (f^n)'(x_0), \tag{8.43}$$

where f^n designates the map of the nth generation.

3. Show that:

$$(f^n)'(x_0) = \prod_{i=0}^{n-1} |f'(x_i)|. \tag{8.44}$$

Hint: use recursive formulae.

4. Deduce that the Lyapunov exponent takes the form:

$$\nu = \lim_{n\to\infty} \frac{1}{n} \sum_{i=0}^{n-1} \ln |f'(x_i)|. \tag{8.45}$$

8.7

Let us consider the tent map, written as $f = ax$, is $0 \leqslant x \leqslant 1/2$ and $f = a - ax$ if $1/2 \leqslant x \leqslant 1$, with $0 \leqslant a \leqslant 2$ and $0 \leqslant x \leqslant 1$.

1. Determine the Lyapunov exponent (see exercise 8.6).

2. For which value of a does the tent map exhibit chaos? Does this result comply with that given by the map defined by equation (8.32)?

Chapter 9

Pattern Formation in One Dimension

Abstract *This chapter introduces the concepts behind the spontaneous emergence of spatial order, such as the ripple patterns in sand or the spots on the fur of animals like jaguars and leopards. Two problems will be discussed in detail: the Turing chemical instability, and hydrodynamic convection. We will discuss the relevance of this chemical instability evidenced in diverse forms as it appears in nature, such as in the patterns of seashells, tropical fishes, and animal furs.*

9.1. Introduction

In the last few chapters, we have focused on the temporal dynamics of point particle systems. However, to study patterns that occur in nature we must consider dynamics not just over time, but across an area of space. More generally, a complete description of dynamics in real systems must take into account the fact that systems have a spatial extent.

The importance of a spatial dimension is evident in one of the examples we encounter regularly: the atmosphere. To establish predictive meteorological models, we need to know some spatial facts such as pressure, temperature, and wind velocity – not only as a function of time, but also throughout a region of points in space. For an understanding of meteorological events, this spatial information is important not just locally but also at a macroscopic scale. For example, certain structures, such as cumulonimbus[1] clouds, house

---

1. Clouds which extend vertically for 6 or 13 kilometers.

© Springer Science+Business Media B.V. 2017
C. Misbah, *Complex Dynamics and Morphogenesis*,
DOI 10.1007/978-94-024-1020-4_9

major atmospheric phenomena (tornadoes or storms); they themselves also show substructures, such as mammatus clouds[2] (see figure 1.4). What is the origin and mechanism for these structures? How does morphogenesis mediate between order and disorder?

Nature offers an overabundance of examples of spatially ordered structures of various categories. Examples include sand ripples, spots in animal fur, the form of sedimentation in geology, the organization of microtubules[3] in biological cells (see section 15.5), etc.

One fascinating aspect of morphogenesis is that often the spatial order emerges in almost every system which is *out of equilibrium*. Generally, the system must be at a critical distance from its thermal equilibrium (to simplify, we can describe this distance as a function of the amount of energy injected into the system). The transition from an initial, unstructured state, called the homogeneous state, to an ordered state, corresponds to – as you may have guessed – a bifurcation!

To begin our study of morphogenesis, we will first describe several examples, propose qualitative interpretations, and then develop explicit calculations with which we can establish the necessary conditions for the spontaneous emergence of order. In the final section, we will look at some contextualized examples in order to pick out, as much as possible, the universal conditions for spatial structure.

We begin our study with a famous example: the Turing instability, named after Alan Turing (1912–1954). Creator of the first programmable computer and father of the domain of artificial intelligence, Turing was also, toward the end of his life, interested in studying models of morphogenesis as found in living beings (Turing structures).

9.2. The Turing System

When two or more soluble chemical substances are mixed, the substances will react with each other to create new products. Often the initial substances and

---

2. The scientific term mammatus comes from the Latin origin *mamma*, meaning udder or breast, in this case used by meteorologists to describe clouds with a pouch or rounded form, often found at the base of different kinds of clouds such as cumulonimbus, cirrus, altocumulus. This structure is also seen in clouds of ash ejected by volcanoes into the atmosphere and in airplane vapor trails (the "contrails" that an airplane forms when crossing certain layers of the atmosphere). The formation of such clouds remains one of the mysteries in atmospheric fluid dynamics and the physics of clouds.

3. Microtubules are long molecules found in the cytoplasm of biological cells. These molecules play an important role in the process of cell division; under certain conditions, these molecules even organize themselves in something like a star formation.

their products mix homogeneously in the medium holding the reaction, like syrup in water: the syrup diffuses homogeneously, slowly coloring the water. However, homogeneous solutions are not always stable; under certain conditions, the distribution of chemical species can become non-homogeneous and evolve into a spatial pattern, sometimes entirely distinct from anything homogeneous. This kind of chemical morphogenesis is without doubt the most famous of the phenomena which can lead to the emergence of patterns in nature. In 1952, Turing published a famous article, "The chemical basis of morphogenesis" [104]. Besides proposing the relatively simple mathematical equations which describe the mechanism for the instability and the origin of the chemical morphogenesis, Turing also makes reference to the more general impact that this instability might have in nature, especially in living creatures.

The genius of the work realized by Turing resides in its disconcerting simplicity, which nonetheless leads to general conclusions of universal importance. We can summarize these results as follows. Suppose we have two chemical reactants where u is an activator (i.e. a chemical that stimulates its own growth) and v an inhibitor (i.e. which acts against the growth of the activator). If we then mix the two we can be certain, without needing to further specify the details of the reaction, that if the inhibitor v diffuses quickly enough with respect to the activator u, the homogeneous mixing of the two reactants will become unstable and we will observe the emergence of a spontaneous spatial chemical order (a pattern of regular alternation between regions rich in u, and regions rich in v). This order can appear as alternating bands, like the white and black stripes of a zebra, but more complicated patterns, such as the hexagonal structures found in bees' nests, are equally possible.

The generality of the Turing model is an indication that chemical order is a robust and inevitable phenomenon. Nonetheless, the reaction is tricky enough that nearly 40 years elapsed after the publication of this article before a laboratory experiment confirming the Turing instability was carried out (see [12, 15, 81]).

The Turing model corresponds to an autocatalytic chemical reaction, but is otherwise independent of the details of the reaction itself. Two phenomena are in play: (i) the reaction between two chemical agents; and (ii) the diffusion of the chemical substances in the solution. To describe the evolution of each concentration, we can write the evolution equations as:

$$\partial_t u = f(u, v) + D_u \partial_{x^2} u,$$
$$\partial_t v = g(v, u) + D_v \partial_{x^2} v, \tag{9.1}$$

where $u(x, t)$ (and respectively $v(x, t)$), dependent on space x and time t, denotes the concentration of the reactant u (respectively, v), $\partial_t u$ and $\partial_t v$ designate the variation (or partial derivative) in the concentration of each chemical species over time at a given point, and $f(u, v)$ (respectively $g(v, u)$) is the function describing the chemical reaction of substance v (respectively, u), ∂_{x^2} is the second derivative with respect to x, and finally, D_u and D_v are the diffusion coefficients of reactants u and v, respectively. The specific process by which the reaction takes place, or the precise form of the reaction equation, is not important.

Common sense would have led us to believe that in the presence of a local heterogeneous distribution of substances, diffusion would slowly create homogeneity, as with a drop of syrup in water. However, on the contrary, the mutual diffusion of two substances can favor, under certain conditions, the segregation, and thus instability, of the homogeneous state.

9.2.1. A Qualitative Picture of the Turing Instability

To think qualitatively about the Turing instability, we can begin by considering the homogeneous mixing of two substances. In real systems, there are always small fluctuations, like thermal agitation, etc. Suppose that, due to some minor fluctuation, a small volume of the solution briefly has a higher concentration of the activator agent (see figure 9.1). Since activator u stimulates its own production, an excess of the activator always creates even more. However, we must not forget that it produces, at the same time, more of its antagonist v. Thus an excess concentration of the activator also indicates an increased production of the inhibitor. In principle, the two substances would have the tendency to homogenize, in line with the normal effect of diffusion. However, and herein lies the originality distinguishing the Turing model: if the inhibitor diffuses significantly more rapidly than the activator, it will diffuse from the high concentration region before the activator, leaving it once again to create a local excess (see figure 9.1) – until an equilibrium is reached between the activator's diffusive and self-catalytic behavior.

It is possible for several disconnected concentration peaks of the activator to form at the same time within a system. There is nevertheless a higher likelihood for a new concentration peak to appear in the neighborhood of the first. The inhomogeneity created by the first creates favorable conditions for the emergence of another. However, note that each chemical substance has a specific lifetime, diffusing for some time before disappearing in an interaction with another substance.

We can denote the lifetimes of the two species by τ_u and τ_v, and specify their diffusion lengths[4] $l_u \sim \sqrt{D_u \tau_u}$ and $l_v \sim \sqrt{D_v \tau_v}$. The distance between two concentration peaks must be of at least one diffusion length: molecules u,

---

4. The diffusion lengths are defined by the average length a molecule travels before undergoing a chemical reaction.

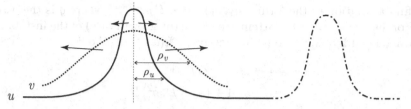

Figure 9.1 – The solid line: the activator concentration u, a self-catalyst, increases; it also produces the inhibitor (dotted line) v. If the inhibitor diffuses more quickly, its concentration peak will spread out more quickly, allowing the activator to continue to increase locally. This is the basic principle behind the Turing instability. Another peak can appear – in principle, anywhere, but there is of course a higher likelihood of it being in the neighborhood of the first, because of the fluctuations which tend to give way to the instability. The second peak concentration of the activator (dot-dash line) has the possibility to grow if it is situated beyond the reach of the excess v created by the first peak. This is typically a distance of the order of the diffusion length of v, l_v. Likewise, the molecules u which come from the first peak react after a distance of the order l_u. The wavelength of the Turing structure is of order $\sqrt{l_u l_v}$. The arrows in the figure schematize the diffusion. $\rho_{u,v}$ is a typical width of the concentration profile.

found in excess around the first peak, must travel a distance of order l_u before they can participate in a reaction again. At the same time, the inhibitor, which diffuses more quickly, finds itself in abundance (compared to the activator) in the surrounding region. The physical wavelength of the pattern is thus a compromise between the two lengths l_u and l_v. For example, the characteristic wavelength can be given by the geometric average of the two lengths, $\lambda \sim \sqrt{l_u l_v}$ (see below).

9.3. Linear Stability Analysis for the Turing System

A stationary, homogeneous state of the Turing system, denoted by (u_0, v_0), would need to satisfy the following conditions: $f(u_0, v_0) = 0$ and $g(u_0, v_0) = 0$ (see equation (9.1)). We can assume the existence of this homogeneous solution, as we will later confirm in many concrete examples. To make a linear stability analysis of the system, we want to look for solutions of the form $u = u_0 + u_1(x, t)$ and $v = v_0 + v_1(x, t)$ and substitute these expressions into the evolution equation for the system (9.1). We first make a Taylor expansion of functions f and g around the homogeneous state (u_0, v_0), to first order (in u_1 and v_1) to give us two linear equations for the pair (u_1, v_1). Then we want to find a solution of the form[5] $(u_1, v_1) = (A, B)e^{iqx+\omega t}$ where q is the wavevector and ω describes the attenuation rate (or growth rate) of the instability, characterized by $\Re(\omega) < 0$ (or respectively, $\Re(\omega) > 0$). We find:

$$\omega A = (f_u - D_u q^2)A + f_v B,$$
$$\omega B = g_u A + (g_v - D_v q^2)B, \tag{9.2}$$

where we have defined

$$f_u = [\partial f/\partial u]_{(u_0, v_0)}, \quad f_v = [\partial f/\partial v]_{(u_0, v_0)},$$
$$g_u = [\partial g/\partial u]_{(u_0, v_0)}, \quad g_v = [\partial g/\partial v]_{(u_0, v_0)}. \tag{9.3}$$

This system is linear and homogeneous, which means the solution is nontrivial only if the determinant is zero. Accordingly, we can write down the following dispersion relation to relate wavevector q with the instability's growth rate, ω:

$$\omega^2 - S\omega + P = 0, \tag{9.4}$$

where:

$$S = f_u + g_v - (D_u + D_v)q^2,$$
$$P = f_u g_v - f_v g_u - (D_u g_v + D_v f_u)q^2 + D_u D_v q^4. \tag{9.5}$$

The factor S is the sum of eigenvalues ω_1 and ω_2 from the dispersion relation (equation (9.4)) (S is the trace of the stability matrix), and the factor P is equal to the product of the eigenvalues, $P = \omega_1 \omega_2$.

9.4. Definition of the Turing Instability

By definition, a Turing instability is stationary and corresponds to a non-zero critical wavevector q_c. We must determine the condition (or conditions) for which the instability growth rate ω becomes zero (both the real and imaginary parts). When the homogeneous solution is stable, any perturbation in the system will attenuate (i.e. $\omega < 0$) for any value of wavevector q (if $\mu < \mu_c$, where μ is a control parameter of the system, see figure 9.2). At the critical point ($\mu = \mu_c$), the perturbation's growth rate ω becomes zero at the critical wavevector q_c, but remains negative for any other value. This point thus defines the maximum of the curve. As we pass the critical point of the instability threshold, ω becomes positive not only at the point $q = q_c$, but for a whole band of wavevectors. Thus, the maximum of $\omega(q)$ (and eventually, its neighborhood) is of particular interest.

---

5. As long as we stay in the linear regime, the different Fourier modes $e^{iqx+\omega t}$ are not coupled, which is what justifies our considering only one mode.

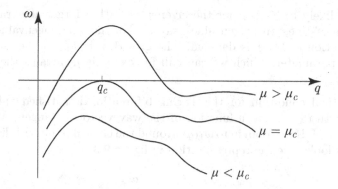

Figure 9.2 – General evolution of $\omega(q)$ in the case of a homogeneous solution which is stable ($\mu < \mu_c$), neutral (or critical, $\mu = \mu_c$), or unstable ($\mu > \mu_c$).

We continue with the following considerations:

1. The necessary conditions for the Turing instability must be characterized by a real eigenvalue[6] ω. This contrasts with many other instabilities such as the Hopf bifurcation, where a transition from a stationary solution toward a time-oscillating solution occurs for a complex ω (see chapter 5).

2. Furthermore, we are looking for an instability which produces spatial order, meaning it can be characterized by a finite critical wavelength λ_c. This implies that the wavevector (inversely proportional to the wavelength, i.e. $q_c = 2\pi/\lambda_c$) must be non-zero (i.e. $q_c \neq 0$). In other words, the growth rate of the instability at a zero valued wavenumber must be negative ($\omega(q = 0) < 0$) to avoid the development of this instability for $q = 0$. Thus, at $q = 0$, the two eigenvalues of the dispersion relation must be negative. The sum of the two eigenvalues is therefore also negative (i.e. $S = \omega_1 + \omega_2 < 0$) and their product, positive (i.e. $P = \omega_1\omega_2 > 0$). From the expression for factors S and P (see equation (9.5)), and for $q = 0$, we get the following conditions:

$$f_u + g_u < 0, \quad f_u g_v - f_v g_u > 0. \tag{9.6}$$

3. The homogeneous state becomes unstable when the system, whose behavior is determined by the control parameters, reacts by amplifying the perturbation rather than damping it. This occurs when the instability growth rate ω becomes positive. This means the Turing bifurcation takes place when one of the two eigenvalues passes zero, becoming positive, which

---

6. Recall that a stationary instability is defined by the evolution of the initially perturbed system from one stationary state (i.e. time-independent, here the homogeneous state) toward another stationary state (an ordered state, like a spatial pattern of waves).

is most likely to happen for the eigenvalue with a larger algebraic value (i.e. the least negative eigenvalue), say ω_1. For the other eigenvalue ω_2 the perturbation $e^{iqx+\omega_2 t}$ is damped. The growth rate approaches zero as a control parameter, which we can call for example μ, attains the critical value μ_c.

4. The critical condition for the Turing bifurcation is satisfied only when the growth rate ω_1, as a function of the wavevector, is tangent at $q = q_c$ (for $\mu = \mu_c$). If we zoom in to $\omega_1(q)$ around the critical point, the dispersion relation looks like the representation in figure 9.3.

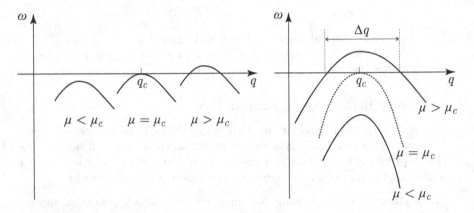

Figure 9.3 – Left: the generic dispersion relation for three situations (with μ the control parameter). (i) Below the critical point ($\mu < \mu_c$); (ii) at the critical point ($\mu = \mu_c$); and (iii) beyond the critical point ($\mu > \mu_c$). Right: the same relation, close to the critical point (this corresponds to a zoom in the neighborhood of the instability threshold) where the dispersion curve is centered around the same q_c. There is a band of wavevectors Δq which correspond to unstable modes.

9.4.1. Emergence of Order

At the critical point, not only does the eigenvalue of the instability growth rate $\omega_1(q_c)$ go to zero, but so does its derivative with respect to the wavevector (see figure 9.3). In other words, at the bifurcation point, the two following conditions are met simultaneously: (i) $\omega_1(q_c = 0)$; and (ii) $(\partial \omega_1(q)/\partial q)_{q_c} = 0$. The first condition determines the value of critical wavevector q_c as a function of the control parameter, and the second imposes a relation between control parameters (or fixes the value of the parameter if there is a single control parameter). This relation represents the frontier in parameter space which delimits regions of stable and unstable solutions. It can define the line, surface, or volume, of a catastrophe (see chapter 4 on catastrophe theory).

All solutions to the linearized Turing system are of the form $a_q e^{iqx+\omega t}$, where a_q is the amplitude associated with the mode having q as a wavevector. The fully general expression is given by the linear combination of all the different modes with wavevector q, as follows:

$$u_1 = \sum_q a_q e^{iqx+\omega t}. \tag{9.7}$$

At the bifurcation point, the growth rate is zero for the critical wavevector ($q = q_c$, $\omega = 0$), but negative for other values (i.e. $q \neq q_c$, $\omega < 0$). These modes will die out with time leaving only the critical mode $q = q_c$. Accordingly, we could describe the stationary solution of the linearized Turing system by:

$$u_1 = C e^{iq_c x} + c.c., \tag{9.8}$$

where $c.c.$ signifies "complex conjugate" and C is the amplitude of the mode associated with vector q_c (this amplitude cannot be determined in the linear regime). This expression already evidences the emergence of order; the Fourier mode corresponds to a spatially periodic solution in the neighborhood of the bifurcation.

9.4.2. Necessary Condition for the Turing Instability, Emergence of Order

Let us now examine the conditions under which the linear Turing system develops the instability. In the last section we saw that a bifurcation ($q = q_c$) occurs when the two following conditions are met: $\omega_1(q_c) = 0$ and $\partial\omega_1(q_c)/\partial q = 0$. These two conditions, along with the system's dispersion relation (see equation (9.4)), place two constraints on factor P: $P = 0$ and $\partial P/\partial q|_{q=q_c} = 0$. Using these, the explicit expression for P, equation (9.5), can be algebraically manipulated to give:

$$(D_v f_u + D_u g_v)^2 - 4D_u D_v (f_u g_v - f_v g_u) = 0 \tag{9.9}$$

and

$$q_c^2 = \frac{D_v f_u + D_u g_v}{2D_u D_v}. \tag{9.10}$$

Since the two diffusion coefficients D_u and D_v are positive (i.e. $D_u > 0$ and $D_v > 0$), we can use the above condition to describe the relation between the coefficients as:

$$D_v f_u + D_u g_v > 0. \tag{9.11}$$

These constraints, placed on reaction functions f and g (equation (9.6)) and on the diffusion coefficients (equation (9.11)), formulate the general necessary conditions for the emergence of a chemical pattern.

We can further simplify these expressions by exploiting the self-catalytic nature of the Turing reaction. With u designating the activator, the derivative with respect to u of the term describing the reaction between u and v must be positive (i.e. $f_u > 0$); necessarily, the term contributes to the growth of u. For example if we choose $f \sim u^2$ for the Turing system (equation (9.1)), the activator u ensures its own growth ($\partial u/\partial t \sim +u^2$). In contrast, inhibitor v must be such that $g_v < 0$, since the term must inhibit the growth of u.

It is easy to show that all four conditions placed on the reaction functions (equations (9.6) and (9.11)) can only be simultaneously satisfied if:

$$f_u > 0, \quad f_v < 0, \quad g_u > 0, \quad g_v < 0 \tag{9.12}$$

or

$$f_u > 0, \quad f_v > 0, \quad g_u < 0, \quad g_v < 0. \tag{9.13}$$

Notice that $f_u > 0$ is always paired with $g_v < 0$.

Combining inequalities (9.6) with (9.11), we can sum up the necessary conditions for the Turing instability by:

$$\frac{D_v}{D_u} > r > 1, \quad \text{with} \quad r \equiv -\frac{g_v}{f_u} > 1. \tag{9.14}$$

This is a formal and explicit description of Turing's result: when the inhibitor diffuses more rapidly than the activator, or precisely r times more quickly with $r > 1$, the homogeneous state becomes unstable in favor of the chemical order described by a sinusoidal oscillation of wavelength $\lambda_c = 2\pi/q_c$. Note that this result is very general and does not make any explicit reference to the kinetic reaction functions f and g above. As we go further past the bifurcation point, the equality (equation (9.9)) becomes an inequality:

$$(D_u f_u + D_u g_v)^2 - 4D_u D_v(f_u g_v - f_v g_u) > 0. \tag{9.15}$$

In this case, as depicted in figure 9.3, the instability growth rate ω_1 becomes zero for two wavevector values q, denoted q_- and q_+, given by:

$$q_{\pm}^2 = \frac{(D_v f_u + D_u g_v) \pm \sqrt{(D_v f_u + D_u g_v)^2 - 4D_u D_v(f_u g_v - f_u g_u)}}{2D_u D_v}. \tag{9.16}$$

The band of wavevectors $\Delta q \equiv q_+ - q_-$ designates the active modes (unstable modes, see figure 9.3) that grow exponentially with time (see equation (9.7)).

In the neighborhood of, and a little beyond, the critical point $\mu = \mu_c$, the derivative of the growth rate as a function of the wavevector is zero ($\partial \omega_1/\partial q|_{q_c,\mu_c} = 0$), and we can write (making an expansion around the threshold values):

$$\omega_1(q) = \alpha(\mu - \mu_c) - \beta(q - q_c)^2 \tag{9.17}$$

where

$$\alpha = \left.\frac{\partial \omega_1}{\partial \mu}\right|_{q_c,\mu_c} , \quad \beta = \left.\frac{1}{2}\frac{\partial^2 \omega_1}{\partial q^2}\right|_{q_c,\mu_c} , \tag{9.18}$$

and $(\mu - \mu_c)$ measures the distance from the bifurcation threshold.

9.5. Introduction of a Few Models Giving Rise to Turing Patterns

Numerous models are susceptible to the Turing instability. For example, if we add a diffusion term to the chemical reaction model known as the Brusselator system (previously studied in the chapter on Hopf bifurcation, see subsection 5.3.4), the model turns into a reaction-diffusion system that can be described by the Turing system (see equation (9.1)) and thus can now develop the said Turing instability.

However, we will turn to two other chemical reaction models to find examples of the Turing instability: the Schnackenberg model, proposed in 1979 [92] and the Lengyel–Epstein model proposed in 1991 [60].

9.5.1. The Schnackenberg Model

The Schnackenberg chemical reaction model is actually a simplified Brusselator model (see subsection 5.3.4), in which the chemical reactions have four distinct stages:

$$A \longrightarrow X \quad \text{constant } k_1 \tag{9.19}$$

$$X \longrightarrow Products \quad \text{constant } k_2 \tag{9.20}$$

$$2X + Y \longrightarrow 3X \quad \text{constant } k_3 \tag{9.21}$$

$$B \longrightarrow Y \quad \text{constant } k_4. \tag{9.22}$$

This model differs from the Brusselator (equation (5.46)) in the last stage: instead of $B + X \to Y$ we have the simpler $B \to Y$. Designating the four constants of the reaction by k_i $(i = 1, 4)$, and using the same approach we used for analyzing the Brusselator model (see subsection 5.3.4), we write the evolution equations for chemical species X and Y:

$$\dot{X} = k_1 A - k_2 X - k_4 B X + k_3 X^2 Y + D_X \nabla^2 X, \tag{9.23}$$

$$\dot{Y} = k_4 B - k_3 X^2 Y + D_Y \nabla^2 Y, \tag{9.24}$$

where D_X and D_Y are the diffusion constants for X and Y. After writing the adimensional[7] versions of these equations, we can use the chemical reaction functions $f(u,v)$ and $g(u,v)$ to write the evolution equations[8] as a general Turing model's system of equations (9.1):

$$\partial_t u = f(u,v) + \partial_{x^2}^2 u,$$

$$\partial_t v = g(u,v) + d\partial_{x^2}^2 v, \qquad (9.28)$$

with the reaction functions and the diffusion coefficients defined as:

$$f(u,v) = \gamma(a-u+u^2 v), \quad g(u,v) = \gamma(b-u^2 v), \quad D_u = 1, \quad D_v = d. \quad (9.29)$$

From this we can establish the criterion for the Turing instability using the general conditions required of any Turing system (equations (9.12) and (9.13)). We can determine the minimum value of the diffusion coefficient[9] D_v for the inhibitor v: $D_v = d > 3 + 2\sqrt{2} \simeq 5.8$. Thus, for the Schnackenberg model, it is necessary that the inhibitor diffuses at least six times more quickly than the activator for the Turing instability to emerge.

9.5.2. The Lengyel–Epstein Model

The Lengyel–Epstein model [60] was proposed in 1991 to characterize the results of an experiment developed in 1990 which showed Turing patterns

---

7. We define the new variables as follows:

$$t^* = \frac{D_X t}{L^2}, \quad r^* = \frac{r}{L}, \quad d = \frac{D_Y}{D_X}, \quad \gamma = \frac{L^2 K_2}{D_X}, \qquad (9.25)$$

$$a = \frac{K_1}{K_2}\left(\frac{K_3}{K_2}\right)^{1/2} A, \quad b = \frac{K_4}{K_2}\left(\frac{K_3}{K_2}\right)^{1/2} B, \qquad (9.26)$$

$$u = \left(\frac{K_3}{K_2}\right)^{1/2} X, \quad v = \left(\frac{K_3}{K_2}\right)^{1/2} Y, \qquad (9.27)$$

where L is the characteristic length of the periodic chemical pattern, the Turing pattern.

8. We conserve the same notation for variables t and x.

9. The stationary and homogeneous solution of (9.28) obeys $f(u,v) = g(u,v) = 0$, which leads to $u_0 = a + b$ et $v_0 = b/(a+b)^2$. The two derivatives of f and g at this point are given by $f_u = \gamma(b-a)/(b+a)$ and $g_u = -2\gamma b/(b+a)$. The first condition in (9.6) imposes that $f_u + g_v = b - a - (a+b)^3 < 0$, while the second is automatically satisfied by virtue of the fact that $f_u g_v - f_v g_u = \gamma^2(a+b)^2 > 0$. The Turing condition (9.14) imposes $d(b-a) - (a+b)^3 > 0$. Adding the left-hand side to the inequality $(-b+a+(a+b)^3 > 0)$ we obtain $(d-1)(b-a) > 0$. Taking into account $b - a > 0$ (which comes from $f_u > 0$, a result of u being an activator; see (9.12) and (9.13)), we obtain $d > 1$: the inhibitor must diffuse more quickly than the activator to assure that the instability condition is met. Such a condition is necessary but not sufficient. The bifurcation requires, in addition, $[d(b-a) - (a+b)^3]^2 > 4d(a+b)^4$, obtained from (9.15) after a few algebraic manipulations. This condition is rewritten, after taking the square root of the inequality: $\alpha\xi^2 - 2\xi\beta^2 - \beta^3 > 0$, where $\alpha = b - a$; $\beta = a + b$; $\xi = \sqrt{d}$. Taking into account the relations established here, we can show that the smallest value of d needed for the Turing instability to occur is given by $d \equiv D_v = 3 + 2\sqrt{2} \simeq 6$.

for the first time, 40 years after their prediction. Based on coupled reaction-diffusion phenomena, the Lengyel–Epstein model proposes the following system of equations to describe the evolution of concentrations:

$$u_t = a - u - 4\frac{uv}{1 + u^2} + D_u \nabla^2 u,$$

$$v_t = b\left(u - \frac{uv}{1 + u^2}\right) + D_v \nabla^2 u, \tag{9.30}$$

where a and b are control parameters (recall that subscript t refers to the time derivative).

We recognize, in the above system (equation (9.30)), the general expression of a Turing system (equations (9.1)) with these reaction functions:

$$f(u, v) = a - u - 4\frac{uv}{1 + u^2},$$

$$g(u, v) = b\left(u - \frac{uv}{1 + u^2}\right). \tag{9.31}$$

From these we can find the expression for the stationary homogeneous state (u_0, v_0) of the Turing system, obtained when the reaction functions become zero at those points, (i.e. $f(u_0, v_0) = 0$ and $g(u_0, v_0) = 0$), as follows:

$$u_0 = \frac{a}{5}, \quad v_0 = 1 + \frac{a^2}{25}, \tag{9.32}$$

and evaluate the derivatives of the reaction functions with respect to reactant concentrations u and v:

$$f_u = \frac{3a^2 - 125}{25 + a^2}, \quad f_v = -\frac{20a}{25 + a^2},$$

$$g_u = \frac{2ba^2}{25 + a^2}, \quad g_v = -\frac{5ba}{25 + a^2}, \tag{9.33}$$

to verify that the general Turing conditions (equations (9.12) and (9.13)) are satisfied. From our previous analysis, we can adapt the conditions to establish the following inequality, made up of the derivatives of the reaction functions and diffusion coefficients (equation (9.15)). For the specific Lengyel–Epstein model, the inequality is satisfied for a control parameter b smaller than the critical value b_c:

$$b < b_c \equiv \frac{D_v}{5D_u a}\left(125 + 13a^2 - 4a\sqrt{10(25 + a^2)}\right). \tag{9.34}$$

The critical value of parameter b (b_c) is proportional to the ratio between diffusion coefficients (equation (9.34)). Thus, fixing the value of b_c is equivalent to placing constraints on D_v/D_u.

Finally, using the general expression (equation (9.4)), we can write the dispersion relation[10] associated with the Lengyel–Epstein model as follows:

$$\omega = (b - b_c) \left(\frac{\partial \omega_1}{\partial b} \right)_{b=b_c, q=q_c} + \frac{(q - q_c)^2}{2} \left(\frac{\partial^2 \omega_1}{\partial q^2} \right)_{b=b_c, q=q_c}. \qquad (9.35)$$

Note that the expression is of the same form as equation (9.18); once again, this is a general result that does not depend on the detail of the system.

9.6. What Are the Conditions for an Inhibitor to Diffuse Quickly Enough with respect to the Activator?

It is important to note that the reactants in the Turing system are molecules of a comparable size. Thus their diffusion coefficients are also comparable and cannot differ by a factor of more than two or three, and very rarely, or almost never, six or more, as is required, for example, by the Schnackenberg model discussed above. This was exactly the major obstacle which kept laboratories from creating an experimental observation of a Turing system until 1990 [15], despite the prediction by Turing in 1952 [104].

Before going into the details of the first successful experimental observation of this instability, let us think through some chemical kinetics.

Autocatalytic reactions, like numerous chemical reactions, take many steps and not only two or three, as one might think from the above models. Luckily, only a few chemical species are actually relevant, based on whether they have slow or fast dynamics. Remember that slow variables dominate the dynamics of a system, while fast variables are adiabatically eliminated (see chapter 2). Thus, when analyzing a model based on the reaction of two or three variables, we must keep in mind that we are looking at the problem after the adiabatic elimination of fast variables in favor of the slow variables[11].

---

10. Expanded to the lowest order in the neighborhood of the bifurcation point.

11. Concretely, this adiabatic elimination occurs by annulling the time derivative of the fast variables. Thus, suppose the problem has three variables, two slow variables u, v and a fast variable w. In zeroing out the time derivative of the fast variable (i.e. $\partial w/\partial t = 0$) we are able to express w as a function of the other variables u and v. Substituting the expression into the equations for the slow variables, we find a coupled system which makes use only of variables u and v. Thus we can say that the coefficients of the system of u and v also hold "hidden" information about w.

The famous chemical reaction which first demonstrated a Turing instability was chlorite–iodide–malonic acid (CIMA)[12] [15]. One of the main difficulties in studying this reaction arises from a natural convection phenomenon which occurs at the same time as diffusion. This convection strongly perturbs the analysis; in particular, convection currents tend to homogenize the mixture and thus hide the Turing patterns. To counter this effect, the reaction was carried out in a water-saturated gel [15], which efficiently blocks the convection motion. One must also introduce a colored indicator to produce a chemical contrast, so as to distinguish between regions rich in u and regions rich in v.

The chosen indicator in the experiment cited above, a large molecule of non-negligible mass, turns out to form a certain chemical complex with the activator. The diffusion of the activator, trapped by the complex, is reduced. As a result, the inhibitor is sufficiently mobile in comparison to satisfy the conditions for the Turing instability. The addition of this indicator was key for the first creation of the Turing instability [15] (see figure 9.4). This first observation was closely followed by another similar discovery (see [81]).

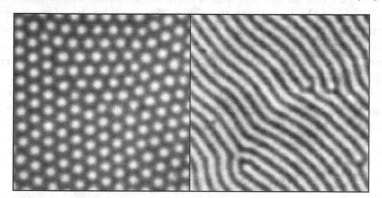

Figure 9.4 – Typical chemical patterns observed experimentally [15]. We observe honeycomb patterns (left) and patterns in a band (right). Bi-dimensional structures will be discussed in chapter 13. The structures in bands are not perfectly periodic: you can observe some defects in the figures. [With the kind authorization of Patrick de Kepper]

---

12. The CIMA reaction derives from the chlorite–iodide reaction. It is often decomposed into two stages described by the following equations:

$$ClO_2^- + 4I^- + 4H^+ \rightarrow 2I_2 + Cl^- + 2H_2O,$$

$$5ClO_2^- + 2I_2 + 2H_2O \rightarrow 5Cl^- + 4IO_3^- + 4H^+.$$

Kinetic studies [55, 68] show that the first stage is autocatalytic in I_2 (an intermediary species of the reaction) and inhibited by I^- (original inhibitor). Several experimental measurements support a description of the reaction by equations containing only two variables (for more detail, see [60]).

9.7. Beyond the Linear Turing Instability

Throughout the chapter, we have seen that under certain general conditions, a homogeneous solution (i.e. a homogeneous mix of two chemical species) can become unstable and evolve spontaneously into a structured state (or spatially ordered) defined by a characteristic wavelength. We can study the stability of the solution using a linear analysis to start with, but need to make a nonlinear evaluation to see the whole picture. The linear study tells us that the perturbation is, at first, infinitesimal[13], and then grows exponentially with time. This means that the perturbation finishes by acquiring a large amplitude over the course of time, large enough that the linear hypothesis becomes invalid. Hence, nonlinear terms must be taken into consideration.

Then, there are two complementary nonlinear studies we can make. The first type of study (already covered in chapter 6), looks at the weakly nonlinear regime. This means making a limited expansion of equations in the neighborhood of the instability threshold. The second type of study is a direct numerical integration of the complete system of equations. This could be done with the Schnackenberg (equation (9.28)) or the Lengyel–Epstein model (equation (9.30)), among others. Numerical investigation is a general rule in the study of nonlinear systems. However, having analytic information, which is generally valid close to the instability threshold, provides an interesting basis and guide for the full nonlinear evolution. The analytic study will be seen in the next chapter.

9.8. The Diverse Turing Patterns as Found in Nature

A fascinating aspect of the Turing model is its potential for engendering many similar forms to those observed in nature, from relatively simple structures such as honeycombs or stripes (see figure 9.4), to the more complex and fascinating structures such as: (i) shell patterns (see the shells in [11]; these patterns are found by combining Turing-type equations with other chemical equations); (ii) the patterns of tropical fish (see [4]); (iii) the patterns in the fur of many animals, such as the leopard or the jaguar (see figure 9.5). These different structures emerge from a modified Turing model, and in some

---

13. The hypothesis of a small amplitude in the perturbation is necessary in order to legitimize the linearization of equations.

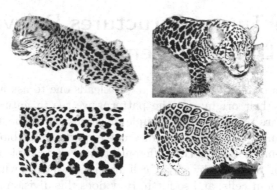

Figure 9.5 – Chemical reproductions of leopard and jaguar patterns, created with some variant of the Turing model. The top figures are real and the bottom ones were obtained with numerical simulations. The real leopard fur (top left) and the real jaguar fur (top right) present spot patterns early on in their development (of the same type as honeycomb patterns). A Turing-type model reproduces the different motifs for adults: leopard spots (bottom left) and polygons for the jaguar (bottom right). [From [62] R.T. Liu, S.S. Liaw & P.K. Maini. Two-stage Turing model for generating pigment patterns on the leopard and the jaguar, *Physical Review E*, **74**(1): 011914, 2006, © American Physical Society]

cases with the adoption of more complex geometries for the substrate (such as a curved geometry) on which the reaction takes place. The Turing model which creates the structures associated with the shell pattern or animal fur has been resolved numerically with methods that are now taught as classics in the elementary courses on numerical simulations.

Close to the instability's threshold, all the well-ordered structures – stripes, honeycombs, squares, etc. – are possible. As we go farther from the threshold, the structures can become more complex. Perfect order becomes a combination of order and disorder (or order with defects). Balancing between parameters and initial conditions allows one to reproduce images such as the ones shown in figure 9.5 (see [62]). Generally, if we begin with a Turing system, an appropriate combination of parameters (and sometimes adding a third concentration field) can also produce the different structures seen in figure 9.5, and more – such as the patterns observed on shells or tropical fish. For more details, we suggest the works which have become the inequivocal references in the domain [69, 70, 76].

9.9. Are Turing Structures Relevant for Living Systems?

The observation of many patterns in nature leads one to ask a fundamental question about their origin: are the patterns we see in nature determined by genetics, or are there relatively simple mechanisms which can explain the diversity of forms? A related topic of great importance in biology, or more specifically in embryology, is cell differentiation. For example a human cell, from fecundation becomes, by way of cell division, two cells, which in turn divide to become four cells, and so forth. How does this division produce, after a certain size of cell aggregate, a complex morphogenesis with a diversity of organs (e.g. liver, lungs, heart) and with distinct functions? For this type of morphogenesis, with cellular differentiation, Turing imagined the following scenario. Cellular aggregation, formed initially from identical cells, produces chemical substances, "morphogens", which, from an interplay of reaction and diffusion such as the autocatalytic reactions, create an alternating chemical composition at the surface of the aggregate (chemical pattern). Since, between two points on the aggregate, the cells do not have the same chemical environment, they could also act differently. For example, some cells may activate one gene while another does not; in this way, the cells can begin to differentiate. Recall, as seen above, that despite the apparent simplicity of the Turing model, complex chemical structures can emerge[14].

The current theories on biological development[15] stipulate that the amount of information necessary to describe a living organism is beyond what can be encoded in the genome. This supports the idea that an important part of this information could be generated in chemical processes such as the Turing reaction. The genome would thus play the role of a "very talented scientist", capable of repairing, for each specific structure, the deficiencies or excesses in concentration that may be crucial for development – that which will, in the end, distinguish individuals from each other. Alongside this essential work assured by the genome, the basic chemical process takes charge of morphogenesis. Of course, there is an intimate interaction between morphogens and the genome since, as mentioned above, the cell must detect the chemical abundance (or deficiency) so that it can then activate some genes and thus influence the chemical dynamics responsible for the emergence of patterns.

---

14. To this end, we invite the reader to consult an excerpt of the text by René Thom, in section 4.10.

15. The biology of development is interested in the evolution of the organism from conception and notably, the formation of the embryo.

Much work remains to be done in order to experimentally verify that such a scenario is indeed the way nature works. The concept of morphogenesis is beginning, however, to show up clearly. One group of scientists in 2001 announced, for the first time, the identification of morphogens in the tropical zebra fish [19]; nonetheless, it is not yet clear how the cells interpret the profiles of morphogens in order to activate genes.

To conclude this section, we sum up the following points. (i) The Turing model, though simple, can produce a great variety of patterns, some of which resemble those seen in certain organisms. (ii) Even if the similarity between patterns seen on some animal furs and solutions of reaction-diffusion models is obvious, a simple comparison is not proof that these mechanisms are the perpetuators: two different processes can give rise to similar patterns. (iii) The hypothesis of morphogens seems to be confirmed, but at present, the scenario based on reaction-diffusion processes to explain the emergence of patterns is not proved. The next few decades promise to be rich in discoveries due especially to understanding the nature of links between certain proteins, their mutual interactions, and their interaction with cells and tissues. This knowledge would permit a better understanding of morphogenesis in the living world.

9.10. Rayleigh–Bénard Convection

In this section we will briefly describe one of the famous hydrodynamic instabilities; if the reader would like to see a detailed study of numerous fascinating instabilities, we recommend the work of Charru [17].

The Rayleigh–Bénard convection, named after Rayleigh (1842–1919) and Bénard (1874–1939), is one of the most studied phenomena in the context of emergence of order. The reasons are many: (i) the experiment is a relatively simple one to carry out; (ii) it is not expensive; (iii) the hydrodynamic and thermal conduction equations which describe it are well established[16], which allows for a quantitative study of the phenomena. Rayleigh–Bénard convection is seen in all natural environments which can be modeled by a layer of fluid between two rigid walls of different temperature, all the while subject to a gravitational field parallel to the thermal gradient. The qualitative and quantitative comparisons between theory and experiment have

---

16. The law of heat conduction, the Fourier law, was established by the French mathematician and physicist Joseph Fourier (1786–1830). The equations describing the fluid dynamics (hydrodynamics), called the Navier–Stokes equations, were established between the end of the eighteenth and beginning of the nineteenth centuries, by the French physicist Navier (1785–1836) and Irish mathematician Stokes (1819–1903).

often been very good. As a central example in morphogenesis, we will present the different stages of the model and its essential results.

Generally, convection is a major process in energy transport, induced by the presence of a thermal gradient. Whenever a fluid is subject to a thermal gradient which is parallel to the gravitational field, this instability may occur, making it a major and omnipresent phenomenon in nature and everyday life. For example, we observe it every time we heat a pot of water. At first, the water heats but remains immobile; the water simply conducts the heat, in a state known as conduction. At the next stage, different parts of the fluid go into motion; this state is known as convection. If one observes the water closely, perhaps by following a particle of dust in the water, one would remark that the movement of any one section of the water is circular. Studied in a precise manner in a laboratory, that is, with strict temperature controls, etc., the convection pattern is well ordered.

On a completely different scale, Earth also experiences the convection phenomenon, due to the changes in temperature of water with depth. On the geological scale, this phenomenon has crucial importance because it is responsible for the recycling of water on the surface of the Earth, known as the lithosphere, a few kilometers thick[17].

Convection is also abundant in the atmosphere, where there exists a thermal gradient between the surface of the Earth, heated by the Sun, and the atmosphere at higher altitude (which is cooler). This temperature gradient can give way to convection. When the rising air carries enough water vapor, the vapor will, in turn, condense, letting off heat (called latent heat), which further augments the temperature gradient and thus the convective current. If the energy of the current is great enough, it can become a strong wind, a tornado, hail, or lightning.

Before proceeding further, let us set straight an error which is often made in the treatment of convection: it is sometimes claimed that the hexagonal honeycomb structure (see figure 1.6) is related to Rayleigh–Bénard convection cells; this is not true. This comes from a different phenomenon known as the Bénard–Marangoni convection. This type of convection is due to surface tension. If you heat a thin layer at the base of a fluid while the top surface is free to deform itself (in the case of the Rayleigh–Bénard mode, the fluid is between two rigid surfaces), the surface tension will modify locally due to

---

17. According to the tectonic plate model, also known as the continental drift model, the lithosphere is composed of rigid plates which move with time, at the surface of the Earth. This movement is the result of the presence of convection in the internal regions of the Earth. Thus, more than 90% of the internal energy of the Earth is brought to the surface by great convection currents, which drive the tectonic plates. The first model of tectonic plates was proposed by the German astronomer Alfred Wegener (1880–1930), based on cartographic, structural, paleontological, and paleoclimatic considerations.

inhomogeneities in temperature. The gradient in surface tension creates a motor force which must be compensated by a flow in the underlying fluid, which creates *in fine* regular motion, visible at the surface of the liquid, creating a pattern of hexagonal cells. This is the Bénard–Marangoni convection, named after the same French physicist Henri Bénard, and the Italian physicist Carlo Marangoni (1840–1925).

The laboratory study of Rayleigh–Bénard (RB) convection is done by analyzing the movements of a layer of fluid placed between two rigid plates, with the system subject to Earth's gravity (see figure 9.6). The lower plate is heated and the difference in temperature, with the top plate maintained at temperature T, is positive (i.e. $\Delta T > 0$). If we designate the spacing between the two plates by d, the temperature gradient $\partial T / \partial z$ is given (in the absence of convection, i.e., in the regime where we have only thermal conduction) by $\Delta T / d$. If the gradient is weak enough, the fluid remains at rest, while the heat diffuses from the bottom toward the top: this is the purely diffusive (or conductive) regime. There is, however, a thermal expansion in the lower and warmer layers. This thermal expansion, or dilation, describes the increase in volume of a constant amount of matter (associated to a decrease in density of the same fluid), due to the increased temperature. The warmer liquid, less dense, thus has the tendency to rise a little, due to the buoyancy described by Archimedes' principle F_A (a force) such that:

$$F_A = V \delta \rho g, \qquad (9.36)$$

where V is the volume of a small element of fluid, $\delta \rho$ the difference in density of the fluid due to thermal expansion, and g the acceleration due to gravity.

If the temperature gradient $\Delta T / d$ is too small, it will not induce the heated liquid to rise. In fact, the effects of the fluid's viscosity create a friction which oppose any movement of the fluid. If we make the temperature gradient high enough, the buoyancy force will eventually become larger than the viscous one, and the fluid will go into motion, entering the convective regime. We will see how convection corresponds to a bifurcation which is associated with the loss of stability of the diffusive regime.

9.10.1. Heuristic Argument for Determining the Threshold Value for Rayleigh–Bénard Convection

In this section, we use a resolutely heuristic approach[18] in order to intuit the physical constraints necessary for the Rayleigh–Bénard instability to take place (see figure 9.6, describing the experimental setup and notations). Though it requires an already good physical intuition, note that this kind of

---

18. In other words, exploratory. Heuristic is an adjective which means "enabling discovery", notably in scientific research; one can speak of heuristic methods.

approach is essential to physicists who, via simplified calculations, are able to establish results which though also simplified, hold some information with which to decipher the physical situation so as to establish a theoretical framework for more precise, but also more mathematically complex, calculations. Note that although this phenomenon was identified at the beginning of the twentieth century, the Rayleigh–Bénard convection continues to be a popular experiment to observe the triggering of convection and the development of turbulence as a function of various physical parameters.

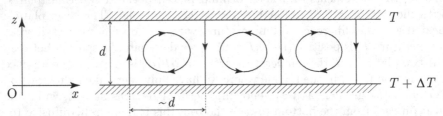

Figure 9.6 – Schematic of a Rayleigh–Bénard cell.

Overall, Rayleigh–Bénard convection arises from the competition between two forces: buoyancy described by Archimedes principle, created by the thermal expansion, and the viscous forces creating friction within the fluid. Even if thermal expansion of a fluid is relatively small (it is associated, at best, to a relative change of density of the order $\Delta\rho/\rho = 10^{-4}$), it is nonetheless enough for the buoyancy to overcome the viscous forces. The heated fluid thus rises to the surface as the colder fluid descends to the bottom, and we have convection. The thermal dilation coefficient α is defined as the relative rate of change of density with respect to temperature:

$$\alpha = -\frac{1}{\rho}\frac{\partial\rho}{\partial T}. \tag{9.37}$$

The negative sign indicates that an increase in temperature contributes to a decrease in density. Since we know that the density of the fluid does not vary radically (of relative order 10^{-4}), we can make a first order Taylor expansion of density ρ in the neighborhood of the initial state, characterized by ρ_0: $\rho = \rho_0 + (\partial\rho/\partial T)\delta T$ where δT is the local temperature variation. Taking into account the expression for the thermal expansion coefficients (equation (9.37)), we could also write the expansion of density as a function of these coefficients:

$$\rho = \rho_0(1 - \alpha\delta T). \tag{9.38}$$

Let us designate one parcel of fluid moving upwards (a spherical volume, with radius R) by δz (see figure 9.7), and v is the velocity of its upward movement. The temperature difference as a function of height is given, to first order, by:

$$\delta T = \delta z \frac{\partial T}{\partial z}. \tag{9.39}$$

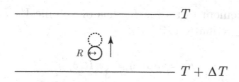

Figure 9.7 – Due to a thermal expansion, a parcel of fluid (continuous circle) rises in the cell (broken circle) as it becomes less dense. Opposing this rising action is viscosity. Convection takes place when the first effect dominates.

For a characteristic time δt, the position of the fluid element changes by $\delta z = v\delta t$ toward the lower temperature regions and the temperature gradient is of order $\Delta T/d$. The fluid parcel is exposed to a temperature variation of order:

$$\delta T = v\delta t\frac{\Delta T}{d}. \tag{9.40}$$

We are assuming that this element does not have enough time to make any significant heat exchange while it is rising[19]. The absence of this local energy exchange is possible only if the time it takes for a full ascension is small with respect to the diffusion time τ_{diff}, which is also the characteristic time for the attainment of thermal equilibrium,

$$\delta t \sim \tau_{\text{diff}}. \tag{9.41}$$

The diffusion time[20] is of the order:

$$\tau_{\text{diff}} \sim \frac{R^2}{\kappa}, \tag{9.42}$$

where κ is the thermal diffusivity of the fluid. Since the typical characteristic time is the diffusion time, we can rewrite the change in temperature (equation (9.40)) experienced by the fluid parcel as:

$$\delta T \sim v\frac{R^2}{\kappa}\frac{\Delta T}{d}. \tag{9.43}$$

---

19. In the opposite case, if the fluid parcel did have a transfer of energy through diffusion with its environment, the local thermal equilibration would inhibit convective movement.

20. The diffusion time is compared with the ascension time of order $\tau_{\text{rise}} \sim R/v$. The condition requiring weak thermal exchange is satisfied if $\tau_{\text{rise}} \ll \tau_{\text{diff}}$, which amounts to $R \gg \kappa/v$. In normal situations, v is of the order 1 cm/s, and $\kappa \sim 10^{-2}\,\text{cm}^2$. We shall see that radius $R \sim d$, and this last condition, are equivalent to a constraint on the distance d between two plates which must be greater than $10^{-2}\,\text{cm}$. In most experimental setups for Rayleigh–Bénard convection, this condition is satisfied.

A spherical fluid element of radius R, thus of volume $V = 4\pi R^3/3$, will feel a buoyant force F_A (equation (9.36)):

$$F_A = \delta\rho V g = \frac{4}{3}\pi R^3 \delta\rho g, \tag{9.44}$$

where the change in density $\delta\rho$ can be expressed as $\delta\rho = \alpha\delta T\rho_0 g$ (see equation (9.38)), giving us:

$$F_A = \frac{4}{3}\pi R^3 \alpha\delta T\rho_0 g. \tag{9.45}$$

If we substitute the change in temperature δT by the approximation in equation (9.43), we have:

$$F_A = \frac{4}{3}\pi \frac{R^5 \alpha\Delta T\rho_0 g\upsilon}{\kappa d}. \tag{9.46}$$

Finally, we also need to remember that every movement that takes place in a fluid encounters a resistance, as we know from our experiences of riding a bike or swimming. The relative movement of a body submerged in water with respect to its immediate environment creates a viscous frictional force. Therefore, a fluid parcel of volume V can only rise efficiently if the buoyancy is greater than the frictional forces it experiences.

Movement in a fluid also creates movement of neighboring layers of the fluid with respect to each other due to friction. For a solid object in the form of a sphere, which has a velocity v relative to the fluid at rest, the friction from the viscous force F_S is given by the classic Stokes' law (see [42]):

$$F_S = 6\pi R\eta v, \tag{9.47}$$

where η is the dynamic viscosity of the fluid. For one parcel of fluid within a larger body of the same fluid, the calculation of the frictional force it experiences is a little more complicated but it has the same functional dependence[21] through the physical parameters of R and η. We thus consider, in our qualitative approach, frictional force F_S (equation (9.47)) as a fair approximation for the friction experienced by the fluid parcel.

Convection begins when the buoyancy overcomes the Stokes force, that is $F_A > F_S$. By using the expressions above for each force (equations (9.46) and (9.47)), we find the following criterion:

$$\Lambda\frac{R^4 \alpha\rho_0 g\Delta T}{\eta\kappa d} \geqslant 1, \tag{9.48}$$

---

21. Using dimensional analysis, it is possible to show that the resistance created by the viscosity on a body of any form with characteristic size R (understood for example as the cube root of the volume of the body) is proportional to $R\eta v$, the only unknown remaining being the dimensionless prefactor.

where Λ is an adimensional number. It is equal to $2/9$ if we use the approximate expressions of the two forces (equations (9.46) and (9.47)). However, remember that our reasoning is purely qualitative, and the values of the numerical prefactors must be treated with a certain caution. Notably, the factor of 6π, which is found in the friction force, corresponds to a rigid spherical body and changes in the case of a fluid. Furthermore, the substitution of time δt in the expression for buoyancy by the thermal diffusion time is yet another approximation. Thus the dimensionless number Λ is a priori of order one or ten, or a fraction of ten (if our argument is sound), but it can only be exactly determined by a precise calculation, as we will see later.

To conclude, note that the expression for the buoyancy force F_A (equation (9.46)) shows that its intensity grows with the radius of the fluid element. This means that the parcels of fluid which jump into motion (i.e. begin convection) are of a relatively large size. In the Rayleigh–Bénard setup, the biggest possible size for radius R of our fluid parcel is of the order $R = d/2$, where, remember, d is the thickness of the fluid layer between the two plates. If we include $R = d/2$ in our criterion for the onset of convection (equation (9.48)), we can write the following as the necessary condition for Rayleigh–Bénard convection:

$$\frac{\alpha \Delta T g d^3}{\nu \kappa} > \frac{16}{\Lambda}, \tag{9.49}$$

where the coefficient ν is the kinematic viscosity of the fluid, related to the dynamic viscosity η by $\nu \equiv \eta/\rho_0$. The two sides in this inequality are dimensionless quantities. The following expression:

$$Ra \equiv \frac{\alpha \Delta T g d^3}{\nu \kappa} \tag{9.50}$$

defines the dimensionless Rayleigh number, a number well known in hydrodynamics for characterizing convection in a fluid. The Rayleigh number Ra depends on geometric parameters of the problem (such as thickness of the fluid layer) and physical parameters characterizing the fluid (such as viscosity ν, thermal diffusivity κ and the thermal dilation coefficient α). The onset of convection (equation (9.49)) can be rewritten as:

$$Ra > Ra_c, \tag{9.51}$$

where we have assumed $Ra_c = 16/\Lambda$, representing the critical Rayleigh number for the onset of convection.

Looking at the functional dependence of the Rayleigh number Ra with respect to the physical parameters (equation (9.50)) reveals that convection occurs more easily (equation (9.51)) when the temperature difference ΔT is large, or when viscosity ν is weak.

Though we have determined this heuristically, the criteria for the onset of convection (9.51) already contain the essential elements needed for a precise calculation, which we will see below. In the meantime, note that the only new information that the precise calculation (sometimes a very laborious calculation[22] for certain geometries) will furnish is the precise value of the critical Rayleigh number Ra_c. We will see how to find the exact value of $Ra_c = 1708$ (which means, in turn, that $\Lambda \sim 0.01$).

9.11. Dispersion Relation for the Rayleigh–Bénard Convection

Mathematically, the linear stability analysis of the convective Rayleigh–Bénard system is not difficult. The temperature obeys the following diffusion-advection equation[23]:

$$\frac{\partial T}{\partial t} + \boldsymbol{v} \cdot \nabla T = \kappa \nabla^2 T, \tag{9.52}$$

where t is the time and $\boldsymbol{v}$ is the velocity field of the fluid. The left-hand side $(\partial T/\partial t) + \boldsymbol{v} \cdot \nabla T$ of equation (9.52) above, can also be written as a total derivative dT/dt, where d/dt is the material derivative (also called the particular or Lagrangian derivative) "felt" by one particle movement. The Lagrangian description thus adopts the reference frame of a moving particle of the fluid. On the other hand, in the Euler frame of description, the reference frame is fixed and the evolution of the fluid is described from a fixed point in space. In the Eulerian description, the partial derivative (or temporal local derivative) is $\partial/\partial t$ (this is the derivative we call Eulerian "seen" by an observer at a fixed point).

Let us take a look at the physics made transparent by the different formulations, Lagrangian and Eulerian. The temperature of a fluid parcel varies as a function of time, but the fluid parcel is simultaneously transported by the hydrodynamical flow, like a particle of dust on the surface of a flowing river. In other words, temperature $T(\boldsymbol{r}, t)$ of a fluid parcel, at time t, becomes $T(\boldsymbol{r} + \boldsymbol{v}\delta t, t + \delta t)$ at time $t + \delta t$, because during the time that has passed, δt, the fluid parcel in question has moved a distance of $\boldsymbol{v}\delta t$. The total variation of temperature T in the limit $\delta t \to 0$ is given by: $dT/dt = [T(\boldsymbol{r} + \boldsymbol{v}\delta t, t + \delta t) - T(\boldsymbol{r}, t)]/\delta t = \partial T/\partial t + \boldsymbol{v} \cdot \nabla T$. The term $\boldsymbol{v} \cdot \nabla T$ is the advection term. The term "material derivative" comes from the fact that

---

22. This is why the qualitative procedure, though calling for a good physical intuition, is so useful, as it allows us to make general conclusions of great relevance. For example, the fact that the Rayleigh number, and thus the instability threshold, are proportional to volume V which is related to the thickness of the fluid layer d (i.e. $Ra \propto d^3$) is a result which is not, a priori, obvious.

23. In addition to diffusion (or conduction), heat is transported by hydrodynamical flow, so we can speak of the transport of heat by "advection".

the evolution of the fluid parcel is followed over the course of its movement, in line with the image of dynamics for a material point, in point particle mechanics.

The velocity field v obeys the Navier–Stokes[24] equation [42]:

$$\frac{\partial v}{\partial t} + v \cdot \nabla v = \nu \nabla^2 v - \frac{1}{\rho} \nabla p + g, \qquad (9.53)$$

where $\nu = \eta/\rho$ is the kinetic viscosity, η the dynamic viscosity, ρ the density of the fluid, p the pressure of the fluid, and g the gravitational field. The left-hand side of the above Navier–Stokes equation (9.53) describes the acceleration of a parcel of fluid. It can also be expressed using the material derivative dv/dt. Meanwhile the right-hand side of the expression is equal to the sum of forces, per unit volume, which are acting upon the fluid parcel, namely: (i) friction[25], created by a non-homogeneous velocity field ($\nabla^2 v$); (ii) pressure ($-\nabla p$), which is always present even for a fluid at rest; and (iii) gravitational force per unit volume g. The two equations (9.52) and (9.53) describe the dynamics of thermal convection.

As we will demonstrate in this paragraph, the purely diffusive and stationary solution (regime of diffusion, also called conductive regime) can become unstable. The diffusive solution is characterized (i) by a fluid at rest (i.e. $\partial v/\partial t$ and $v = 0$); and (ii) by a stationary temperature profile (i.e. $\partial T/\partial t = 0$). Since the two plates that border the fluid (see figure 9.6) are supposed to extend infinitely long in both directions (by "infinite", we mean a large enough extension with respect to the distance between the two plates), thus we have translation invariance along the horizontal axis Ox. Consequently, temperature T and pressure p depend only on vertical spatial variable z. Using the heat transport equation (9.52) along with the Navier–Stokes equation (9.53), we have the following system: $d^2 T_0(z)/dz^2 = 0$ and $\partial p_0/\partial z + \rho_0 g = 0$, where index "0" refers to a stationary solution. Integrating we get the temperature profile: $T_0 = a + bz$, where a and b are two integration constants, whose values are constrained by the boundary conditions: $T_0(z = 0) = T_\downarrow$ and $T_0(z = d) = T_\uparrow$. We find the following temperature profile:

$$T_0(z) = T_\downarrow - \frac{\Delta T}{d} z \qquad (9.54)$$

where ΔT is the difference in temperature between the two plates (i.e. $\Delta T = T_\downarrow - T_\uparrow$). The pressure profile associated with the stationary diffusive solution takes the form:

$$p_0 = c - \rho_0 g z, \qquad (9.55)$$

---

24. We have already seen this equation in subsection 4.1.2.
25. All movements of inhomogeneous velocities imply the relative movement of layers of fluid with respect to one another; this creates friction.

where c is an arbitrary constant, since pressure is always defined with respect to a reference[26]. In principle, density ρ_0 is determined from pressure and temperature in the thermal equation of state[27]. Even in the presence of a hydrodynamical flow or of a temperature gradient, continuum mechanics (such as thermal convection studied here) relies on a hypothesis of local thermodynamic equilibrium. This means that for a small parcel of fluid, we can locally use thermodynamical principles, such as the first and second laws of thermodynamics.

9.11.1. Linear Stability Analysis

The nonlinearity of the thermal convection equations (9.52) and (9.53) can be seen in the two advection terms: $v \cdot \nabla T$ (in the heat transport equation (9.52)) and $v \cdot \nabla v$ (in the Navier–Stokes equation (9.53)). A full solution to these equations can only be found numerically. Nevertheless, we can make a useful study of the linear stability of the diffusive solution that we found previously (equations (9.54) and (9.55)) in order to perform an analytic calculation. We lightly perturb the system in its purely diffusive stationary state and introduce a small parameter ϵ:

$$T(r,t) = T_0(z) + \epsilon T_1(r,t). \tag{9.56}$$

We can do the same for the velocity field v, pressure p and density ρ.

We then substitute these expressions into the thermal convection system of equations (9.52) and (9.53), and keep only the terms which are linear in ϵ (as befits a linear stability analysis).

We use the original heat transport equation (9.52) and the temperature profile associated with the purely diffusive solution (equation (9.54)) to get:

$$\frac{\partial T_1}{\partial t} + v_{z1}\frac{\Delta T}{d} = \kappa\nabla^2 T_1. \tag{9.57}$$

Using the pressure profile of the diffusive solution (equation (9.55)) and the expansion to first order for density (equation (9.38)), we can write $\rho_1 = -\rho_0\alpha\delta T = -\epsilon\rho_0\alpha T_1$. Finally, the projection of the Navier–Stokes

---

26. Note that only the pressure gradient, and not pressure itself, plays a role in the dynamical equations, as is the case for point particle dynamics which depend on the difference in the potential energy and not the potential itself.

27. An equation of state for a system in equilibrium is a relation between different physical parameters, state variables, which determine the exact details of the state. For a fluid of a single component, this equation of state relies on density, temperature, and pressure. If we use the characteristic state equation of a physical system, it is possible to determine all the thermodynamic quantities that describe the system and, accordingly, predict its properties.

equation (9.53) onto vertical axis Oz gives us the following:

$$\frac{\partial v_{z1}}{\partial t} = \nu \nabla^2 v_{z1} + \frac{\rho_1}{\rho_0^2}\frac{\partial p_0}{\partial z} - \frac{1}{\rho_0}\frac{\partial p_1}{\partial z}, \tag{9.58}$$

$$= \nu \nabla^2 v_{z1} + \alpha g T_1 - \frac{1}{\rho_0}\frac{\partial p_1}{\partial z}. \tag{9.59}$$

The projection of the Navier–Stokes equation (9.53) on the Ox-axis gives:

$$\frac{\partial v_{x1}}{\partial t} = \nu \nabla^2 v_{x1} - \frac{1}{\rho_0}\frac{\partial p_1}{\partial x}. \tag{9.60}$$

Now we assume the temperature T, pressure p, and velocity v only depend on the two spatial variables z and x and are independent of y. In other words, we only look at the development of patterns along the Ox-axis (unidimensional order). In chapter 13 we will look at bidimensional order.

We have three equations: heat transfer (equation (9.57)) and the two projections of the Navier–Stokes equations along the Oz and Ox axes (equations (9.59) and (9.60)), but, we have four unknowns: v_{1x}, v_{1z}, T_1, p_1. The additional equation needed to resolve this set of equations is the mass conservation equation (also called the continuity equation): $\partial \rho/\partial t + \nabla \cdot (\rho v) = 0$, which can be rewritten as $\partial \rho/\partial t + v \cdot \nabla \rho + \rho \nabla \cdot v = d\rho/dt + \rho \nabla \cdot v = 0$.

For an incompressible fluid[28], we have $d\rho/dt = 0$, and we can express the conservation of mass simply:

$$\nabla \cdot v = 0. \tag{9.61}$$

To linear order in ϵ, we have:

$$\frac{\partial v_{1x}}{\partial x} + \frac{\partial v_{1z}}{\partial z} = 0. \tag{9.62}$$

We have now four linear and differential equations: (9.57), (9.59), (9.60), and (9.62), which are also autonomous[29] with respect to t and x. The solution has the form $e^{iqx+\omega t}$, where q is the wavevector and ω the growth, or attenuation, rate ω (a priori complex) of the instability, while q is real, since otherwise we would find spatially divergent solutions which is contradictory

---

28. An incompressible fluid does not mean that ρ is constant everywhere, but rather that, in following one parcel of fluid in its movement, the density of the parcel remains constant, $d\rho/dt = \partial \rho/\partial t + v \cdot \nabla \rho = 0$. This hypothesis is valid for all liquids, unless the velocity of the flow is close to the speed of sound. In particular, we know that sound waves correspond to a contraction/dilation of fluid layers – the sound would not be able to propagate without this compression and dilation. For the majority of hydrodynamical problems, the effects of compressibility are nearly zero given the small velocities in play (small velocities with respect to that of sound).

29. Remember that an equation is called autonomous with respect to a variable (time, or space) if its coefficients are independent of the variable in question.

to our search for spatial order on a finite scale[30]. The solution for temperature is given by:

$$T_1 = f(z)e^{iqx+\omega t} + c.c., \tag{9.63}$$

where c.c. is again complex conjugate. Similar expressions hold for the other quantities (v_{1z}, v_{x1} and p_1). It is useful to eliminate v_{1x} and p_1 from the equations in order to focus attention on T_1 and v_{z1} only. To do so, we express v_{1x} as a function of v_{1z} using the continuity equation for an incompressible fluid (9.62):

$$v_{1x} = -\frac{1}{iq}\frac{\partial v_{1z}}{\partial z}. \tag{9.64}$$

We then use this in the Navier–Stokes equation as projected onto the Ox-axis (equation (9.60)) to obtain:

$$p_1 = \frac{\rho_0}{q^2}\left[\nu\nabla^2 - \frac{\partial}{\partial t}\right]\frac{\partial v_{1z}}{\partial z}. \tag{9.65}$$

Substituting the expression for the perturbation in pressure p_1 as a function of v_{1z} into the Navier–Stokes as projected onto the Oz-axis (equation (9.59)), we finally write the equation with the sole variables T_1 and v_{z1}:

$$\omega(q^2 - \mathcal{D}^2)v_{1z} - \alpha g q^2 T_1 + \nu(\mathcal{D}^2 - q^2)^2 v_1 z = 0, \tag{9.66}$$

while the perturbed equation for heat transport (9.57) is written as:

$$\omega T_1 - \kappa\left[\mathcal{D}^2 - q^2\right]T_1 - \frac{\Delta T}{d}v_{1z} = 0, \tag{9.67}$$

where, to simplify the expressions of the last two equations (9.66) and (9.67), we have defined:

$$\mathcal{D} \equiv \frac{d}{dz}. \tag{9.68}$$

In order to solve the above set of differential equations (and finally obtain the dispersion relation which relates the instability growth rate ω to wavenumber q), we need to specify the boundary conditions.

Boundary Conditions

The temperature is fixed on both plates, so we must have $T(z = 0) = T\downarrow$ and $T(z = d) = T\uparrow$. Since the temperature at equilibrium T_0 already satisfies these conditions, we must impose that the perturbations in temperature at the base and top of the fluid be equal to zero:

$$T_1(z = 0) = 0, \quad T_1(z = d) = 0. \tag{9.69}$$

---

30. We will see in a later chapter that it can be useful, for mathematical reasons, to formally treat the wavevector q as a complex quantity.

The boundary conditions for velocity necessitate more attention. Equation (9.66) is a fourth order equation, as operator $\mathcal{D}^4$ confirms – which means we need to specify four boundary conditions before we can solve it.

The two rigid plates which confine the liquid impose that velocity field v of the fluid be zero at the fluid/plate interface. This provides two boundary conditions. However, it turns out that these boundary conditions make a full analytic solution challenging (we would have to perform a numerical study instead). A situation which *does* lend itself to a full analytic solution corresponds to the case where the surfaces which border the fluid are free. We will see that the principal result of our study is the determination of the critical Rayleigh number. In other words, changing the boundary conditions simply means a change in the critical Rayleigh number, and will not affect the qualitative physical aspects of the system's solutions.

If we assume that our top and bottom surfaces remain flat (though they are free), this means that the vertical component of the velocity is zero both at the bottom and top, $z = 0$ and $z = d$:

$$v_{1z}(z = 0) = 0, \quad v_{1z}(z = d) = 0. \tag{9.70}$$

The other two boundary conditions follow from the fact that the free surface[31] is, by definition, a force-free surface, both at $z = 0$ and $z = d$.

The frictional forces of the fluid create hydrodynamical stress, proportional to $\eta \partial v_{1z}/\partial x$ and $\eta \partial v_{1x}/\partial z$, respectively, which must vanish at the surfaces. Since $v_{1z} = 0$ along the plates, the first term is zero[32]. Using equation (9.64), we can write the condition of vanishing $\eta \partial v_{1x}/\partial z$ at both surfaces as:

$$\mathcal{D}^2 v_{1z}\big|_{z=0} = 0 \quad \text{and} \quad \mathcal{D}^2 v_{1z}\big|_{z=d} = 0. \tag{9.71}$$

Having written the six boundary conditions (equations (9.69), (9.70) and (9.71)), each associated with the differential equations describing the evolution of the system, (9.66) and (9.67), it is easy to see that the z-dependence

---

31. Experimentally, one way of approaching this situation would be to use oil for the fluid and place it between a layer of mercury and gaseous helium (see [39]). Since the oil has a high viscosity with respect to mercury and helium, it is possible to show that the surface of the oil will behave as if it is a free surface. The fluctuation of temperature T_1 at the oil/mercury interface is almost zero because of the high conduction of heat in mercury, while the weak heat conductor helium also creates a nearly zero heat flux at the helium/oil interface because of its near-zero heat conduction. This gives us a mixed boundary condition for heat.

32. Since $v_{1z} = 0$ along Ox, it is constant with respect to variable x everywhere on the surface. Consequently, the derivative of v_{1z} with respect to x becomes automatically zero at the surface.

of the solution is sinusoidal[33] and, ignoring for now the dependence of x that we saw in e^{iqx}, can be written as:

$$v_{1z}(z) = A_1 \sin\left(\pi\frac{z}{d}\right) \quad \text{and} \quad T_1(z) = B_1 \sin\left(\pi\frac{z}{d}\right), \tag{9.72}$$

where A_1 and B_1 are two integration constants. These solutions satisfy the boundary conditions (equations (9.69), (9.70), and (9.71)). Now we must just see under what conditions these solutions also satisfy the differential equations of the fluid. Looking at equations (9.66) and (9.67), we find two linear and homogeneous algebraic equations whose solution is non-trivial only if the determinant is zero. Thus, we have a final condition, the dispersion relation ω, which satisfies:

$$\omega^2 - S\omega + P = 0, \tag{9.73}$$

where

$$S = -\Lambda_q(\nu + \kappa), \quad P = \nu\kappa\left(\Lambda_q^3 - \frac{Ra\,q^2}{d^4}\right), \tag{9.74}$$

with

$$\Lambda_q = q^2 + \frac{\pi^2}{d^2}. \tag{9.75}$$

This dispersion relation is formally identical to that obtained for the Turing system (equation (9.4)).

It is easy to verify that the discriminant of this dispersion relation (equation (9.73)) is always positive, indicating that the two solutions for the instability growth rate ω are real; we will call these solutions ω_1 and ω_2 (with $\omega_1 > \omega_2$).

According to the definition of S, which is equal to the sum of both eigenvalues, we have $S = \omega_1 + \omega_2 < 0$. This means that at least one of the two roots must be negative. Only the larger value – in algebraic measure – of the two roots is of interest. As we pointed out in section 9.3 describing the linear stability analysis of the Turing model, this is the first root which is likely to become positive: the signature of an instability. The product of the two roots $P = \omega_1\omega_2$ becomes zero when:

$$\Lambda_q^3 - \frac{Ra\,q^2}{d^4} = 0. \tag{9.76}$$

---

33. In reality, it is easy to see that every function of type $\sin(nz/d)$, with integer n could be a solution. As an exercise, show how the sinusoidal dependence leads to a greater critical value of thermal gradient $\Delta T/d$ (still a critical Rayleigh number) at which convection takes place. Since we are interested in the minimum value of the thermal gradient which can give rise to convection, the values of $n > 1$ are not taken into consideration in our calculations.

This condition can be rewritten as:

$$Ra = Ra_c(q) = \pi^4 \frac{\left[1 + (dq/\pi)^2\right]^3}{(dq/\pi)^2}, \tag{9.77}$$

where $Ra_c(q)$ indicates the critical value of the Rayleigh number for which the first root becomes zero (i.e. $\omega_1 = 0$). The critical Rayleigh number $Ra_c(q)$ is called the neutral (stability) curve (see figure 9.8).

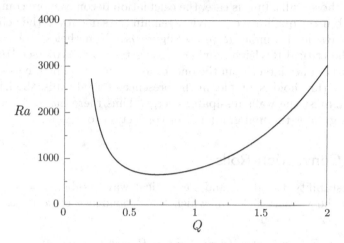

Figure 9.8 – Neutral convection curve for Rayleigh–Bénard convection where we plot Ra as a function of $Q = qd/\pi$.

This curve possesses a minimum at q_c:

$$q_c = \frac{\pi}{d\sqrt{2}}. \tag{9.78}$$

At $q = q_c$ the numerical value of the Rayleigh number is given by:

$$Ra_c = \frac{27\pi^4}{4} \simeq 657.5. \tag{9.79}$$

This is the minimum value of Rayleigh number Ra, corresponding to the smallest possible thermal gradient, for which convection will take place (see equation (9.50)). For this value, periodic velocity fields and temperature gradients will emerge, of a spatial periodicity $\lambda_c = 2\sqrt{2}d$. Thus, for a convective system (respectively, conductive), which is unstable (respectively, stable), with $\omega_1 > 0$ (respectively, $\omega_1 < 0$), the Rayleigh number must be greater than (respectively, inferior to) this minimal critical Rayleigh number, $Ra > Ra_c$ (respectively, $Ra < Ra_c$).

Now we discuss the orders of magnitude of the thermal gradients which are necessary for the emergence of the convective instability. Consider the case of a Rayleigh–Bénard system which is characterized by a thickness of $d = 10^{-2}$ m, made of a silicon oil with viscosity $\nu \simeq 10^{-4}$ m^{2}s^{-1}, with a thermal diffusion of $\kappa \simeq 10^{-7}$ m^{2}s^{-1} and a thermal dilation coefficient $\alpha \simeq 10^{-3}$ K^{-1}. Using the definition of the Rayleigh number (equation (9.50)) and its critical value (equation (9.79)), we conclude that the minimal temperature difference (for $d \sim 1$ cm) needed for convection to take place is $\Delta T \simeq 1.7$ K.

If the surfaces of the fluid are confined by a rigid surface, the velocity will go to zero on those walls. In this case, the calculation becomes more complicated (see [16]), but the qualitative behaviors remain the same (even the behavior of the critical Rayleigh number $Ra(q)$; see figure 9.8). Nonetheless, the numerical value of the critical Rayleigh number changes to $Ra_c \simeq 1707$ (see [16]). This value is three times larger than the one for a free surface. The physical reason for a higher threshold is simple: in the presence of rigid walls, the friction of the fluid against the walls dissipates energy. Thus there has to be a greater force, or a greater thermal gradient, for convection to begin.

9.11.2. Convection Rolls

At the instability threshold and for critical wavenumber $q = q_c$, the solution for the velocity field (in which the dependence on x is seen in the solution (9.72)) is given by:

$$v_{z1} = B_{c1} \sin(q_c z) \cos(q_c x), \quad v_{x1} = B_1 \cos(q_c z) \cos(q_c x). \tag{9.80}$$

The vertical and horizontal velocities are periodic functions in x. Using $v_{x1} = dx/dt$ and $v_{z1} = dz/dt$, we can link the two expressions, writing the following differential equation: $dz/dx = \sin(q_c z)/\cos(q_c z)$, whose solution is a circle of radius B_1. Since, furthermore, the period is given by $\lambda_c = 2\sqrt{2}d$, the currents (trajectories $z(x)$) make circular movements (see figure 9.9).

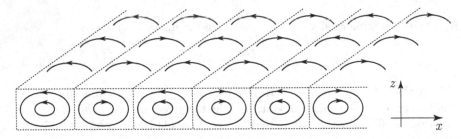

Figure 9.9 – Schematic of the structure of rolls in the velocity field.

Since we are also interested in the case where velocity $v_y = 0$ (invariance along Oy), the velocity field can be seen as convection rolls, which are also called Bénard rolls, observed in experiments. We will see later (in chapter 13) why these structures are unidimensional (with $v_y = 0$), and not bidimensional.

Note once more that amplitude B_1 is undetermined in the linear regime (since a multiplication of a system of linear and homogeneous equations by any number leaves the system of equations unchanged). A nonlinear analysis is needed to determine the amplitude (discussed in the next chapter).

Generic Form of the Dispersion Relation Near the Critical Point

At the instability threshold, the value of the instability growth rate ω_1 (from here on, we will simply call it ω) is zero. We can verify that $\partial\omega/\partial q \equiv \omega'(q) = 0$. In the neighborhood of the critical point, the dispersion relation takes the same form as in the Turing model (equation (9.4)). If we differentiate the dispersion relation from the Rayleigh–Bénard instability (equation (9.73)) with respect to q and set $\omega = 0$ at the critical point, we find $\partial\omega(q)/\partial q|_{q_c} = 0$. Thus at the critical point, the instability growth rate ω must satisfy two conditions:

$$\omega(q_c) = 0, \qquad \left.\frac{\partial\omega}{\partial q}\right|_{q_c} = 0. \tag{9.81}$$

The instability growth rate ω, a function of q and Ra, is thus tangent to the axis of q at $q = q_c$. These two conditions determine both the critical Rayleigh number (equation (9.79)) and the critical wavevector q_c (equation (9.78)). Using these two conditions, we will find a result identical to that from the Turing instability (see especially figure 9.3). A Taylor expansion to second order of the instability growth rate ω gives us:

$$\omega(q, Ra) = \omega(q_c, Ra_c) + (q - q_c)\frac{\partial\omega}{\partial q} + (Ra - Ra_c)\frac{\partial\omega}{\partial Ra} + \frac{1}{2}(q - q_c)^2\frac{\partial^2\omega}{\partial q^2}. \tag{9.82}$$

The derivatives are evaluated at the critical point (q_c, Ra_c). At this point, the first and second terms above (equation (9.82)) go to zero at the critical point (equation (9.81)).

If we take the derivative of the dispersion equation (9.73) once with respect to Rayleigh number Ra, and twice with respect to wavenumber q, and evaluate the results at the critical point, the expression of ω near the instability (equation (9.82)) becomes

$$\omega = \frac{\nu + \kappa}{\nu\kappa}\left\{\frac{3}{2}\frac{\pi^2}{d^2}\frac{Ra - Ra_c}{Ra_c} - 2(q - q_c)^2\right\}. \tag{9.83}$$

The first term $\epsilon \equiv \left[(\nu + \kappa) \times 3 \times \pi^2 \times (Ra - Ra_c)\right]/\nu\kappa 2d^2 Ra_c$ measures the distance from the instability threshold. The second term tells us of the existence of a band of wavevectors of order $|q - q_c|$ for which the instability growth rate is positive, corresponding to the active (unstable) modes.

Generically, the dispersion relation found for any instability which has at its base the transition from a homogeneous state toward an ordered state, with q_c as its critical wavevector, is of the form:

$$\omega = \epsilon + \omega''(q - q_c)^2/2, \tag{9.84}$$

where $\omega'' = \partial^2\omega/\partial q^2$ is the concavity (of negative value) of the instability growth rate ω, and ϵ is a small parameter measuring the distance from the instability threshold. The dispersion relation for the Rayleigh–Bénard instability will thus resemble that of the Turing instability, equation (9.18) (with the substitution $\mu \to \epsilon$ and $\mu_c \equiv 0$).

When the instability threshold is passed, the linear analysis tells us that perturbations in the system will grow exponentially with time. The perturbation, though initially small, becomes a large enough amplitude that the nonlinear terms of the system can no longer be ignored; we will look at this in the next chapter.

9.12. Exercises

9.1

Consider the so-called "Brusselator" model that reads:

$$\partial_t u = a - (b + 1)u + u^2 v + D_u \partial_{x^2} u, \tag{9.85}$$

$$\partial_t v = bu - u^2 v + D_v \partial_{x^2} v, \tag{9.86}$$

where $u(x, t)$ (respectively $v(x, t)$), depends on the space variable x and on time t. u (respectively v) denotes the concentration of the reactant u (respectively v), $\partial_t u$ and $\partial_t v$ refer to time derivatives. a and b are two control parameters.

1. Is this an autocatalytic reaction? If so, which is the activator and which is the inhibitor?

2. Write down the necessary condition for the occurrence of a Turing instability. Write down the sufficient and necessary condition for this instability. What can you conclude?

3. Determine the critical value of b beyond which the Turing instability takes place.

4. Determine the critical wavevector.

9.2

Consider the general study of the Turing instability seen in this chapter and show that beyond the instability threshold, the most unstable mode has a wavevector which is, in general, different from that corresponding to the critical mode, q_c.

9.3

We consider the following reaction-diffusion model:

$$\partial_t u = au - (u^2 + v^2)(u + cv) + \partial_{x^2} u - b\partial_{x^2} v, \tag{9.87}$$

$$\partial_t v = av - (u^2 + v^2)(v - cu) + b\partial_{x^2} u + \partial_{x^2} v. \tag{9.88}$$

1. Specify the simplest stationary solution.

2. Perform the linear stability analysis. Discuss different instability scenarios (stationary instability, oscillatory instability – Hopf bifurcation, with zero and non-zero wavevectors).

3. Can the general condition for the Turing instability be satisfied? Justify your answer.

9.4

Consider the following nonlinear equation:

$$\partial_t u = f(u) + D\partial_{x^2} u, \tag{9.89}$$

where f is a certain nonlinear function of u.

1. We assume that $f(u) = 0$ for $u = u_0$ (constant). Show that the fastest mode corresponds to $q = 0$.

2. Which condition should be satisfied by f in order to have an instability of the fixed point?

3. Explain why the results of questions 1 and 2 remain valid even if u depends on two space variables x and y, so that the equation becomes:

$$\partial_t u = f(u) + D\nabla^2 u, \tag{9.90}$$

where $\nabla^2 = \partial_{x^2} + \partial_{y^2}$ is the Laplacian operator.

9.5

Consider the following nonlinear equation (see also section 10.1 to learn about the context of this equation):

$$\frac{\partial \phi}{\partial t} = \left[\epsilon - \left(1 + \frac{\partial^2}{\partial x^2} \right)^2 \right] \phi - \phi^3, \tag{9.91}$$

where ϕ is a scalar function of two variables x and t. We consider a finite domain $x \in [-L/2, L/2]$ where L is the domain extent along Ox. At the two boundaries of the domain we impose the following boundary conditions:

$$\phi = \phi_x = 0, \tag{9.92}$$

where ϕ_x is an abbreviation of the space derivative.

1. Linearize the equation about $\phi = 0$ (u denotes the perturbation about $\phi = 0$), and write the resulting equation for u and the corresponding boundary conditions.

2. We seek a solution in the form $u = e^{\omega t} f(x)$. Show that f obeys (at the instability threshold):

$$\epsilon f = \left(1 + \frac{\partial^2}{\partial x^2}\right)^2 f, \tag{9.93}$$

with $f = f_x = 0$ as a boundary condition.

3. We seek solutions in the form $f = ae^{iqx}$. Show that q can only take the following values: $q_\pm = \sqrt{1 \pm \sqrt{\epsilon}}$.

4. Write the general solution for f; the integration factors (complex numbers) will be denoted by a and b.

5. Using solution to question 4 combined with boundary conditions, deduce the following relation between L and ϵ:

$$q_+ \tan\left(\frac{Lq_+}{2}\right) = q_- \tan\left(\frac{Lq_-}{2}\right) \tag{9.94}$$

6. For a given extent of the domain, L, the instability threshold, is reached provided that the above equation has at least one positive solution. Show that if L is large enough, we have the following expression:

$$\epsilon_c \simeq \left(\frac{2\pi}{L}\right)^2. \tag{9.95}$$

We have seen in this chapter that the instability threshold for an infinitely extended system is equal to zero ($\epsilon_c = 0$). The present result shows that the finite size delays the instability (it requires $\epsilon_c > 0$).

Chapter 10

Universality of Pattern Description near Threshold

Abstract *In this chapter we unearth the universal equation describing every bifurcation from which spatial order can emerge. Any system which can lead to this spatial order can be described by this universal equation (by definition, an equation independent of the specific system under consideration). We will first derive this equation from a concrete example, and then use symmetry to argue for its universal application. We will see that even for fixed control parameters, there exist an infinite number of possible stable solutions, distinguishable by the resulting pattern's wavelength and amplitude. We will study what is called a secondary instability – an instability which the periodic pattern, itself resulting from an instability (primary instability) of a homogeneous state, experiences. Known as the "Eckhaus instability", it leads to a change in the pattern's wavelength by the creation or destruction of a few repetitions of the pattern motif. Though the Eckhaus instability reduces the range of possible wavelengths for the periodic solutions, we will still be left with a band of possible wavelengths, all of which can be a priori realized within a given system depending on initial conditions. This will lead us naturally to the notion of wavelength selection, which will be addressed later in the book.*

In the last chapter we studied the Turing and Rayleigh–Bénard instabilities, which are triggered at a critical value of the control parameters. Each time, the system loses its original homogeneity along the Ox-axis in favor of some spatial structure, in a move which we can also describe as the emergence of spatial order, corresponding to a *symmetry breaking*. The continuous translation symmetry along the Ox-axis is broken for a weaker, discrete symmetry, characterized by repetition of the same motif or pattern at distances corresponding to any integer value of the spatial periodicity. Most of the nonlinear problems in this book will focus on the regime near the instability's threshold. A brief introduction to dynamics at greater distances from the instability threshold will be presented in the exercises of chapter 14.

When we described the emergence of spatial order in the last chapter, we began with a homogeneous state along the Ox-axis and examined the evolution of a perturbation to the homogeneous state, if the perturbation was of the form:

$$e^{iqx+\omega t},\tag{10.1}$$

where q is the wavevector of the perturbation and ω is the growth (or attenuation) rate of the perturbation. Remember that a negative growth rate (which may also be called an attenuation rate) indicates a stable solution. In the last few chapters we have looked at examples in which ω is real[1]. There, when $\omega < 0$ for all wavevectors q, the perturbation (equation (10.1)) decreases over time, and the homogeneous state is called linearly stable. Otherwise, when $\omega > 0$ for at least one wavevector q, the perturbation grows exponentially and the homogeneous state is unstable.

In the neighborhood of the instability, the dispersion relation $\omega(q)$ (see figure 9.3 for the Turing instability's dispersion relation) takes the generic form we have seen before (equation (9.84)), with the control parameter's critical value μ_c playing a pivotal role. The control parameter μ could be, for example, the Rayleigh number (see equation (9.50)) in the convective system we studied in the previous chapter (Rayleigh–Bénard convection).

For $\mu < \mu_c$, the growth rate is negative ($\omega < 0$), and the perturbation is attenuated over time. For $\mu > \mu_c$, there is an interval of values Δq of wavevectors q (see figure 9.3), called the active band, for which the growth rate is positive ($\omega > 0$). In this case, the perturbation grows exponentially. An initially small perturbation will grow with time, until it becomes so large that nonlinear effects can no longer be ignored.

When $\mu = \mu_c$, the system is in the critical state with a null growth rate. The corresponding wavevector q is exactly equal to the critical value q_c. The growth rate is negative ($\omega < 0$) for all other wavevectors. At this threshold, the perturbation creates a spatial order characterized by the wavelength $\lambda \simeq \lambda_c = 2\pi/q_c$.

In the last chapter, we investigated two systems described by nonlinear evolution equations: the chemical system of the autocatalytic Turing reaction, and the physical system of Rayleigh–Bénard convection. The nonlinear nature of these systems makes a direct analytic solution difficult to find; generally, we have to resort to numerical calculations for a full resolution of equations. However, the analytic study remains useful, in particular, in the vicin-

---

1. The case where ω is complex is treated in chapters 5 and 12.

ity of the instability, where a power law expansion of the small parameter describing the distance from the threshold provides much insight. Even if (as for the Turing problem) a numerical solution is easily obtained, an analytic treatment offers several advantages. On one hand, it saves us from a fastidious investigation of all possible values of parameters in a nonlinear system of many control parameters; on the other hand, it gives us some basic insights, through the approximation close to the instability threshold, to nonlinear behaviors. From this will emerge the universal character of nonlinear phenomena (near to the instability threshold) that we seek.

10.1. Universal Amplitude Equation

Though every model of a nonlinear system has its particular details, the ensemble of these models can be described by a single equation, accordingly called, a universal equation. It is also known as an *amplitude equation*, as will soon become clear. To derive it we could begin from either the Turing or the Rayleigh–Bénard system or, in this case, we will opt for a third system in order to have simpler algebraic calculations, since we would prefer to focus on an overall analysis and not so much the specific algebra of any one system. The universality of the problem gracefully allows us to use this simplified point of departure without compromising the generality of our results.

With this in mind, we choose the relatively simple system described by the following nonlinear equation:

$$\frac{\partial \phi}{\partial t} = \left[\epsilon - \left(1 + \frac{\partial^2}{\partial x^2} \right)^2 \right] \phi + \alpha \phi^2 - \phi^3, \tag{10.2}$$

where ϕ is a scalar function of two variables, x and t, used to describe the evolution of the nonlinear dynamical system (for example, a component of velocity in convection), and α and ϵ are two parameters of the system. This model equation has two nonlinearities, one quadratic and the other cubic[2]. If we were to leave out the quadratic term, we would find what is known as the Swift–Hohenberg (SH) equation (see [24]). For reasons which will become clear later in this chapter, we have deliberately modified the original SH equation by adding the quadratic nonlinearity. Despite its additional term, we will still refer to equation (10.2) as the SH equation throughout this chapter.

---

2. Note that for now, the choice of an arbitrary nonlinear function leads to the same conclusion with respect to the final form of the evolution equation. In effect, this equation can be modified at will, for example we could add derivatives of a higher order or higher powers of function ϕ without affecting the form of the amplitude equation which we will derive.

10.1.1. The Homogeneous Solution

The Swift–Hohenberg (SH) equation was first introduced as a simple model of Rayleigh–Bénard convection (ϕ is proportional to the velocity field in Rayleigh–Bénard convection), in which $\epsilon = (Ra - Ra_c)/Ra$ is the analog of the relative Rayleigh number, Ra the analog of the proper Rayleigh number, and Ra_c the analog of the critical Rayleigh number, which defines the value at which the liquid begins to undergo motion. Equation (10.2) has a homogeneous solution $\phi = 0$ (corresponding to a null velocity in the convection problem).

10.1.2. Linear Stability Analysis

Our first step in studying this dynamical system will be a study of the linear stability of its homogeneous solution. To this end, we introduce a perturbation in the homogeneous solution of the form $\phi = \nu\phi_1(x,t)$ (ν is a small parameter). We substitute this into the time-dependent equation (10.2) and keep only the first order terms in ν, to obtain the following:

$$\frac{\partial \phi_1}{\partial t} = \left[\epsilon - \left(1 + \frac{\partial^2}{\partial x^2} \right)^2 \right] \phi_1. \tag{10.3}$$

Since we are working with a linear and autonomous equation (i.e. the coefficients in the equation are independent of time), we can look for a solution of the form $e^{iqx+\omega t}$, yielding the following dispersion relation:

$$\omega = \epsilon - \left(q^2 - 1 \right)^2. \tag{10.4}$$

Thus, the perturbation's growth rate ω is negative for all negative values of the relative Rayleigh number (ϵ). When the Rayleigh number Ra is equal to the critical Rayleigh number Ra_c, parameter ϵ becomes zero and the curve $\omega(q)$ becomes tangential to horizontal axis q at the critical wavevector $q_c = 1$ (see figure 10.1). A null value of the parameter ($\epsilon = 0$) represents the critical condition needed for the instability to emerge. As soon as the parameter becomes positive ($\epsilon > 0$), all modes corresponding to the active band Δq (for which $\omega > 0$, see figure 10.1) grow exponentially with time, rendering the linear analysis invalid. Far from the instability threshold (ϵ being large), the nonlinear effects generally prohibit us from finding an analytic solution to the full model equation (10.2) and we turn to a numerical calculation. However, if we remain near the instability threshold (ϵ being small), an analytic study becomes possible, as we shall now see.

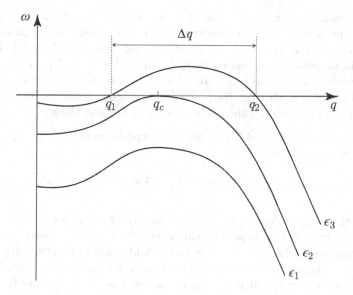

Figure 10.1 – Dispersion relation for three values of ϵ: $\epsilon_1 < 0$ (stable), $\epsilon_2 = 0$ (critical) and $\epsilon_3 > 0$ (unstable).

10.1.3. Introduction of Multiscales

A powerful method in the study of nonlinear systems is that of multiscales, which we already encountered in our study of purely temporal dynamics (i.e. without taking spatial variables into account, see chapter 6). The objective now is to see how we can generalize the concept of multiscales to include spatial variables as well.

For small values of ϵ, the band of wavevectors corresponding to the unstable modes (which can be found by solving $\omega = 0 = \epsilon - (q^2 - 1)^2$) is also small. To see this, consider the two wavevectors q_1 and q_2 (see figure 10.1), the solutions of this equation, given by $q_2 = \sqrt{1 + \sqrt{\epsilon}} \simeq 1 + \sqrt{\epsilon}/2$ and $q_1 = \sqrt{1 - \sqrt{\epsilon}} \simeq 1 - \sqrt{\epsilon}/2$. The active band can thus be approximated as:

$$\Delta q = q_2 - q_1 \simeq \sqrt{\epsilon}. \tag{10.5}$$

We can see that the band of wavevectors describing unstable modes is narrow. In the linear regime, all functions of e^{iqx} (with q an arbitrary real number) are a possible solution, and the general solution can be written as a superposition of all the modes as follows:

$$\phi_1(x, t) = \sum_q a_q(t) e^{iqx} \simeq \sum_{\Delta q} a_q(t) e^{iqx}, \tag{10.6}$$

where $a_q(t)$ is the amplitude of the Fourier mode which depends on time t (in the linear regime the expression is given by $e^{\omega t}$, but we keep, for now,

a_q as our notation, since our goal is to extend the result to the nonlinear regime). The first expression (involving $\sum_q$) above corresponds to a sum of all possible wavevectors. The second expression (involving $\sum_{\Delta q}$) is limited to the sum of wavevectors belonging to the unstable band Δq. In effect, only the wavevectors inside this band are active (i.e. unstable), and modes outside of the band relax with time toward a homogeneous solution. This justifies our decision to retain only the active modes in our calculations.

Introducing critical wavevector q_c into the expression $\phi_1(x,t)$ (the perturbation equation (10.6)) we get:

$$\phi_1(x,t) = e^{iq_c x} \sum_{\Delta q} a_q(t) e^{i(q-q_c)x}. \tag{10.7}$$

Given that $q - q_c \sim \Delta q \sim \sqrt{\epsilon} \ll 1$, we can say that ϕ_1 describes a narrow wavepacket centered around critical wavevector q_c in Fourier space (in the extreme case where $\epsilon \to 0$, we would find a Dirac function $\delta(q - q_c)$). The Fourier transform of a very narrow function in Fourier space is one which is widely spread out in real space (for example, the Fourier transform of a Dirac function $\delta(q - q_c)$ is a constant). We define a complex amplitude A as:

$$A(x,t) \equiv \sum_{\Delta q} a_q(t) e^{i(q-q_c)x}. \tag{10.8}$$

Since $q - q_c \sim \sqrt{\epsilon}$ is small, the complex amplitude A varies on the spatial scale of order $1/\sqrt{\epsilon}$. Indeed, if we make the following change of variables: $Q = (q - q_c)/\sqrt{\epsilon}$, variable Q is of order one and the amplitude $A(x,t)$ looks like:

$$A(x,t) \equiv \sum_{\Delta Q} a_Q(t) e^{iQ\sqrt{\epsilon}x}. \tag{10.9}$$

This formulation makes it evident that A is, in fact, a function of product $\sqrt{\epsilon}x$ and not simply x. We can rewrite solution $\phi_1(x,t)$ as:

$$\phi_1(x,t) = A(\sqrt{\epsilon}x,t)e^{iq_c x} + c.c., \tag{10.10}$$

where $c.c.$ stands for "complex conjugate". Furthermore, the dispersion relation (equation (10.4)) tells us that in the neighborhood of the instability threshold ($q \simeq 1$), growth rate ω is of the order ϵ (i.e. $\omega \simeq \epsilon$). Thus, by writing $\Omega = \omega/\epsilon$ (with Ω of order one), we can express the time dependence of perturbation $e^{\omega t}$ as $e^{\Omega \epsilon t}$, associating it to the characteristic evolution time $1/\epsilon$. This shows the dynamics are slow in the neighborhood of the instability (see the concept of slow dynamics and critical slowing down in subsection 2.3.2). The perturbation ϕ_1 is thus composed of a fast spatial oscillation ($e^{iq_c x}$) and a slowly evolving amplitude $A(\sqrt{\epsilon}x, \epsilon t)$. To bring out this slow evolution of A, it is common practice to introduce slow variables X and T (see subsection 2.3.3), defined by:

$$X = \sqrt{\epsilon}x, \quad T = \epsilon t. \tag{10.11}$$

The amplitude A is only a function of slow variables $A(X, T)$, whereas function ϕ in the SH equation depends on three variables $\phi(x, X, T)$. We are thus using the same multiscales in our approach as seen earlier in this book (in particular see subsection 2.3.3 and chapter 6). Though a temporal treatment with multiscales has been taken up many times, this is the first time that we have also seen it in a spatial variable, with the introduction of slow variable X.

While X and T are originally not independent variables from x and t, the spirit of multiscale analysis lies precisely in treating the three variables x, X, T as if they were independent. We must modify the derivative operators accordingly:

$$\frac{\partial}{\partial x} \longrightarrow \frac{\partial}{\partial x} + \sqrt{\epsilon} \frac{\partial}{\partial X}, \tag{10.12}$$

$$\frac{\partial}{\partial t} \longrightarrow \epsilon \frac{\partial}{\partial T}. \tag{10.13}$$

Substituting these into the SH equation (10.2), we get

$$\epsilon \frac{\partial \phi}{\partial T} = \left[\epsilon - \left(1 + \frac{\partial^2}{\partial x^2} + 2\sqrt{\epsilon} \frac{\partial^2}{\partial x \partial X} + \epsilon \frac{\partial^2}{\partial X^2} \right)^2 \right] \phi + \alpha \phi^2 - \phi^3. \tag{10.14}$$

By introducing multiscales in our analysis we make the function's dependence on parameter ϵ explicit. This makes the task of counting powers of ϵ automatic, facilitating the perturbative calculation based on power expansion of the solutions in terms of the small parameter ϵ.

10.1.4. Derivation of the Amplitude Equation

As stated above, once the instability threshold is crossed ($\epsilon > 0$), a linear analysis tells us that perturbations which were initially small will grow exponentially in the course of time, so that nonlinear terms can no longer be disregarded. In order to deduce the higher order solutions, we now expand function ϕ into a power series of small parameter ϵ. The lowest power appearing in the evolution equation (10.14) is $1/2$, so we might guess ϕ has the following expansion[3]:

$$\phi(x, X, T) = \sqrt{\epsilon} \phi_1 + \epsilon \phi_2 + \epsilon^{3/2} \phi_3 + \cdots. \tag{10.15}$$

We proceed by carrying this expression into the system's evolution equation (10.14) to deduce the contributions order by order.

---

3. This ansatz is based on intuition, and if it is not correct we will end up with some contradiction in the course of the calculations, leading us to correct our guess. Another way to determine which power to use would have been to postulate a general power, writing the solution as a power series of ϵ^α, with α unknown at this stage, and to be determined a posteriori.

Order $\sqrt{\epsilon}$

At this order, $\sqrt{\epsilon}$, the evolution equation of the system (10.14) is written as:

$$\left(1 + \frac{\partial^2}{\partial x^2}\right)^2 \phi_1 \equiv L\phi_1 = 0, \tag{10.16}$$

where we have introduced the operator $L = (1 + \partial^2/\partial x^2)^2$. This is a partial differential equation since function ϕ_1 depends on the three variables x, X, T (i.e. $\phi_1 = \phi_1(x, X, T)$), so that in all generality, the integration constants depend on variables X and T. The acceptable solution[4] is of the form:

$$\phi_1 = A(X, T)e^{ix} + c.c. \tag{10.17}$$

This solution is, in fact, the same linear stationary solution seen in the linear analysis section, except that now the integration factor A depends on X and T.

Order ϵ

At this order, we obtain:

$$L\phi_2 = \alpha\phi_1^2 - 4\left(1 + \frac{\partial^2}{\partial x^2}\right)\frac{\partial^2}{\partial x \partial X}\phi_1 = \alpha\left(A^2 e^{2ix} + AA^* + c.c.\right), \tag{10.18}$$

where A^* designates the complex conjugate of A. The solution to equation (10.18) is the sum of the solution to the homogeneous equation ($B(X, T)e^{ix} + c.c.$), and a particular solution of the complete equation:

$$\phi_2 = B(X, T)e^{ix} + \alpha\frac{A^2}{9}e^{2ix} + \alpha AA^* + c.c. \tag{10.19}$$

Note that the response of the system at this order includes a second harmonic e^{2ix}, and that coefficients A and B are not determined; we need to go one order further to find the complete result.

Order $\epsilon^{3/2}$

At this order, the equation of the system is written:

$$L\phi_3 = \phi_1 - 4\frac{\partial^4\phi_1}{\partial x^2\partial X^2} - \frac{\partial\phi_1}{\partial T} - 4\left(1 + \frac{\partial^2}{\partial x^2}\right)\frac{\partial^2\phi_2}{\partial x \partial X} + 2\alpha\phi_1\phi_2 - \phi_1^3, \tag{10.20}$$

---

4. By acceptable, we mean a solution which behaves as it should at infinity (i.e. for $x \gg 1$); in other words, a solution that does not diverge at infinity. In full generality, the complete solution is composed of two individual solutions e^{ix} and xe^{ix}, for instance in the form $\phi_1 = A(X, T)e^{ix} + D(X, T)xe^{ix} + c.c.$, but since the full solution must not diverge at infinity, we eliminate the second contribution by imposing $D = 0$.

which can be rewritten as:

$$L\phi_3 = \left[A + 4A_{XX} - A_T - \left(3 - \frac{38}{9}\alpha^2\right) A^2 A^*\right] e^{ix}$$

$$+ F(X, T)e^{3ix} + G(X, T)e^{2ix} + c.c., \tag{10.21}$$

where F and G are functions dependent on coefficients A and B, though we do not need to specify this for reasons which will become clear. To simplify the written expression, we have used index notation for the derivatives (i.e. $\partial A/\partial T = A_T$ and $\partial^2 A/\partial X^2 = A_{XX}$).

The evolution equation of the system at this order, $\epsilon^{3/2}$ (equation (10.21)), is inhomogeneous[5]: the complete solution is composed of the solution to the homogeneous equation $(E(X, T)e^{ix} + c.c.)$ plus a particular solution to the full equation. The thing is, the right-hand side of the equation includes function e^{ix}, which is a *resonant* term; that is, e^{ix} is an eigenfunction of operator L, of eigenvalue zero. In response to this function the system diverges, just like in the case of a frictionless harmonic oscillator in the presence of an external force of a frequency equal to the system's natural frequency. The only way, then, to ensure an acceptable behavior (i.e. non-divergent behavior) is to zero out the factor in front of e^{ix} in the evolution equation (10.21). In other words, we are defining a solvability condition akin to the one introduced in the study of the amplitude near a Hopf bifurcation (chapter 6). We could equally have used the Fredholm alternative theorem (see subsection 6.1.1) to find the same result.

Canceling out the divergent contribution of function e^{ix} boils down to setting its prefactor $[A + 4A_{XX} - A_T - (3 - (38/9)\alpha^2)A^2 A^*]$ to zero, revealing the amplitude equation we are searching for:

$$A_T = A + 4A_{XX} - \left(3 - \frac{38}{9}\alpha^2\right) |A|^2 A. \tag{10.22}$$

This equation can be called an *amplitude equation*, the *Landau–Ginzburg equation* or also the *Newell–Whitehead–Segel equation*[6]; the five surnames come from scientists who derived it in different contexts. Note that, since functions F and G are not factors of e^{ix}, we do not need to place any constraints on them, and we do not need to worry about them in our calculation[7].

---

5. Remember that an equation is called homogeneous when it reads $L\phi = 0$ (where L is some operator) and inhomogeneous otherwise (i.e. the equation reads as $L\phi(x, X, T) = F(X, T)$).

6. This equation was derived by studying the problem of Rayleigh–Bénard convection (see [24]).

7. This remains true as long as we restrict our analysis to the lowest order of the amplitude equation.

We shall assume hereafter that $3 - (38/9)\alpha^2 > 0$, that is to say the nonlinear term plays a saturating role. This assumption is fulfilled for the pure SH equation for which $\alpha = 0$.

10.2. Some Properties of the Amplitude Equation

In this section we look at some of the general properties of the amplitude equation.

10.2.1. The Form of the Amplitude Equation Using Symmetries

Perhaps you remember that our interest lies in the *universality* of the form of the amplitude equation when in the vicinity of the instability threshold. We are not interested in the originating equation (in this case, equation (10.2)) because the amplitude equation will always have the same form as long as the system (here the one associated with equation (10.2)) is invariant through translation and inversion. The amplitude equation (10.22) thus has symmetry $X \to -X$ corresponding to invariance by inversion, as does any originating equation. The equations are also invariant by translation in x.

Recall the form of the general solution: amplitude A is a prefactor of harmonic e^{ix} (see equation (10.17)), A^2 is in front of second harmonic e^{2ix} (see equation (10.19)), etc. If we make the transformation $x \to x + x_0$, modifying the origin of the Ox-axis by moving it to the arbitrary point x_0, we are changing $Ae^{ix} \to Ae^{i(x+x_0)} = Ae^{ix_0}e^{ix}$ in the general solution. In other words, the translation operation is the same as changing amplitude A like:

$$A \to Ae^{ix_0}. \tag{10.23}$$

This operation leaves the amplitude equation (10.22) invariant. However, if the amplitude equation contained quadratic terms, such as A^2 or $|A|^2$, the equation would not remain invariant under $A \to Ae^{ix_0}$. Hence there are no quadratic terms, nor power three terms such as A^3, A^{*3} (except $|A|^2A$) allowed in the amplitude equation. Earlier, we added a quadratic term to the original SH equation (10.2) in order to show that the presence of a quadratic nonlinearity in the original model will not imply the presence of a quadratic term in the amplitude equation.

10.2.2. The Coefficients Are Real: Consequences of Inversion Symmetry

Let us examine the amplitude equation (10.22) in its most general form:

$$A_T = \alpha_1 A + \alpha_2 A_{XX} + \alpha_3 |A|^2 A. \qquad (10.24)$$

Here α_ℓ ($\ell = 1, 2, 3$) are coefficients which are a priori complex numbers. By making the inversion $X \to -X$, we are applying the transformation $A \to A^*$ to the expression of field ϕ describing the dynamics (notably, to expressions ϕ_1, ϕ_2 in (10.10) and (10.19)). The result is that the amplitude equations associated with A and A^* are equivalent. If we take the complex conjugate of the generic amplitude equation (10.24), and impose equivalence between the two equations (that obeyed by A and that obeyed by A^*), we find $\alpha_\ell = \alpha_{\ell *}$ ($\ell = 1, 2, 3$). In other words, the coefficients α_ℓ must all be real. We will see that this is no longer true in the presence of a Hopf bifurcation, even if the original equation has inversion symmetry. In that case, even though the equation itself is symmetric under inversion, the solution of the equation is not. The symmetry breaking is observed in the solutions to the equations, not in the evolution equations themselves. For example, an equation with inversion symmetry can have a solution without it; this is why the expression "spontaneous symmetry breaking" is often used.

10.2.3. Canonical Form and Other Equivalent Forms of the Amplitude Equation

The amplitude equation can always be reduced to canonical form, where all the coefficients are equal to one, through some changes of variables. If we write the equation in the form:

$$A_T = aA + bA_{XX} - c|A|^2 A, \qquad (10.25)$$

where a, b, c are real constants, and then make the following transformation of variables: $A \to A\sqrt{c/a}$, $X \to X\sqrt{a/b}$ and $T \to aT$; the amplitude equation takes the canonical form:

$$A_T = A + A_{XX} - |A|^2 A. \qquad (10.26)$$

It is sometimes useful for small parameter ϵ to appear explicitly in the amplitude equation. The amplitude, as introduced before, is of order one, and we know, since ϵ appears explicitly in the expansion of function ϕ (see equation (10.15)), that the physical amplitude is of order $\sqrt{\epsilon}$. This suggests a useful change of variable: $\tilde{A} = A\sqrt{\epsilon}$ where $\tilde{A}$ is the physical amplitude. Using the relationship between the fast and slow variables $X = x\sqrt{\epsilon}$ and $T = \epsilon t$, the previous amplitude equation (10.26) can be rewritten as

$$\tilde{A}_t = \epsilon \tilde{A} + \tilde{A}_{xx} - |\tilde{A}|^2 \tilde{A}. \qquad (10.27)$$

In what follows, we omit the "tilde". Looking at a homogeneous solution to this equation, we can easily see that the phase is constant, and can be defined (arbitrarily) as equal to zero. Taking $A = \rho e^{i\psi}$, where ρ is the amplitude and ψ is the phase, the previous amplitude equation (10.27) tells us: $d\psi/dt = 0$, so that ψ is constant (and can be set to zero). We can thus take A to be real. To summarize, the amplitude equation (10.27) has the following equilibrium (fixed point) solutions:

$$A_0 = 0, \quad A_0 = \pm\sqrt{\epsilon}. \tag{10.28}$$

The first solution always exists, but the second solution only exists for values of parameter ϵ that are strictly positive (i.e. $\epsilon > 0$). In the plot of amplitude A as a function of parameter ϵ (see figure 10.2), we can see the equilibrium solutions (or fixed points) and the presence of a *pitchfork* bifurcation[8].

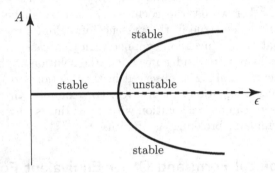

Figure 10.2 – Equilibrium solution. The dotted branch corresponds to the unstable solution.

10.2.4. Variational Dynamics of the Amplitude Equation

When A is independent of x, equation (10.27) can be written as (by omitting tilde):

$$A_t = -\frac{\partial}{\partial A}\left[-\epsilon\frac{A^2}{2} + \frac{A^4}{4}\right] = -\frac{\partial V}{\partial A}, \tag{10.29}$$

with a real physical amplitude A.

This equation is analogous to that of a massless particle subject to friction, in motion along the potential $V(A) = -\epsilon A^2/2 + A^4/4$ (see the shape of the potential in figure 10.3). For negative values of ϵ ($\epsilon < 0$), the potential V has a single minimum at $A = 0$. By analogy with mechanics, the solution $A_0 = 0$ is stable (all these concepts are described in chapters 2 and 3). For $\epsilon > 0$,

---

8. It is called also a direct bifurcation – in contrast to the inverted bifurcation, known as the subcritical bifurcation; this bifurcation is analogous to a second order transition. All of these concepts have been introduced before (in particular, see chapters 2 and 3).

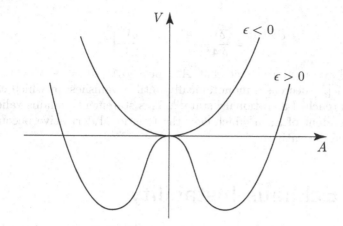

Figure 10.3 – The potential $V = -\epsilon A^2/2 + A^4/4$ as a function of A.
For $\epsilon > 0$, solution $A = 0$ becomes unstable.

the potential minimum at $A = 0$ becomes a maximum, while two minima appear at $A = A_0 = \pm\sqrt{\epsilon}$. The solution $A_0 = 0$ loses its stability in favor of the two solutions $A = A_0 = \pm\sqrt{\epsilon}$; we say it has bifurcated toward a new stable solution. The stable branches in figure 10.2 are represented by the continuous line, the unstable branches by the dotted line.

Before concluding the section, it is important to note that in the general case, where the amplitude is inhomogeneous, the amplitude equation (10.27) is derived from a functional[9] $V[A]$ (we use notation $V[A]$ in order to stress that V is a not an ordinary function, but a functional), such that:

$$A_t = -\frac{\delta}{\delta A^*}V[A] = -\frac{\delta}{\delta A^*}\int dx \left(-\epsilon\frac{|A|^2}{2} + \frac{|A|^4}{4} + \frac{|A_x|^2}{2}\right), \qquad (10.30)$$

where $\delta/\delta A^*$ designates the functional derivative with respect to A^* (A's complex conjugate). This functional derivative is expressed with total and partial derivatives as follows:

$$\frac{\delta}{\delta A^*} \longrightarrow \frac{\partial}{\partial A^*} - \frac{d}{dx}\frac{\partial}{\partial A_x^*}. \qquad (10.31)$$

This is the usual expression of the Euler–Lagrange derivative, found in the principle of least action of mechanics (see equation (3.53)). In the case where amplitude is homogeneous, we recover – by analogy with point mechanics – the amplitude equation as derived from a potential (equation (10.29)). Quantity $V[A]$ is known as the *Lyapunov functional* and has the following

---

9. Refer to subsection 3.9.3 (the Euler problem of a soap film) for more on the concept of a functional.

property:

$$\frac{\partial V[A]}{\partial t} = \frac{\delta V}{\delta A^*}\frac{\partial A^*}{\partial t} = -\left|\frac{\delta V}{\delta A^*}\right|^2 \leqslant 0, \qquad (10.32)$$

where we have used the fact that A^* obeys $\partial A^*/\partial t = -\delta V[A]/\delta A$. The functional $V[A]$ decreases monotonically until it vanishes, in which case the system has reached a stationary state[10]. This statement remains valid when A is independent of x, in which case the functional derivative becomes the usual partial derivative.

10.3. Eckhaus Instability

In physical situations, amplitude A is generally a function of the space it inhabits[11]. Imperfections and boundaries (for example, walls) always exist which lead to modulations of what might otherwise be perfect order, and thus too, the amplitude depends on the spatial variable. In fact, we will see that these modulations can be what destabilize order, making these inhomogeneous aspects of the problem even more important to look at.

Consider the canonical form of the amplitude equation (10.26). This equation has a stationary solution:

$$A_0(X) = \sqrt{1 - q_0^2}\,e^{iq_0 X}, \quad |q_0| < 1, \qquad (10.33)$$

where q_0 is any real number smaller than one. We can write the linear solution of the SH equation (10.10) as:

$$\phi_1 = \sqrt{\epsilon}A_0(X)e^{ix} + c.c. = \sqrt{\epsilon}\sqrt{1 - q_0^2}\,e^{iqx} + c.c., \qquad (10.34)$$

with $q = 1 + q_0\sqrt{\epsilon}$. We have used the fact that $X = x\sqrt{\epsilon}$. The spatial oscillation frequency q_c which is equal to 1 in the linear regime becomes $q = 1 + q_0\sqrt{\epsilon}$ because of nonlinear effects. The condition $|q_0| < 1$ becomes

$$\epsilon > (q - 1)^2. \qquad (10.35)$$

Consider again the canonical form of the amplitude equation (10.26). Before rescaling amplitude equation (10.22), the prefactor of the above periodic solution (equation (10.34)) would have been $\sqrt{1 - 4q_0^2}/\sqrt{3 - 38\alpha^2/9}$, and not

---

10. The problem of the dynamics close to the instability threshold is similar to the problem of approaching thermodynamical equilibrium, where A would be the thermodynamical variable and $V[A]$ the appropriate potential – for example, the free energy in the case of an evolution at constant volume and temperature.

11. When amplitude A is reduced to a constant, it is because we see perfect order, which is, to say the least, rare in nature.

simply $\sqrt{1 - q_0{}^2}$, as shown in equation (10.34). It follows (keeping the same notations as in our study of the canonical form) that the stationary solutions exist only for $q_0^2 < 1/4$, and condition (10.35) becomes

$$\epsilon > 4(q - 1)^2. \tag{10.36}$$

The dispersion relation of the Swift–Hohenberg equation (10.4) reveals that the trivial solution $\phi = 0$ is unstable (i.e. $\omega > 0$) when $\epsilon > (q^2 - 1)^2$, or (since q is close to one, ϵ being small) when $\epsilon > 4(q - 1)^2$. Thus, the condition for the existence of a stationary state (inequality (10.36)) coincides with the region of linear instability for solution $\phi = 0$. We will continue our analysis of the canonical amplitude equation (10.26), recalling that the condition for stability of state $\phi = 0$ in this form corresponds to $\epsilon > (q - 1)^2$, and not $\epsilon > 4(q - 1)^2$.

The stationary periodic states in x defined above are not all stable. Some are unstable due to a modulation of the phase (or, equivalently, modulation of the wavelength). If we explicitly write the real and imaginary parts of the amplitude, we have:

$$A = \rho e^{i\psi}, \tag{10.37}$$

where ρ is a real variable describing the intensity of the real amplitude and ψ, describing the phase. The canonical amplitude equation (10.26) is divided into two equations corresponding to the real and imaginary parts of the amplitude:

$$\rho_T = \left(1 - \psi_X^2\right)\rho - \rho^3 + \rho_{XX},$$

$$\psi_T = \psi_{XX} + \frac{2}{\rho}\psi_X \rho_X. \tag{10.38}$$

We investigate the linear stability of the stationary solution (equation (10.33)) by superposing a small perturbation to the amplitude $\sqrt{1 - q_0^2}$ and phase $(q_0 x)$:

$$\rho = \sqrt{1 - q_0^2} + \rho_1, \quad \psi = q_0 x + \psi_1. \tag{10.39}$$

We keep only the linear terms of amplitude equation (10.38) to find a linear and homogeneous system for variables ρ_1 and ψ_1. Searching for a solution of the form $(\rho_1, \psi_1) = (a, b)e^{iQx + \sigma t}$, with a and b two arbitrary constants, we establish a system with two unknowns, (a, b):

$$\left[\sigma + 2\rho_0^2 + Q^2\right]a + 2i\rho_0 q_0 Q b = 0,$$

$$-\frac{2iq_0 Q}{\rho_0}a + \left[\sigma + Q^2\right]b = 0. \tag{10.40}$$

We set the determinant to zero to find the dispersion relation, whose solution is:

$$\sigma_\pm = q_0^2 - Q^2 \pm \sqrt{(q_0^2 - 1)^2 + 4q_0^2 Q^2}. \tag{10.41}$$

σ_- is always negative. On the other hand, the branch σ_+ can become positive. We can easily verify this for $q_0^2 > 1/3$ (see the next section): the stationary solution which exists for parameters $|q_0| < 1$, becomes unstable when $|q_0| > 1/\sqrt{3}$. The relation between q and q_0 (see equation (10.34)) shows that this instability, called the "Eckhaus instability" [30], occurs for:

$$\epsilon < \frac{(q-1)^2}{3}. \tag{10.42}$$

This instability reduces the initial band of possible states (i.e. $|q - 1| < \sqrt{\epsilon}$, equation (10.35)) to the narrower band defined by $|q - 1| < \sqrt{\epsilon}/\sqrt{3}$ (see figure 10.4).

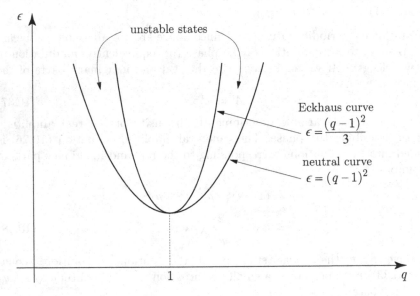

Figure 10.4 – The neutral curve and the curve delimiting the Eckhaus instability.

10.3.1. Eckhaus Instability: a Phase Instability

To better identify the nature of this instability, we continue our analysis by looking at the eigenvectors of eigenvalues $\sigma_\pm$. The branch σ_+ first becomes positive when the values of parameter Q are small enough (i.e. for large enough wavelengths). Indeed, if we rewrite the expression for σ_+ as:

$$\sigma_+ = 2Q^2 \frac{3q_0^2 - 1 - Q^2/2}{\rho_0^2 + Q^2 + \sqrt{(q_0^2 - 1)^2 + 4q_0Q^2}}, \tag{10.43}$$

we see that the sign of σ_+ is determined by the numerator. As soon as the inequality $q_0^2 > 1/3$ is satisfied, branch σ_+ becomes positive. This first occurs for small values of Q.

For these small values of Q, we can easily show that:

$$\sigma_+ \simeq \frac{3q_0^2 - 1}{1 - q_0^2} Q^2, \quad \sigma_- \simeq -2\rho_0^2. \tag{10.44}$$

We carry each expression into one of the two equations[12] of linear system (10.40) to obtain the following conditions:

$$\frac{a}{b} \simeq \frac{-iq_0}{\rho_0} Q \ll 1, \quad \frac{a}{b} \simeq \frac{i\rho_0^3}{Qq_0} \gg 1, \tag{10.45}$$

where we have assumed a small value of Q. The first inequality, associated with the unstable mode for eigenvalue σ_+, indicates that the constants a and b are such that the value of a must be much less than b (i.e. $a \ll b$); consequently, the dynamics are dominated by the phase[13]. The second inequality qualifies the stable mode of the amplitude associated with σ_-. The amplitude relaxes quickly with time (stable mode), and the unstable mode – sometimes called the dangerous mode – is associated with the phase. In other words, the amplitude can be considered to be adiabatically slaved to the phase. This comes down to saying $\psi_T = 0$. If we ignore ρ_{XX} (see below), this leaves us $\rho_0 \simeq \sqrt{1 - K^2}$ with $K = \psi_X$ the local wavevector. Through iteration we see that $\rho_{XX} \sim \psi_{XXX}$ produces a higher order derivative in ψ than is present in the phase's diffusion equation, which we will derive next; the term can be justifiably eliminated.

Qualitatively, the first amplitude equation of system (10.38) is "steep" enough, in the sense that the small perturbation relaxes immediately to intensity ρ_0 (the linear part of the right-hand side of the equation shows the exponential relaxation). The term ρ_{XX} on the right-hand side of the equation is only saying that the perturbation which is propagating by diffusion cannot make any quick modifications to the amplitude at any given point. This means we can assume ρ to be fairly uniform between two points, and determined by the local wavevector. In fact, we can deduce that the amplitude is approximately given by its *adiabatic* value:

$$\rho \simeq \sqrt{1 - K^2}, \tag{10.46}$$

where $K \equiv \psi_x$ is the local wavevector. If we substitute this expression into the second equation of system (10.38) to determine the evolution of variable ψ,

---

12. They are equivalent, since the determinant of the system (10.40) is zero.

13. Remember that a and b are, respectively, the amplitudes of the perturbation ρ and the phase ψ, see equation (10.37).

we get:

$$\psi_T = D\psi_{XX}, \quad D = \frac{1 - 3K^2}{1 - K^2}. \tag{10.47}$$

This is the typical form of an equation describing phase diffusion. It shows how any change or perturbation of the local wavevector leads to a readjustment of the entire system through a process of phase diffusion. A negative diffusion coefficient means there is an instability. Recall that, at the lowest order, the local wavevector is related to the overall wavevector q by $q \simeq 1 + K\sqrt{\epsilon}$ (in this case $K \sim q_0$). The denominator is positive for $1 > K^2 \simeq (q-1)^2/\epsilon$, which is the condition of existence for stationary states. The numerator is negative when $3\epsilon < (q-1)^2$. We recover the same Eckhaus instability as above (figure 10.4).

10.3.2. Eckhaus Instability: Destruction and Creation of Cells

Consider a system composed of an ensemble of cells (in the form of curves, waves, etc.) of a regular size and representing a periodic motif. Forcing the system into the unstable band with respect to the Eckhaus instability (for example, by giving it a wavelength which is too small), causes the system to react by eliminating a cell (or several) in order to augment its wavelength and make the structure "enter" into the stable band (see figure 10.5). After the local elimination of the wavelength, the system readjusts through a process of phase diffusion. If instead we force the system by imposing a too-large wavelength, the system creates, by division of one (or many) cells, new cells (see figure 10.5), this time reducing the wavelength until the final periodic structure has a wavelength within the stable band.

Wavelength Selection

The Eckhaus instability reduces the band of wavevectors associated with periodic modes. Still, any wavevector within the stable Eckhaus band is a potential solution; the solutions make up a continuous family, as parameterized by the wavevectors. In short, we still have an infinite ensemble of possible periodic solutions. Naturally, the question arises: given a system which undergoes a bifurcation leading to the emergence of a spatial pattern, is the final choice of wavelength due to some robust property, or does it depend on details of the system's preparation? This is a complex problem, referred to as the problem of *wavelength selection*. We will see in chapter 14 that there are many mechanisms by which wavelength selection can occur, some of which are robust, and others less.

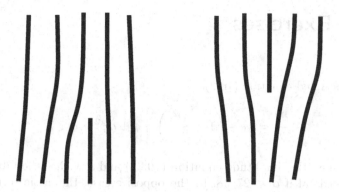

Figure 10.5 – The bars represent the position of, for example, the maxima of the periodic structure, and time reads from bottom to top. Left: starting from a periodic solution with six cells, but unstable, because of a too-small wavelength. The system eliminates cells to augment its wavelength. After elimination, the readjustment of the wavelength is done through a diffusion of the phase. Right: starting with a four cell solution, also unstable, because of the existence of a too-large wavelength. The system creates cells to diminish its wavelength.

10.4. Some Remarks on the Instabilities of Unidimensional Structures

In this chapter the study has been based on the following hypotheses: (i) we are limited to the unidimensional case; (ii) we have considered that the primary instability (that gives way to order) is temporally non-oscillatory; and (iii) we have placed ourselves very close to the threshold of the primary instability. If we discard one or more of these hypotheses, there is a greater range of behaviors, such as the "zig-zag" instability which crops up in the study of bidimensional instability; see exercises in chapter 13. Furthermore, if we place ourselves far enough way from the instability threshold, new instabilities can occur, such as parity violation (where a structure spontaneously loses the symmetry $x \to -x$); see exercises in chapter 14. There is a wide field concerning secondary instabilities (such as the Eckhaus instability), but we will only investigate this briefly.

A detailed study of the nonlinear instability in convection (and in other systems) can be found in important books such as those of Manneville [67], Cross & Greenside [23], and Hoyle [46].

10.5. Exercises

10.1

When we studied equation (10.2):

$$\frac{\partial \phi}{\partial t} = \left[\epsilon - \left(1 + \frac{\partial^2}{\partial x^2} \right)^2 \right] \phi + \alpha \phi^2 - \phi^3, \tag{10.48}$$

we determined the amplitude equation (10.22) and saw that the bifurcation was supercritical if $\alpha < 27/38$. In the opposite case, the cubic term has a global positive sign, meaning that the first nonlinear contribution amplifies the linear instability, instead of saturating it. In this case, we have to expand the solution to an order higher than three.

1. Without calculation, can you guess how to write the next order term in the amplitude equation. Can you give a general argument for your choice?

2. By pushing the expansion performed in this chapter to next orders, derive the amplitude equation up to the next relevant order. For simplicity, we consider the steady-state version of equation (10.48), and set $\alpha = 0$ and call cubic coefficient b_3.

10.2

Consider the following nonlinear amplitude equation:

$$A_T = \epsilon A + A_{XX} + |A|^2 A - |A|^4 A. \tag{10.49}$$

1. Show that equation (10.49) admits a family of steady and spatially periodic solutions, $A_0(x) = Re^{iqx}$. Determine R as a function of q and ϵ.

2. Specify the condition required for the existence of such solutions.

3. Introducing a modulus ρ and a phase ψ, $A = \rho(X, T)e^{i\psi(X,T)}$, write down the coupled nonlinear equations satisfied by the modulus and the phase. Express the solution A_0 in terms of ρ_0 and ψ_0 (the modulus and the phase of the stationary solution A_0).

4. Setting $\rho = \rho_0 + \rho_1$ and $\psi = \psi_0 + \psi_1$, provide the linearized version (with respect to ρ_1 and ψ_1) of the equations obtained in question 3.

5. By seeking a solution in the form $(\rho_1, \psi_1) = (a, b)e^{iQX + \Omega T}$, with a and b real constants, determine the dispersion relation $\Omega(Q)$. Discuss the possibility of instability and the associated conditions. Which type of instability do we have? Justify your answer.

6. Following the same reasoning as that adopted in subsection 10.3.1, determine the phase equation. What can you conclude?

10.3

We have seen in this chapter that the form of the amplitude equation can be obtained on the basis of symmetry considerations (see subsection 10.2.1).

Discuss all possibles cases regarding amplitude equations containing terms of the form $A^m(A^*)^n$ (we limit the discussion to the eighth order, i.e. $n+m \leqslant 8$). Justify your answer.

10.4

1. Can equation (10.48) be written in a variational form? If so, write down the functional.

2. The same question for equation:

$$u_t = -u_{xx} - u_{xxxx} + u_x^2. \tag{10.50}$$

If this equation has a variational form, provide the functional; if not, explain which are the terms that do not allow an expression in terms of a functional.

10.5

Consider the following nonlinear equation:

$$\frac{\partial \phi}{\partial t} = \left[\epsilon - \left(1 + \frac{\partial^2}{\partial x^2} \right)^2 \right] \phi + \left(\frac{\partial}{\partial x} \phi \right)^2 \frac{\partial^2}{\partial x^2} \phi. \tag{10.51}$$

Using the method developed in this chapter, determine the amplitude equation to cubic order.

10.6

Assume that A satisfies the following amplitude equation:

$$A_t = A + A_{xx} - |A|^2 A. \tag{10.52}$$

1. Show that for the steady-state solutions we have two invariants, denoted I_1 and I_2, and that they obey the following relations:

$$I_1 = \rho^2 \psi_x, \quad I_2 = \frac{1}{2}\rho_x^2 + \frac{I_1^2}{2\rho^2} + \frac{\rho^2}{2} - \frac{\rho^4}{4}, \tag{10.53}$$

where $A = \rho e^{i\psi}$.

2. Use this result to show that if $\rho = 0$ is any arbitrary point, then the phase must remain constant everywhere in x, and that the wavevector should be equal to q_c, the critical value following from linear stability analysis.

10.7

Consider the following amplitude equation

$$A_t = \nu A + A_{xx} - |A|^2 A, \tag{10.54}$$

where ν is real and measures the distance from the instability threshold.

1. Write the amplitude equation (10.54) as a function of a phase ϕ and an amplitude ρ.

2. We recall that the amplitude equation has $\rho_0 e^{iq_0 x}$ as a steady solution with $\rho_0 = \sqrt{\nu - q_0^2}$. We have also seen in this chapter that the phase diffusion coefficient vanishes for $\nu - 3q_0^2 = 0$, and becomes negative when $\nu - 3q_0^2 < 0$. In the vicinity of the phase instability threshold higher order terms must be taken into account. For that purpose we set $\epsilon^2 = \nu - 3q_0^2$ to be measure of the distance from the Eckhaus instability threshold. The small ϵ parameter will serve as an expansion parameter. Show from the dispersion relation of the Eckhaus instability found in this chapter that the growth rate can be written as $\sigma \sim \epsilon^2 Q^2 - aQ^4$ where a and b are of order one.

3. Deduce that dynamics can be described in terms of a slow spatial and a temporal scale, and determine them as a function of t, x and ϵ.

4. Write the phase and amplitude equation in terms of the new slow variables.

5. Set $\phi = q_0 x + \epsilon \psi$ (with ψ of order one) and rewrite the phase and amplitude equations in terms of ρ and ψ.

6. We seek a solution in the form

$$\rho = \rho_0 + \epsilon \rho_1 + \epsilon^2 \rho_2 + \cdots, \quad \psi = \psi_0 + \epsilon \psi_1 + \epsilon^2 \psi_2 + \cdots \tag{10.55}$$

Show by expanding the equations for ψ and ρ that to order ϵ^0 we obtain

$$\rho_0 = 0, \quad \rho_0 = \pm\sqrt{\nu - q_0^2}. \tag{10.56}$$

7. Show that to next order we obtain $\rho_1 = 0$.

8. Show that to order ϵ^2 we obtain

$$\rho_2 = -\frac{q_0 \psi_{0X}}{\rho_0}, \tag{10.57}$$

to order ϵ^3 we obtain

$$\rho_3 = -\frac{q_0 \psi_{1X}}{\rho_0} \tag{10.58}$$

and to order ϵ^4 we obtain

$$\rho_4 = -\frac{q_0 \rho_2 \psi_{0X}}{\rho_0^2} - \frac{\psi_{0X}^2}{2\rho_0} + \frac{\rho_{2XX}}{2\rho_0^2} - \frac{q_0 \psi_{2X}}{\rho_0}. \tag{10.59}$$

9. Show that to order ϵ^5 we obtain

$$\psi_T = D\psi_{XX} - \alpha\psi_{XXXX} + \beta\psi_X\psi_{XX}. \qquad (10.60)$$

10. Provide the expressions for α and β.

11. Show (using an adequate change of variables) that equation (10.60) can take the following canonical form:

$$\psi_t = -\psi_{xx} - \psi_{xxxx} + \psi_x\psi_{xx}, \qquad (10.61)$$

where here we have kept the same original notation ψ.

12. Can you explain the form of the nonlinear term on the basis of symmetries?

Chapter 11

Fronts between Domains and Invasion of One State by Another

Abstract *Once we reach the threshold of the instability, the way in which order overtakes a system may be spontaneous and throughout all parts of the system or, instead, begin as a localized pattern which then propagates until it has taken over the rest of the system, rather like a forest fire. We will see how the latter (and more usual) case boasts of an invasion described by an infinitude of possible propagation speeds, and show how the system settles on a unique value for this speed. In other words, we will introduce the general concept of a front, and explain the problem of selecting a unique propagation speed from among an infinite manifold of possibilities. We will conclude with a brief introduction to the concept of marginal stability, in which a system chooses to exist at the frontier between stable and unstable solutions.*

In the previous chapter we saw that spatial order emerges when the homogeneous solution, a solution without any particular order or spatial pattern, loses its stability. Having reached this instability threshold, the solution with a spatial pattern corresponding to spatial order takes over. However, the mechanism through which the spatial order manifests itself is not a priori obvious. One possibility would be for the order to appear as a pattern throughout the entire system simultaneously, though that would reflect an idealized situation – since the imperfections of any real system would favor the emergence of the ordered pattern at one or a few particular points, which would then bit by bit spread to the rest of the system. As an example, consider the case of Rayleigh–Bénard convection (section 9.10). We have seen that Bénard rolls occur when the temperature difference between the two plates above and below the fluid is large enough. Once the system reaches or surpasses this critical temperature difference, the conductive solution gives way to the convective solution. But, what does the transition from conductive to convective *look* like? Since the plates are made of a naturally imperfect material, the temperature difference between them will have a slight variation at some points, and thus one or two rolls will begin to appear at a specific locality before progressively spreading to the rest of the fluid. The onset of this instability could also be provoked by artificially creating one or two rolls at a specific location (for example, through localized heating). In either case, we can say that the solution without any rolls (the conductive solution) is invaded by the solution with rolls (the convective solution).

To approach this kind of problem we will look again in the neighborhood of the instability corresponding to the emergence of order, where the systems can be described by a universal equation, called the *amplitude equation*, which we recall for you here (it was also equation (10.27)):

$$A_t = \epsilon A + A_{xx} - |A|^2 A. \tag{11.1}$$

We readily find the following homogeneous and stationary solutions to this equation: $A = 0$ and $A = \pm\sqrt{\epsilon}$. Solution $A = 0$ corresponds to a solution without any patterns, whereas solutions $A = \pm\sqrt{\epsilon}$ refer to solutions displaying a coherent structure; amplitude A describes the amplitude of the structure. Thus, we have the emergence of order for positive values of parameter ϵ (i.e. $\epsilon > 0$, see figure 10.2). The parameter's critical value, $\epsilon = 0$, corresponds to the bifurcation threshold of the homogeneous (patternless) toward the ordered state.

We will now focus on the ordered solution associated with $\epsilon > 0$, and consider an initial condition defined by a localized function of space, such as a Gaussian or Dirac function (see figure 11.1), denoted by $A_0(x)$. For $\epsilon > 0$, solution $A = 0$ is unstable, whereas $A \neq 0$ is stable. A numerical simulation of equation (11.1) reveals that the initial solution $A_0(x)$ invades the system leaving an amplitude $A = \pm\sqrt{\epsilon}$ in its wake; the final choice of + or − depends on the initial conditions. Since the amplitude (or envelope) A is describing the ordered state, we are looking at a problem where the unstable state $A = 0$ is invaded by the stable state $\pm\sqrt{\epsilon}$ (see figure 11.1). Recall that the amplitude A corresponds only to the envelope of the periodic pattern, and that the full waveform is described by the product of envelope A and an oscillating function e^{iqx} (see chapter 9).

In this chapter, we will explain the fundamental difference between the problem of invasion of an unstable solution by a stable solution, and the invasion of a metastable solution by a stable solution (see, in particular, section 3.3, in which an example of the passage from a metastable to a stable state is given).

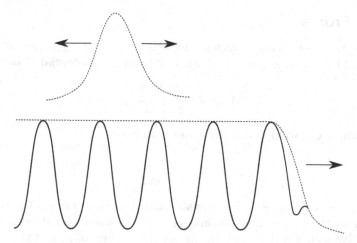

Figure 11.1 – We create an initial local perturbation (top) and choose $\epsilon > 0$ so that solution $A = 0$ is unstable. The fact that $A = 0$ is unstable means that local perturbations will invade the rest of the system with a velocity v (to be determined), resulting in a structure of a nearly constant amplitude, equal to $\sqrt{\epsilon}$ (see the dashed line in the bottom figure). Also shown (bottom), a plot showing the underlying periodic structure not seen in equation (11.1) (this equation describes only the envelope). The full waveform is the product of envelope A and an oscillating function e^{iqx} (see chapter 10).

The essential physical criterion in distinguishing the two situations comes from the invasion velocity with which one state takes over the other. When a metastable state is invaded by the stable state, the invasion velocity is well defined; whereas in the invasion of an unstable state, an infinite number of a priori possible solutions for the choice of the invasion velocity exists.

11.1. Invasion of a Metastable Solution by a Stable Solution

We first quickly review the canonical case, often seen in problems dealing with first order phase transitions: that of coexistence of two states of the same stability. Varying the control parameter of the system changes the stability of one of the states, making it less stable than the other, which we then call metastable. The stable state then invades the metastable state with a well defined velocity, as we shall see, but first, we will introduce the main concepts.

11.1.1. Fronts

Let us begin by studying a system characterized by the generic amplitude equation (11.1), with A a real amplitude. Introducing potential V such that:

$$V = -\epsilon \frac{A^2}{2} + \frac{A^4}{4}, \tag{11.2}$$

the amplitude equation (11.1) can be rewritten as:

$$A_t = A_{xx} - \frac{\partial V(A)}{\partial A}. \tag{11.3}$$

For positive ϵ, solution $A = 0$ is unstable and $A = \pm\sqrt{\epsilon}$ stable. Directing our attention to the two stable solutions $A = \pm\sqrt{\epsilon}$, imagine that we attribute the negative solution $A = -\sqrt{\epsilon}$ to the negative values of x, and the positive solution to positive x values. The two solutions being equivalent, we may intuitively expect a solution in which $A = \pm\sqrt{\epsilon}$ coexist and are connected through a point, say $x = 0$. Let us prove this.

The stationary version (i.e. $A_t = 0$) of the amplitude equation (11.3) is of the form:

$$A_{xx} + \epsilon A - A^3 = 0. \tag{11.4}$$

This equation allows the following solution:

$$A(x) = \sqrt{\epsilon}\tanh\left(x\sqrt{\frac{\epsilon}{2}}\right). \tag{11.5}$$

To convince yourself, just plug the expression for amplitude $A(x)$ into the stationary amplitude equation (11.4).

Given the inversion symmetry ($A \to -A$) of the stationary amplitude equation, we know that $-A(x)$ is also a solution to (11.4).

Notice that the stationary solution $A(x)$ (equation (11.5)) also satisfies $A(x \to \pm\infty) \to \pm\sqrt{\epsilon}$ (see figure 11.2). Thus, the solution indeed links the two solutions $\pm\sqrt{\epsilon}$ of the homogeneous equation ($\epsilon A - A^3 = 0$) at $x = 0$. This stationary solution (equation (11.5)) represents a front (or a *kink*, a topological defect[1]) (again, see figure 11.2).

---

1. This solution is sometimes called a topological defect. Indeed, this kink cannot easily be eliminated, because all translations of the kink to left or right are also viable solutions for the system. More generally, we can show that the general solution is given by $A(x) = \sqrt{\epsilon}\tanh[(x - x_0)\sqrt{\epsilon/2}]$, with x_0 an arbitrary constant. This is a result of the amplitude equation's translation invariance.

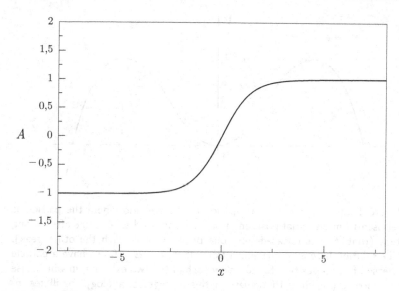

Figure 11.2 – A front solution obtained from equation (11.5) for $\epsilon = 1$.

11.1.2. Analogy with Newtonian Mechanics

We can use an analogy from Newtonian mechanics to paint a qualitative picture of the stationary solution's (equation (11.5)) behavior. Imagine x is a variable representing "time". The amplitude A then becomes a function of temporal variable x, and could be interpreted in a mechanical system as position (or an angle, or any other degree of freedom). In this framework, the amplitude equation (11.3), expressed as a function of potential V, is analogous to that of the Newton equation, where the term A_{xx} is associated with acceleration, and the expression $F(A) = \partial V(A)/\partial A = -\epsilon A + A^3$ would be the "force" (note that the mass of the particle, described by the coefficient of A_{xx}, is equal to one). The stationary amplitude equation (11.4) would be:

$$A_{xx} = F(A) = -\epsilon A + A^3 = -\frac{\partial}{\partial A}\left[\epsilon\frac{A^2}{2} - \frac{A^4}{4}\right] = -\frac{\partial}{\partial A}W(A) \qquad (11.6)$$

with

$$W = \epsilon\frac{A^2}{2} - \frac{A^4}{4} = -V. \qquad (11.7)$$

W is the potential defining the force, responsible for the displacement of the particle (see figure 11.3 which plots potential W for $\epsilon > 0$, the case of interest). Note that potential W – which we will call the fictitious potential, since it is the potential of the mechanical analogy – is the opposite of potential V, the physical potential. Potential W has two maxima, or two fixed points, located at $A = \pm\sqrt{\epsilon}$; these fixed points are stable in the physical counterpoint but unstable in the present mechanics analogy. By analogy

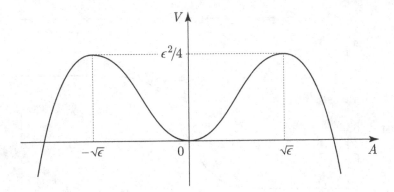

Figure 11.3 – $E = \epsilon^2/4$ corresponds to the scenario where the particle is released from an initial position $A = -\sqrt{\epsilon}$ and reaches $A = \sqrt{\epsilon}$ after a long time (that is, it is released from one peak and will reach the other peak). The solution here is of one single front. If $\epsilon^2/4 < E < 0$, we have a particle released from a position found below each of the two extrema, in which case the particle (which is frictionless in this mechanical analogy) oscillates infinitely between two positions, corresponding to an infinite number of fronts.

with conservative systems' mechanics[2], characterized by the absence of term A_x in the amplitude equation, we can write the following condition:

$$\frac{1}{2}A_x^2 + W = E. \tag{11.8}$$

This defines the law for conservation of total energy in a conservative mechanical system, where the sum of the kinetic and potential energies, which we call E, must remain constant over time x.

In the mechanical analogy, the path between the two stationary solutions $\pm\sqrt{\epsilon}$ (or the solution with a front defined by equation (11.5)) expresses the trajectory of the particle between time $x = -\infty$ where the particle is at position $A = -\sqrt{\epsilon}$ (the left extrema of the potential in figure 11.3) and time $x = \infty$ where it is at $A = \sqrt{\epsilon}$ (the right extrema of the same figure). If we consider a mechanical system without any friction, the situation corresponds to an amplitude equation without an A_x term (i.e. first time derivative in the mechanics analogy). Now imagine that we let go of the particle from the leftmost peak of the potential (see figure 11.3). A perturbation, even infinitesimal, would make the particle fall toward the bottom of the potential[3],

---

2. We can multiply the amplitude equation (11.4) by A_x, and then write $\partial_x[A_x^2/2 + W(A)] = 0$, meaning that the term in brackets is a constant of movement, which we denote E.

3. The particle could also fall to the left, but in this case we have a divergent solution because the particle would never stop during its fall; this situation is not relevant to our discussion.

and thus toward the right side of the potential traced in figure 11.3. However, in the absence of friction, the particle would not stop at the bottom of the potential but rather continue its course toward the right and, thanks to conservation of energy, would in fact finish its trajectory exactly at the top of the right peak of the potential, at the same height as its initial position. This gives a qualitative description of the solution for a front formulated in equation (11.5).

It takes the particle an infinite amount of time to pass from one peak of the potential to the other. At each peak, the velocity of the particle becomes zero, which means the time needed for the particle to reach the other maximum tends to infinity[4].

Setting constant E to a value less than that of the peak of the potential (i.e. $E < \epsilon^2/4$) is equivalent to letting go of the particle from a height somewhere between the potential's minimum and maximum (see figure 11.3). Since there is no friction, the particle will undergo an infinite number of oscillations inside the potential well. Thus, there must be a family of continuous periodic solutions, a solution with an arbitrary number of fronts, for all values of the constant that are positive and less than the maximum value of the potential (i.e. $0 < E < \epsilon^2/4$). We will later see that this solution of multiple fronts is unstable; only a solution with a unique front can be stable.

When we describe a system corresponding to a particle being released from the peak of the potential (i.e. by setting $E = \epsilon^2/4$), we find two solutions $\pm\sqrt{\epsilon}$ which coexist. If we pick a non-symmetrical potential $V(A)$ such that each minimum has a different value, one of the two solutions becomes metastable (see section 3.3 which introduces the notion of metastability). Remember that in the mechanical analogy, the minimum of physical potential V corresponds to the maximum of fictitious potential W, from which the mechanical force is derived. In the mechanical analogy (see figure 11.4) we easily see that by dropping a particle from the lower peak, the particle descends the length of the potential but, due to conservation of energy[5], it will never reach the other side's higher peak. Thus, the two solutions $\pm\sqrt{\epsilon}$ cannot coexist.

---

4. The interval of "time" required for the transition from one peak to the other is given by equation

$$\lambda = \int_{-\lambda/2}^{\lambda/2} dx = \int_{-\sqrt{\epsilon}}^{\sqrt{\epsilon}} dA/A_x \sim \int dA/[(A - \sqrt{\epsilon})(A + \sqrt{\epsilon})],$$

where the constant E is fixed at $E = \epsilon^2/4$, corresponding to the peak of potential V. To obtain this result, we used the conservation law (equation (11.8)) to find the expression for A_x. We can see that the time interval λ diverges logarithmically in the neighborhood of each peak.

5. Of course, the particle will stop at the same potential height as the lower peak, and can return to its initial position. If the motion of the particle stops there, we are looking at a solution with two fronts (figure 11.4). Another possibility is that the particle will go back and forth an arbitrary number of times, making it a multifront solution. However, we will not go into detail about these distinctions here.

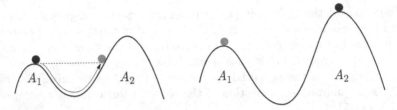

Figure 11.4 – Left: the particle (in black) released with zero velocity from the left-hand peak, climbs and stops at the same height if $v = 0$ (i.e. no friction; see text in subsection 11.1.3). Thus, the two fixed stable points (the peaks; remember the minima of physical potential V correspond to the maxima of the fictitious potential W) A_1 and A_2 cannot coexist. Thus there is an invasion of the metastable solution (A_1) by the stable solution (A_2). Right: we release the particle from A_2. This time the particle cannot stop its trajectory at position A_1 unless the friction v is adjusted so that the particle's potential energy reserve, $W(A_2) - W(A_1)$, is exactly compensated for by the dissipation of energy by friction. The invasion speed v (corresponding to friction; see text in subsection 11.1.3) is thus unique.

11.1.3. Determining the Invasion Velocity of a Metastable Solution by a Stable Solution

To begin, we should make the preliminary remark that it is natural for the stable solution to invade the metastable solution; it is the case analogous to dropping a particle from the higher peak to the lower peak (figure 11.4). Let us denote the invasion speed, which is a positive constant independent of time, by v, and look for a solution of the form $A(x - vt)$. We have thus to make the substitution $A_t \rightarrow -vA_x$. The amplitude equation (11.3) thus becomes:

$$A_{xx} + vA_x - \frac{\partial V}{\partial A} = 0. \tag{11.9}$$

In the mechanical analogy, where variable x represents time, we can think of this equation as analogous to an equation including frictional forces. The value of friction coefficient v is not given, and must be determined later. The uniqueness of the value is intuitive enough; since the particle leaves without any velocity from the top right peak of the potential (at location $A = A_2$), it will not stop at the second peak (found at $A = A_1$, see figure 11.4) unless the value of friction v is such that the difference in potential energy $W(A_2) - W(A_1)$ is exactly equal to the energy dissipated by friction. Multiplying the stationary amplitude equation which includes friction (equation (11.9)) by A_x we can then integrate from $x = -\infty$ to $x = \infty$ to find the expression for the invasion velocity:

$$v = \frac{V(A_2) - V(A_1)}{\int_{-\infty}^{\infty} A_x^2 dx}. \tag{11.10}$$

The numerator represents the motor of the dynamics, describing the reserve of potential energy, while the denominator describes the dissipated energy. The velocity becomes zero when the two extrema of the potential are at the same height, that is, when the two solutions have the same stability. This is known as the coexistence of solutions, or bistability. A small bias in the potential would bring about the invasion of the metastable state by the stable state at a well-defined velocity; in the problem we later propose (subsection 11.5.1), we will carry out an explicit calculation of the invasion velocity.

11.2. Invasion of an Unstable Solution by a Stable Solution

Now we want to look at the situation in which a stable solution invades an unstable solution, adopting the amplitude equation (11.3) as our model equation. As we have emphasized a few times in this chapter, in a system that is characterized by a positive ϵ parameter ($\epsilon > 0$), an initial local perturbation (see figure 11.1) will spread to invade the whole system. We will now make a detailed analysis of the invasion of the unstable solution $A = 0$ by the stable solution $A = \sqrt{\epsilon}$ to determine invasion velocity v. Keep in mind that this problem is conceptually very different from the invasion of a metastable solution by the stable solution discussed in the previous section.

A front solution moving at velocity v means that we have (as above) $A(x,t) = A(x - vt)$. This gives us the following relation between the two partial derivatives of the amplitude: $A_t = -vA_x$. Picking amplitude A such that it is always real, the amplitude equation (10.26) is written as:

$$A_{xx} + vA_x = -\epsilon A + A^3 = -\frac{\partial}{\partial A}\left(\epsilon\frac{A^2}{2} - \frac{A^4}{4}\right). \qquad (11.11)$$

Pursuing the mechanical analogy, the amplitude equation (11.11) is analogous to a particle at position A evolving in time x. The mechanical system is then characterized by the "mass" of the fictitious particle, corresponding to the coefficient of the term in A_{xx}, equal to one, having a friction term A_x with friction coefficient equal to invasion velocity v. This particle is moving in an anharmonic potential W:

$$W = \epsilon\frac{A^2}{2} - \frac{A^4}{4} = -V. \qquad (11.12)$$

Note once more that this potential[6] (see figure 11.5) is the opposite of the potential in figure 10.3.

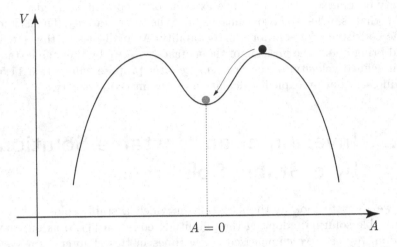

Figure 11.5 – The particle is released at an initial time at position $A = \sqrt{\epsilon}$ and must reach position $A = 0$ after a certain time. Since the particle experiences some friction, it will always end up at the bottom of the well, no matter what the (non-zero) value of the friction v actually is. If the friction is weak enough, there will be some number of oscillations in the potential well before the particle stops its motion (we are speaking of a semi-damped movement). If, on the contrary, the friction is very high, the particle will stop at the bottom of the well without any oscillation (an overdamped movement).

Taking cues from the mechanical analogy (see figure 11.5), we can think of this invasion as determining the position of the particle at instant $x = \infty$ after we drop it from an initial position $A = \sqrt{\epsilon}$ at time $x = -\infty$. Since the friction is non-zero, the particle always ends up at $A = 0$, either after a few oscillations around the same $A = 0$, or without any oscillation, depending on whether friction v is high or low (see figure 11.6). In other words, whatever the value of friction v, the particle always reaches the bottom of the potential well, and there is always a path from $A = \sqrt{\epsilon}$ to $A = 0$. The problem thus has an infinite number of solutions, or rather, there is a continuous family

---

6. Remember that the quadratic form of a potential defines a harmonic potential; in other words, the fact of being harmonic is associated to a potential in the form of a parabola. In many applications of physics, the harmonic potential is justifiably used as an approximation to study the evolution of a system. The basic concept of a harmonic potential plays a fundamental role in physics. Nevertheless, the hypothesis using a harmonic potential is not valid if the amplitude of the oscillating solution becomes too large. To study deviations from harmonic behavior for a real system, we need to look at higher orders of the potential, and then the potential is called anharmonic.

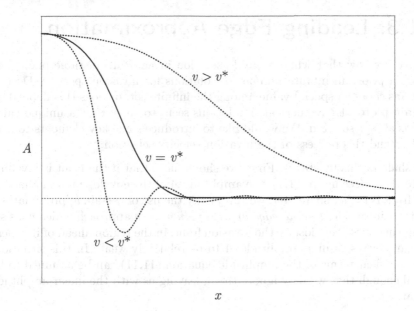

Figure 11.6 – The stationary solution for different values of the velocity.

of solutions (i.e. any positive value of v is a possible solution) which in turn raises the question of velocity selection: does a real system select, or not, a unique velocity among the infinite number of possible solutions?

We would like to stress the conceptual difference between the invasion problem of a metastable solution by a stable solution (studied in the previous section) and the invasion of an unstable solution by a stable solution. In the first scenario, associated with a potential like the one drawn in figure 11.4, we need to find the path between stable solution $A = A_1$ and metastable solution $A = A_2$. Starting from the peak at $A = A_1$ (left side in figure 11.4), we need to find a particular value for the "friction" coefficient in order to make it to position $A = A_2$. If the friction is too low, the particle will go past $A = A_2$ and sink forever into infinity, and if the friction is too high, the particle cannot reach $A = A_2$. In other words the friction must have a precise value so that the particle reaches $A = A_2$ with zero velocity and sits there; the solution for v is unique. As such, if the solutions A_1 and A_2 have the same stability (i.e. are found at the same height of the potential), the friction coefficient must be zero (i.e. $v = 0$) so that the particle, released from one of the two summits, can climb exactly to the other peak: this is a bistability which corresponds, in conventional terminology for phase transitions (such as in the case of the transition between a liquid and a solid state), to the coexistence of two "phases" in equilibrium (in the case of phase transitions, the liquid state can coexist with a solid state if the temperature is maintained at the melting temperature).

11.3. Leading Edge Approximation

Above, we saw that when a stable solution invades an unstable one, there exists a priori an infinite number of solutions for invasion speed v. Do real systems choose a special value among the infinite set, or does this depend on certain protocols? Numerical simulations seem to show that a unique value of v is always selected. We would like to introduce some key elements to help understand this process of the invasion velocity selection.

We shall do this by stages. First, we should note that if the front is invading state $A = 0$ (see figure 11.1), the amplitude A of the envelope found ahead of the front is rather small, and we can justify the use of certain approximations in its vicinity. The *leading edge approximation* is an approach which consists of placing ourselves close to the invasion front, in the region ahead of it, where we can always assume amplitude A to be relatively weak. In this situation, the nonlinear terms of the amplitude equation (11.11) can be assumed to be small enough that we may ignore them, leaving us with the linear amplitude equation:

$$A_{xx} + vA_x + \epsilon A = 0. \tag{11.13}$$

The general solution to this equation has the form $A(x) = ae^{-qx}$, where a is an amplitude and q is a constant that can be determined. If we carry this into equation (11.13), we find:

$$q_{1,2} = \frac{1}{2}\left(v \mp \sqrt{v^{*2} - v^2}\,\right), \tag{11.14}$$

where $v^* \equiv 2\sqrt{\epsilon}$.

In the mechanical analogy, the particular value v^* is the critical friction coefficient which marks the border between an overdamped solution (overcritical damping) and an underdamped solution (subcritical damping). For lower values of the friction coefficient (i.e. $v < v^*$), q becomes a complex number, signaling a spatially oscillatory solution (see figure 11.7 showing the two branches q as a function of v). For higher friction coefficient values, the front is overdamped. Since any value of the friction coefficient v is a priori acceptable, we have a continuum of solutions. The problem of velocity selection for a front is the question of which solution will be chosen, if any, among the infinite possibilities.

Let us first focus our attention on the scenario corresponding to values of parameter v higher than critical value v^* (i.e. $v > v^*$, see figure 11.7). The solution's upper branch (labeled q_2 on figure 11.7) corresponds to larger wavevectors than for the lower branch (q_1). The exponential form of the leading edge solution tells us that a front associated with the upper branch vanishes more quickly than one associated with the lower branch. Consequently, the leading edge is determined by branch q_1. We will analyze the behavior of

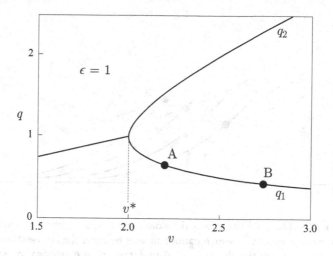

Figure 11.7 – The curve $q(v)$. The branch of the curve $q(v)$ for $v < v^*$ is the branch corresponding to complex values of q. We will prove in the following sections that this branch is unstable. The branch q_2 corresponds to a front which vanishes more quickly than the one corresponding to q_1 (see text). We will show that the dynamics of the front are determined by the branch q_1. We have set $\epsilon = 1$.

the front associated with branch q_1 to determine how the system ends up choosing the smallest value q of that branch. We begin by choosing two solutions of the lower branch, designated by A and B (see figure 11.7). The solutions are chosen such that the wavevector of solution A is greater than that of solution B (i.e. front B is more spread out than A). If we plot the position of each front over successive periods of time, we will notice that the intersection of the two fronts rises in the course of time (see figure 11.8) and that front A determines the global velocity for both fronts! The slower phenomenon is thus governing the dynamics. We can conclude that the leading edge will correspond to the smallest value of q, namely q^*. The question now remains, why does the system not select a value of $v < v^*$ (corresponding to a complex q)? This is the question we will answer in the next section.

Before continuing, we would like to make a small digression on the subject of the dominance of slower fronts. The problem is similar to that of the growth of a faceted crystal (this reasoning is extracted from [105]), for which only the slowest facet survives (see figure 11.9). The fastest disappears with time: the dynamics and form of the crystal are thus determined by the slowest facets. We can conclude that the front must choose the slowest velocity, and that this velocity is linked to critical parameter v^*. It remains for us to prove that the velocities lower than v^* cannot also be solutions. To answer this question, we will use another concept which will tell us that solutions with $v < v^*$ are unstable.

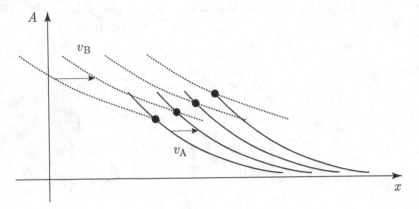

Figure 11.8 – The solid line shows the slower front A, and the dotted line represents the faster front B, which cannot survive because the intersection point steadily rises, leaving the slower front A to determine the propagation speed.

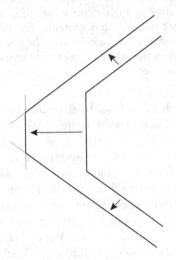

Figure 11.9 – A faceted crystal grows by an advancement of its facets. As one can see in the figure – which shows the position of the facets at two different times (the magnitude of the facet growth velocities are indicated by the arrows) – it is the faster facet which will disappear, leaving the space to the slower facet.

11.3.1. Instability of the Solution Corresponding to $v < v^*$

Our starting point is equation (11.11), and A_0 will now denote the stationary solution with constant velocity $A_0(x - vt)$. We study its stability by setting $A(x, t) = A_0(x - vt) + \eta(x, t)$ and linearizing equation (11.11) with respect to η (linear stability analysis). Keeping the first order terms of η we get:

$$\eta_t = \eta_{xx} + v\eta_x + \epsilon\eta - 3A_0^2\eta. \qquad (11.15)$$

We can describe perturbation η in the following form[7]:

$$\eta = \psi(x)e^{-\omega t}e^{-vx/2}, \qquad (11.16)$$

where $\psi(x)$ is the amplitude of the perturbation and the terms $e^{-\omega t}$ and $e^{-vx/2}$ correspond to temporal and spatial modulations respectively. If we introduce this expression into the linearized amplitude equation (11.15), we obtain:

$$\omega\psi = \left(-\frac{\partial^2}{\partial x^2} + U(z)\right)\psi, \qquad (11.17)$$

with

$$U(z) = \frac{v^2}{4} - \epsilon + 3A_0^2. \qquad (11.18)$$

The equation describing the perturbation's amplitude ψ (equation (11.17)) resembles the Schrödinger equation, where ω plays the role of energy and U that of the potential. It is clear that the linearized amplitude equation to first order (equation (11.15)) has a neutral solution (or stationary, i.e. $\eta_t = 0$, which comes down to saying the solution has $\omega = 0$ as its eigenvalue), given by the function:

$$\eta_G(x) = \frac{dA_0(x)}{dx} \equiv A_{0x}. \qquad (11.19)$$

To prove this, let us start from the stationary amplitude equation (11.4):

$$A_{0xx} + vA_{0x} + \epsilon A_0 - A_0^3 = 0 \qquad (11.20)$$

and differentiate with respect to x, to find the following:

$$(A_{0x})_{xx} + v(A_{0x})_x + \epsilon A_{0x} - 3A_0^2 A_{0x} = 0. \qquad (11.21)$$

Defining $\eta_G(x) = A_{0x}$ returns the equation describing the behavior of perturbation η (equation (11.15)) in the stationary regime (or for $\omega = 0$). This solution, sometimes called the Goldstone mode, is a consequence of translational invariance[8]. The Goldstone mode is a solution to the Schrödinger

---

7. Note that we have chosen $-\omega$ and not ω for our rate of change, as we have always done up to now. This is done only to have a more direct correspondence with an analogy we will make, further on, with the Schrödinger equation.

8. If $A_0(x - vt)$ is a solution, so is $A_0(x - vt + x_0)$, where x_0 is an arbitrary displacement of the front. If x_0 is small, we can write $A_0(x - vt + x_0) = A_0(x - vt) + x_0 A_{0x}$. Thus, mode A_{0x} is associated to an infinitesimal displacement of the front.

equation for eigenvalue $\omega = 0$. In other words, the linearized amplitude equation (11.17) has the following eigenfunction ψ and eigenvalue ω:

$$\psi_G = e^{xv/2}A_{0x}, \quad \omega = 0. \tag{11.22}$$

This is one of many eigenfunctions; the question is, is it a fundamental state or an excited state? From the Schrödinger equation we know that the states – fundamental and excited – of a quantum system can be classified by the number of nodes in the wavefunction: the fundamental state has no nodes, the first excited state (if one exists) has one node, and so on. If you cannot remember this result, just think about the situation of a free particle (respectively, a vibrating rope), where energy (respectively, frequency) behaves like $E_n \sim n^2\pi^2/L^2$ and the associated wavefunction takes the form $\psi(x) \sim \sin(xn\pi/L)$ where n is an integer number telling us which state the system is in, and L is the length of the well (or respectively, the chord). For $n = 0$, the wavefunction does not have any node, for $n = 1$ there is one, and so on. Physically, it is evident that having many nodes would correspond to having many arches in the rope, which is energetically more expensive.

As we have underlined above, the scenario where the invasion speed is less than the critical speed $v < v^*$ (see equation (11.14)) corresponds to the case of low friction in our mechanical analogy and A_0 shows oscillations (see figure 11.6), while for values that are greater than the critical value (i.e. $v > v^*$) result in an overdamping of A_0. Thus, in the first scenario ($v < v^*$) the wavefunction ψ_G would have nodes[9], and we know it is not the fundamental state of the system. Since eigenfunction ψ_G corresponds to a zero value eigenenergy (i.e. $\omega = 0$) we can assert that the fundamental state must have an energy which is less than zero (i.e. $\omega < 0$). However, if we look at the form of the perturbation (equation (11.16)), we can see that a negative energy would lead to an exponential growth of the perturbation over time. Hence we must conclude that the solution is unstable for that domain (i.e. $v < v^*$), thus excluding this domain of invasion velocities from being viable stationary solutions within the dynamics of the system.

In conclusion, we have just seen that all fronts displacing themselves at speeds which are less than v^* are unstable, which excludes the first straight line in figure 11.7 from the domain of possible solutions. We have also seen that the front corresponding to the upper branch q_2 of the same figure vanishes quickly, leaving the front to be determined by the lower branch q_1. From the analysis summarized in figure 11.8, we have seen that front A (with the smaller q) dominates over front B. This led us to pick the smallest possible velocity. The only possible choice left to us is then a front propagating at exactly $v = v^*$. Numerical and detailed mathematical analyses come to the same conclusion (see [106]).

---

9. The oscillation means the function must necessarily pass through zero.

11.3.2. Marginal Stability

When the propagation speed is exactly equal to the critical value (i.e. $v = v^*$), variable q is real, and we can affirm that this situation describes the fundamental state of the linearized equation and has the eigenvalue of $\omega = 0$ (since q is real, the front has no nodes and thus corresponds to the fundamental state). In other words, this selection criterion tells us that the solution is marginal, or neutral: a perturbation neither grows nor decays. This is also equivalent to saying that if the front moves with velocity $v = v^*$, the perturbation spreads through invasion as quickly as it grows locally. If instead the front has a velocity which is faster, the perturbation dies out; and when slower, the perturbation grows locally. The stability criterion's dependence on the frame of reference means it has a *convective* nature, since its stability is not an absolute fact, but depends on the frame in which the perturbation is observed. Most literature agrees on the criterion of marginal stability in determining the invasion velocity, as in our analysis. In certain cases, a more in-depth analysis is necessary, especially when the approximation of the leading front is not valid [106] (see also section 11.4).

In conclusion, let us remember that we briefly, via an analogy with mechanics, verified that the amplitude equation has multifront solutions, called *multikinks* (see subsection 11.1.2). We were able to deduce, from previous considerations, that these solutions are unstable. Using the terminology proper to the Schrödinger equation, a wavefunction is proportional to the derivative of solution A of the amplitude equation (see equation (11.22)), and for the multiple front solution, the derivative of A passes through zero; thus, it has at least one node. We have just now seen that all solutions with nodes must have at least one negative eigenvalue ($\omega < 0$), which means it is also an unstable solution (remember that we have picked the temporal dependence to look like $e^{-\omega t}$; see equation (11.16)). Thus, only a unique monotonic front can be stable.

11.4. Conclusion

In this chapter, we have described the possible scenarios for an invasion of one solution by another by solving the amplitude equation of a physical system. The invasion problem is fairly general; it is rare for a solution to install itself simultaneously throughout an entire system. In general, the solution will emerge locally and, bit by bit, invade the rest of the system. The new solution which we call "invasive" (or propagating) takes over when it becomes more stable than the solution in place before it, that solution having become either metastable or unstable. In this type of problem, the question naturally arises as to how to determine the speed of the invading front. To solve the problem, we looked at the specific case of a one dimensional system described by an amplitude equation. We introduced the notion of a front separating the system's two solutions. We then studied the invasion of one state by another, first that of a metastable state invaded by a stable state, then of an unstable state invaded by a stable state. We saw that the first scenario shows classic behavior, letting us find a unique solution for the invasion velocity easily. However the second case is far from classic and instead we find the following problem: there are, a priori, an infinite number of possible velocities for the propagation speed. Using a combination of heuristic arguments and analysis of linear stability, we have shown that the system chooses one solution among a continuum of solutions, in which v^* is the critical velocity demarcating the ensemble of unstable solutions (associated with the velocities which are less than v^*) and the ensemble of stable solutions (associated with the velocities which are above v^*). The system's choice of solution seems to be dictated by a principle of marginal stability: the system chooses exactly the solution where the velocity corresponds to the frontier between stability and instability, for which reason the solution is said to have a marginal stability.

The principle of marginal stability here is not the result of a rigorous mathematical proof. We simply verified the stability of solutions with a velocity less than v^*, and then to assure ourselves that the system does choose velocity v^* among all the possible stable solutions (corresponding to $v > v^*$), we turned to physical intuition. This physical intuition seems to be confirmed by computational simulations; they show that the solution chosen by the system does in fact move with velocity v^*. A rigorous mathematical proof was presented by Aronson and Weinberger in 1978 [3]. Starting from the universal amplitude equation seen at the beginning of the chapter (see equation (11.1)) and for the class of perturbations which begin localized (see figure 11.1), these authors have shown the emergence of a front solution which little by little establishes itself until finally invading the system with the velocity of marginal stability exactly v^*. This mathematical proof is more elaborate than our purpose requires.

Note that our approach in this chapter is based on linear approximations near the leading edge. These approximations are fully justified because the bifurcation is supercritical (pitchfork bifurcation); the amplitude grows in a continuous manner from a zero amplitude solution (i.e. $A = 0$) to a non-zero amplitude solution (i.e. $A \neq 0$). The velocity of the front is determined by the dynamics downstream of the front, which is why it is called a "pulled" front. If the bifurcation is subcritical, the situation becomes more complex, in particular, when the invasion velocity depends on nonlinear mechanisms. In this case, the invasion velocity is determined by the dynamics upstream from the front (the amplitude is larger; the predominance of this region is all the greater because of the important nonlinear effects). We call this a "pushed" front. We invite the reader who wishes to go deeper into the subject to look at a more complete review in [106].

11.5. Solved Problem

11.5.1. Determining the Spreading Velocity for the Invasion of a Metastable State by a Stable State

Solution. Consider a potential with a bias; for example, we could add an extra term $\nu(A^3/3)$ with positive, and small enough, ν (i.e. $\nu > 0$ and $\nu \ll 1$) to write:

$$V = -\frac{A^2}{2} + \frac{A^4}{4} + \nu\frac{A^3}{3}. \tag{11.23}$$

For $\nu = 0$, there are two stable solutions $A = \pm 1$ (we have picked $\epsilon = 1$) and the invasion velocity v is zero. The non-zero value of ν creates an asymmetry in the potential, as seen in figure 11.4. This asymmetry causes a non-zero invasion velocity, making the stationary solution (11.5) invalid. However, if ν is small enough, the new solution should not be very different: it would be the stationary solution (11.5) but with an extra perturbative term $\nu\tilde{A}(x)$, with $\tilde{A}$ the perturbation (i.e. $A(x) = A_0(x) + \nu\tilde{A}(x)$). This is confirmed by noting that the numerator of the invasion velocity (equation (11.10)) becomes zero for $\nu = 0$ (the potential energy reserve is zero because the values of the two extrema are the same). For the lowest order, we can thus replace the amplitude in the denominator of the invasion velocity (equation (11.10)) by the stationary solution A_0. This lets us rewrite the invasion velocity as:

$$v \simeq \frac{V(A_2) - V(A_1)}{\int_{-\infty}^{\infty} A_{0x}^2 dx} \simeq \frac{2\nu/3}{\int_{-\infty}^{\infty} A_{0x}^2 dx} = \frac{\nu}{\sqrt{2}}, \tag{11.24}$$

where we have made the following approximations in the numerator:

$$V(A_2) \simeq V(A_{20}) = V(1) = \nu/3,$$
$$V(A_1) \simeq V(A_{01}) = V(-1) = -\nu/3.$$

Since $[\partial V/A]_{A=A_{02}} = [\partial V/A]_{A=A_{01}} = 0$, this approximation is valid to order ν^2.

11.6. Exercises

11.1

Consider the following nonlinear equation:

$$A_t = \epsilon A - A^3 + A_{xx}. \tag{11.25}$$

Suppose there is a front solution moving at a constant speed v. Imagine a perturbation of the steady-state solution which is proportional to $e^{\omega t}$. Recall that if the real part of ω is positive the solution is unstable, and is otherwise stable. One may ask the question whether the perturbation (in the case of instability) grows faster or slower than the front motion. By way of example, imagine a river flowing at a certain speed. Consider a fixed observer who is not in the river watching a fixed point on its surface. Suppose that suddenly, at that fixed point, the surface which was initially flat, acquires a very small pump (like a localized perturbation) that will grow with time (due to some instability). Since the river is flowing at a certain speed, that localized wave will be advected by the river while its amplitude grows (since we assumed there is instability). There are two possibilities: (i) the river is flowing fast enough for the perturbation to escape from the initial fixed point before it has time to grow at that point (its amplitude will, by assumption of instability, continue to grow, and a large amplitude will be attained somewhere other than the initial point where the perturbation took place; (ii) the river flows so slowly that the perturbation grows rapidly at its initial point. In the first case the observer, who is always watching the initial fixed point, will see that the perturbation has decayed, while in the second case the opposite conclusion is reached. However, an observer who moves with the speed of the river (and always watches the same fixed point in the moving frame) will always conclude that the initial amplitude of the perturbation has grown. If both the stationary and the mobile observers conclude that the perturbation has grown, we say that the instability is *absolute*. If, on the contrary, only the observer watching the moving frame sees the growth of the perturbation, we then say that the instability is *convective*. This is why, in order to ascertain an instability, we have to analyze perturbations in the moving frame.

From a mathematical point of view the idea is to look for perturbations in the form $e^{(\omega+iqv)t}$ where q can be viewed as a wavevector. In order to understand this expression, the perturbation can be viewed first as a Fourier component, $e^{\omega t + iqx}$ (as we have regularly done), and then we need to focus on a point x moving at the front velocity. Another point to realize is that any perturbation, in general, is made not of a single wavevector, but of many wavevectors, so that a proper representation (useful here) is that the perturbation is a wavepacket in the form $\int_{-\infty}^{\infty} dq a_q e^{(\omega+iqv)t}$ where a_q is the Fourier amplitude of the packet. Finally, another ingredient that is needed is the fact that if t is large, this integral contributes more when the phase of the wavepacket $\omega + iqv$ is maximal. It is known in mathematics that we must treat q and ω as complex numbers (see [77]). This method, which consists of finding the maximum of a Fourier integral, is known as the stationary phase method. The condition of having an extremum is, as usual, written as:

$$\frac{d}{dq}\left(iqv + \omega(q)\right)_{q=q^*} = 0, \tag{11.26}$$

where q^* is the wavevector corresponding to an extremum. We can equivalently write:

$$v = i\frac{d}{dq}\omega(q). \tag{11.27}$$

Recall that ω is the usual dispersion obtained from linear stability analysis.

1. Since the wave is dominated by the mode which gives the maximal phase, the wavepacket can be reduced to:

$$e^{(\omega(q^*)+iq^*v)t}. \tag{11.28}$$

The stationary condition requires that v should correspond to a neutral point (i.e. the real part of the argument in the above equation to vanish). Show that this condition corresponds to a special value of v where $v = v^*$, which satisfies:

$$v^* = \left.\frac{\Re(\omega)}{\Im(q)}\right|_{q=q^*}. \tag{11.29}$$

2. Write the dispersion relation of equation (11.25).

3. Show by using (11.27) and (11.29) that:

$$v^* = 2q_i^* \quad \text{and} \quad v^* = \frac{\epsilon + q_i^{*2}}{q_i^*}. \tag{11.30}$$

Hint: first make sure that q^* is a purely imaginary number (by requiring that the velocity v be real), and denote it as q_i.

4. Deduce that $q^* = i\sqrt{\epsilon}$ and that $v^* = 2\sqrt{\epsilon}$. What can you conclude?

5. Explain why in the case where $v > v^*$, the solution is stable, whereas it is unstable otherwise. The speed $v = v^*$ corresponds to marginal stability.

6. Show that if v^* and q^* are given by (11.27) and (11.29), then the perturbation $e^{(\omega+iqv)t}$ is purely oscillatory and is given by $e^{i\Omega t}$ where $\Omega \equiv \Im(\omega) + v^*\Re(q^*)$. The front problem can be viewed, in general (except if Ω is zero) as an oscillator that leaves behind a periodic structure. It is like a string where, for example, one of its end points oscillates, generating a wave. We know in that problem that the wavelength of the wave is fixed by the pulsation of the end point and the propagation speed. Show that the selected wavevector is given by:

$$q_s = \Re(q^*) + v^{*-1}\Im(\omega(q^*)). \tag{11.31}$$

Please be sure not to confuse the wavevector q_s, which is the wavevector of the structure, with q^*, which is an abstract one that provides the maximum of the phase of the wavepacket (which is a complex number in general).

What are the underlying assumptions that allow one to write (11.31)?

7. What is the value of q_s for equation (11.25)?

11.2

Consider the Swift–Hohenberg equation:

$$\frac{\partial \phi}{\partial t} = \left[\epsilon - \left(1 + \frac{\partial^2}{\partial x^2} \right)^2 \right] \phi - \phi^3. \tag{11.32}$$

1. Show by using the conditions derived above (i.e. stationary phase, marginal stability), that the front speed v^* and the wavevector q^* (with $q^* = q_r^* + i q_i^*$) are given by (consider only the solution that $q_r \neq 0$):

$$v^* = \frac{4}{3\sqrt{3}} \left[\sqrt{1 + 6\epsilon} + 2 \right] \left[\sqrt{1 + 6\epsilon} - 1 \right]^{1/2}, \tag{11.33}$$

$$q_r^* = \frac{1}{2} \left[\sqrt{1 + 6\epsilon} + 3 \right]^{1/2}, \tag{11.34}$$

$$q_i^* = \frac{1}{2\sqrt{3}} \left[\sqrt{1 + 6\epsilon} - 1 \right]^{1/2}. \tag{11.35}$$

2. For small ϵ, check that the results are consistent with those of the amplitude equation (exercise 11.2). We recall that the amplitude equation obtained from the Swift–Hohenberg one is not canonical, which means the coefficients are not equal to unity.

11.3

Determine v^* and q^* for the following nonlinear equation (Benney equation):

$$u_t = -u_{xx} - u_{xxxx} + b u_{xxx} + u_x^2. \tag{11.36}$$

Show that there exists a critical value of parameter b (denoted by b_c) below which we have absolute instability (i.e. $\Re(\omega + iqv)$ is always positive whatever the value of v).

Chapter 12

Order and Disorder both Spatial and Temporal near a Hopf Bifurcation

Abstract *In this chapter we learn to describe the type of bifurcation from which both spatial patterns and temporal oscillations emerge (Hopf bifurcation). Using symmetry, we show that these systems are described by a universal equation (that is, independently of the system under consideration) in the neighborhood of the instability threshold. Unlike the stationary bifurcation, stationary in time, a Hopf bifurcation can lead to a dizzying array of rich behaviors, ranging from order (in the form of bands which move with time – progressive waves), to spatio-temporal chaos.*

In chapter 10, dedicated to the study of the emergence of spatial order, we mainly used the generalized Swift–Hohenberg system (equation (10.2)) in order to derive the amplitude equation and show its universality around a stationary bifurcation point.

In this chapter, we want to consider spatial bifurcations of a different nature (an oscillating or a Hopf bifurcation) and study the different resulting evolutions of the system. We will see that the system can go many ways, either into some sort of ordered state or into chaos both temporal and spatial. How is the Hopf bifurcation nature reflected in the amplitude equation?

In chapter 5 where we studied the homogeneous (i.e. with amplitude independent of the spatial variable) Hopf bifurcation, the amplitude equation was identical to that for the stationary case with the exception of the coefficients, which were complex (they have a non-zero imaginary part). Thus, when the spatial variable is introduced, we might a priori suppose that the amplitude

© Springer Science+Business Media B.V. 2017
C. Misbah, *Complex Dynamics and Morphogenesis*,
DOI 10.1007/978-94-024-1020-4_12

equation will have the same form as the one we derived for the stationary bifurcation (equation (10.24)), but with complex coefficients.

To formally derive the Hopf bifurcation's amplitude equation, we can follow the same procedure as in our study of the stationary bifurcation (see chapter 10). Starting from the given model, we expand the solution in powers of small parameter ϵ (which measures the distance from the instability threshold) to derive the amplitude equation. Since we have already shown this type of derivation several times throughout the book, we will not go into the details of the Hopf bifurcation case here, but will assume the reader may adeptly combine the acquired knowledge to derive it for themselves. We will simply present the equation, motivate its form by symmetries, and focus our attention on the results that follow from this equation, for the unidimensional case.

12.1. The Amplitude Equation with Complex Coefficients

We can justify the form of the amplitude equation in the neighborhood of a Hopf bifurcation by looking at symmetries, as we did previously for a stationary bifurcation. In subsection 10.2.2, we saw that coefficients of a stationary bifurcation must be real due to the inversion symmetry $x \to -x$. For a stationary bifurcation, the inversion is mirrored by A^* ($A \to A^*$), which requires that the coefficient be real (again, see subsection 10.2.2).

12.1.1. Some Preliminaries

Remember that, in a linear regime, the Fourier mode for a perturbation is that of a plane wave described by $e^{iqx+\omega t}$, where q is the wavevector and ω is the growth rate of the perturbation. For a Hopf bifurcation we have a complex growth rate $\omega = \omega_r + i\omega_i$, where ω_r and ω_i represent the real and imaginary components of ω. The instability is characterized by a positive real component ($\omega_r > 0$). Near to the bifurcation, this component ω_r (see equation (9.84)) is written as:

$$\omega_r = \epsilon + \omega_r'' \frac{(q - q_c)^2}{2}, \tag{12.1}$$

where ϵ measures the distance from the instability threshold, q_c is the bifurcation's wavevector and ω_r'' is the second derivative, evaluated at the bifurcation point (i.e. at $q = q_c$ and $\epsilon = 0$). Remember that this expression is the result of a Taylor expansion in the neighborhood of the bifurcation (that is, in the neighborhood of $q = q_c$ for $\epsilon = 0$, see subsection 9.11.2).

The first derivative of the instability's growth rate ω_r at the bifurcation point (equation (12.1)) becomes zero, indicating that the real part of the growth rate ω_r is at a maximum at this point (again, see subsection 9.11.2).

We can also make a Taylor expansion of the imaginary part ω_i around the bifurcation point, as:

$$\omega_i = \omega_{i0} + (q - q_c)\omega_i' + \omega_i''\frac{(q - q_c)^2}{2} + \cdots, \qquad (12.2)$$

where $\omega_{i0} \equiv \omega_i(q = q_c, \epsilon = 0)$ and ω_i', ω_i'' are respectively the first and second derivatives, evaluated at the bifurcation point. Notice that the first derivative of the imaginary component, ω_i', is not zero, unlike ω_r'.

Finally, note that the number of dynamic variables that are really independent from each other in any given problem may be greater than one (see chapter 9). However, the usual hypothesis made at the bifurcation point ($q = q_c$ and $\epsilon = 0$) assumes that the first derivative of the growth rate only becomes zero for one eigenvalue[1] ω. This eigenvalue becomes positive for an ensemble of wavevectors contorting around q_c when $\epsilon > 0$ (see chapters 9 and 10). The eigenmode h_q for eigenvalue ω is written as a linear superposition of different dynamic variables, included in the system of equations which characterizes the evolution of the system. In the linear regime, the Fourier eigenmode h_q obeys:

$$\frac{\partial h_q}{\partial t} = \omega h_q. \qquad (12.3)$$

The solution is given by $h_q(t) = h_q(0)e^{\omega t}$.

For a system having a very large size in the Ox direction (taken theoretically to be an infinite system)[2] there are no restrictions on q, meaning that all solutions with arbitrary q values may be a solution. As such, we can write the general solution to $h(x, t)$ as a linear superposition of all the Fourier modes:

$$h(x, t) = \sum_q h_q(t)e^{iqx} + c.c. = e^{iq_c x}\sum_q h_q(t)e^{i(q-q_c)x} + c.c.$$

$$= e^{iq_c x}\sum_Q h_Q(t)e^{iQX} + c.c., \qquad (12.4)$$

where we have made the following change of variables: $Q = (q - q_c)/\sqrt{\epsilon}$, and introduced a slow (spatial) variable, $X = x\sqrt{\epsilon}$ (see chapter 10). Now, we want to define h_Q through its dependence on slow and fast variables in the linear regime (since our study is focused on the region near a bifurcation, the linear regime suffices in our search for information about these variables).

---

1. We sometimes call this a co-dimension 1 bifurcation.
2. All real systems are finite. The hypothesis of an infinitely extending system in practice corresponds to having a lengthscale (or wavelength when the pattern is ordered) which is small with respect to the extension of the system in the Ox direction.

So, remaining in the linear regime, we can use equations (12.1) and (12.2) to describe the time dependence of the perturbation by:

$$h_q(t) \sim e^{\omega t} = e^{[i\omega_{i0} + \epsilon + \omega''(q-q_c)^2/2 + i(q-q_c)\omega_i']t}$$

$$= e^{i\omega_{i0}t} e^{[1 + \omega''Q^2/2 + i\omega_i'Q/\sqrt{\epsilon}]T}. \tag{12.5}$$

If the bifurcation is stationary, ω is real (i.e. $\omega_i = 0$) and the temporal dynamics is a function only of slow variable T. For a Hopf bifurcation we see from equation (12.5) that we have both the fast variable t and the slow variables ϵt and $t\sqrt{\epsilon}$. Thus, to express the slowly varying amplitude of the wave in all rigor, we can write $a_Q(t\epsilon, t\sqrt{\epsilon})$. We may easily introduce two distinct timescales T and τ, such that $T = \epsilon t$ and $\tau = \sqrt{\epsilon}t$, like in any problem involving multiscales (see chapter 10).

Accordingly, the partial differential operator is written as:

$$\frac{\partial}{\partial t} \longrightarrow \frac{\partial}{\partial t} + \epsilon \frac{\partial}{\partial T} + \sqrt{\epsilon}\frac{\partial}{\partial \tau}. \tag{12.6}$$

Note that the operator is composed of three terms, related to the three timescales in the problem, whereas the analogous operator for the stationary bifurcation depends on a single slow variable (equation (10.13)). To avoid going into a lengthy multiscale analysis for three temporal variables (especially since our objective is to determine the amplitude equation via purely symmetrical considerations), we adopt a hypothesis which somewhat arbitrarily simplifies our problem: if the derivative of the imaginary components of the growth rate (i.e. ω_i') is of the order $\sqrt{\epsilon}$, the development of equation (12.5) only involves slow variable T. We can assume thus $\omega_i' = \bar{\omega}_i'/\sqrt{\epsilon}$ and that $\bar{\omega}_i'$ is of order one.

The physical field $h(x,t)$ is thus composed of two terms, describing a fast variation (via t) and a slow variation (via T). In the weakly nonlinear regime (and by continuity of the linear analysis), the fast part looks like $e^{i\omega_{i0}t}$ while the temporal dependence of the slow variable is encoded in the wave's amplitude (or its envelope[3]), as follows:

$$h(x,t) = e^{iq_c x + i\omega_{i0}t} \sum_Q a_Q(T)e^{iQX} + c.c., \tag{12.7}$$

where we have defined $h_Q(t) = e^{i\omega_{i0}t}a_Q(T)$ to clearly separate the function's dependence on the slow and fast variables, t and T.

---

3. At the bifurcation point, we have: $\omega_i = \omega_{i0}$. In the neighborhood of the bifurcation, for small ϵ, ω_i has correctional terms of order ϵ – but these corrections are already taken into account in the nonlinear amplitude equation $a_Q(T)$. This is why it is legitimate not to explicitly keep the dependence on fast variable $e^{i\omega_{i0}t}$ in the expression for the physical field $h(x,t)$.

Following the same procedure as in chapter 10, on the universality of spatial order, we rewrite the physical field $h(x,t)$ (equation (12.7)) as:

$$h(x,t) = e^{iq_c x + i\omega_{i0} t} A(X,T) + e^{-iq_c x - i\omega_{i0} t} A^*(X,T), \qquad (12.8)$$

where we have defined $A(X,T) = \sum_Q a_Q(T) e^{iQX}$. The sum, which in principle counts all the real wavevectors, can be restricted to the band of wavevectors associated with unstable modes.

Performing a Fourier transform of eigenmode h_q of the linear regime (equation (12.3)) with respect to variable Q, and using the relation $\omega = \omega_r + i\omega_i$ (where ω_r and ω_i are given by equations (12.1) and (12.1)), we obtain the following linear amplitude equation:

$$A_T = A + \bar{\omega}_i' A_X + \frac{1}{2} \left(\omega_r'' + i\omega_i'' \right) A_{XX}. \qquad (12.9)$$

We have used the slow variables and defined $A_T = \partial A / \partial T$, $A_X = \partial A / \partial X$ and $A_{XX} = \partial^2 A / \partial X^2$. If we compare this result to the stationary bifurcation (chapter 10) we see two main differences: (i) the coefficient in front of the second derivative A_{XX} is complex; and (ii) the equation has a new term, $\bar{\omega}_i' A_X$.

Before writing down the nonlinear amplitude equation from symmetry considerations, we would like to point out that the field h of the wave (equation (12.8)) is a product of the fast oscillation $e^{iq_c x + i\omega_{i0} t}$ and a slow envelope $A(X,T)$; in contrast with a stationary bifurcation, for a Hopf bifurcation the fast oscillation looks like a plane wave propagating with phase velocity ω_{i0}/q_c. We must distinguish between two types of waves: a left wave with amplitude A and a right wave[4] with amplitude $B = A^*$, with the wave always propagating at group velocity $\bar{\omega}_i'$.

12.1.2. Deriving the Amplitude Equation from Symmetry

Unlike for a stationary bifurcation (i.e. a bifurcation with a growth rate which is real, $\bar{\omega}_i = 0$), for a Hopf bifurcation the inversion operation x to $-x$ is not equivalent to a switch between amplitude A and its complex conjugate A^* (equation (12.8), describing field $h(x,t)$, is not invariant under that operation). We used the inversion property in subsection 10.2.2 to show that the coefficients of the stationary bifurcation's amplitude equation had to be real. The absence of the same property for a Hopf bifurcation entails that the coefficients are generally complex (i.e. they have a non-zero imaginary component).

---

4. To be completely rigorous, amplitude B of the right wave could happen not to be conjugate to the right amplitude (i.e. $B \neq A^*$), which would introduce new interaction terms between the two waves. For the sake of simplicity, we will focus on the case where $B = A^*$.

On the other hand, we know that for a Hopf bifurcation, we have a wave which is propagating in a certain direction (let us say, increasing X) at the group velocity ω_i'. This means that the form of the amplitude equation must be identical if we apply the inversion operation X to $-X$ simultaneously to both the spatial coordinate X and the wave's group velocity (the simultaneous change in sign of the Ox-axis and wave velocity mean that the physical configuration is preserved), which in turns implies that we are looking to keep terms like γA_X (with γ a real coefficient proportional to the group velocity) – terms which fulfill this requirement.

We can therefore write the amplitude equation as:

$$A_T = (1 + ia)A - (1 + i\beta)|A|^2 A + \gamma A_X + (1 + i\alpha)A_{XX}, \qquad (12.10)$$

where a, α, β, and γ are real constants[5].

The linear part of equation (12.10) is similar to the linear amplitude equation (12.9), except for one term on the right, iaA. In the next section we will see that this term can be absorbed by a simple change of variables, highlighting the equivalence between the two equations.

We have also assumed, by assigning a negative sign to the real part of the cubic coefficient (for a nonlinear saturation), that the bifurcation is supercritical.

The term γA_X can be easily eliminated from equation (12.10) by a Galilean transformation. Calling $X' = X + \gamma T$, and $T' = T$, our new amplitude equation written in terms of these new variables has new derivatives:

$$\frac{\partial}{\partial T} = \frac{\partial}{\partial T'}\frac{\partial T'}{\partial T} + \frac{\partial}{\partial X'}\frac{\partial X'}{\partial T} = \frac{\partial}{\partial T'} + \gamma\frac{\partial}{\partial X'} \qquad (12.11)$$

and

$$\frac{\partial}{\partial X} = \frac{\partial}{\partial T'}\frac{\partial T'}{\partial X} + \frac{\partial}{\partial X'}\frac{\partial X'}{\partial X} = \frac{\partial}{\partial X'}. \qquad (12.12)$$

Note that the second spatial derivatives, like the first, are identical $(\partial^2/\partial X^2 = \partial^2/\partial X'^2)$. Using these derivatives in the amplitude equation (12.10) shows that the change of variables lets us eliminate the term γA_X.

We can further simplify the complex amplitude equation (12.10) by eliminating the term in ia via the operation $A \rightarrow Ae^{iat}$, equivalent to placing ourselves in a rotating frame in the complex plane $(\Re(A)\text{–}\Im(A))$. Through

---

5. We might have imagined that the coefficient in front of A_X could also be complex, like the other coefficients in the equation, but this would violate the parity symmetry of the system which imposes the conservation of the amplitude equation's form under $x \rightarrow -x$ and $\omega_i' \rightarrow -\omega_i'$. Another consideration to further justify the absence of a term like iA_X relies on the linear analysis developed in the last section where we established that the coefficient weighing term A_X is real (as it is given by the group velocity).

this transformation, we get the following complex amplitude equation:

$$A_t = A - (1 + i\beta)|A|^2 A + (1 + i\alpha)A_{xx}, \qquad (12.13)$$

where, having adopted a lowercase notation, x and t still designate the slow variables. Notice that in the preliminary linear analysis at the start of the chapter, we factored out the term $e^{i\omega_{i0}t}$ from our description of the perturbation field (equation (12.5)). As a result, the linear amplitude equation describing the perturbation (equation (12.13)) has no term in $i\omega_{i0}A$. In other words, we see that this choice is equivalent to placing ourselves in the rotating frame.

12.2. Non-variational Dynamics

One important property of equation (12.13) is that it is non-variational: it cannot be written as a gradient of a Lyapunov function (i.e. of a potential). However, the equation *can* still be written as the derivative of a *functional*[6]:

$$A_t = -\frac{\delta V}{\delta A^*}, \qquad (12.14)$$

defining functional V as:

$$V = \int dx \left(-\frac{|A|^2}{2} + (1 + i\beta)\frac{|A|^4}{4} + (1 + i\alpha)\frac{|A_x|^2}{2} \right). \qquad (12.15)$$

Though the complex amplitude equation (12.14) looks like it may come from variational dynamics, we must beware confusion: the complex character of function V, characteristic of a Hopf bifurcation, means that the condition $\partial V/\partial t \leqslant 0$ cannot be satisfied, in contrast to the scenario of a stationary bifurcation with variational dynamics.

We can write:

$$\frac{\partial V[A]}{\partial t} = \frac{\delta V}{\delta A^*}\frac{\partial A^*}{\partial t} = -\frac{\delta V}{\delta A^*}\frac{\delta V}{\delta A}, \qquad (12.16)$$

where we have used the fact that the conjugate amplitude A^* obeys the evolution equation $\partial A^*/\partial t = -\delta V[A]/\delta A$.

For a stationary bifurcation (see subsection 10.2.4), function V is a real function, and we can write $(\delta V/\delta A^*)(\delta V/\delta A) = |\delta V/\delta A|^2$ to deduce a monotonic decrease of function V over time (i.e. $\partial V/\partial t \leqslant 0$). If instead, function V

---

6. See the Euler problem solved in subsection 3.9.3 for more on functional derivatives and subsection 10.2.4 for a discussion on the variational nature of the dynamics of the amplitude equation.

is complex, the product of $(\delta V/\delta A^*)(\delta V/\delta A)$ is no longer definitely positive, so we no longer see a monotonic decrease. This is a fundamental property of non-equilibrium[7] phenomena: the evolution is not monotonic, letting the system's dynamics be very rich.

12.3. Some Properties of the Complex Amplitude Equation

12.3.1. Plane Waves and Stability

The amplitude equation of complex coefficients (12.13) has a simple solution in the form of a plane wave:

$$A_0 = R_0 e^{i\omega t + iqx}. \tag{12.17}$$

If we carry this expression for amplitude into equation (12.13) we find:

$$R_0^2 = 1 - q^2, \quad \omega = -\beta - (\alpha - \beta)q^2, \tag{12.18}$$

where q must satisfy: $|q| \leqslant 1$.

Before studying this solution's stability, let us first split the amplitude equation into its imaginary and real components. We define amplitude A such that:

$$A = R(x,t)e^{i\theta(x,t)}, \tag{12.19}$$

where R is the amplitude and θ is the phase of the wave. Carrying this expression into the amplitude equation (12.13), we find the following system of equations:

$$R_t = R + R_{xx} - R\theta_x^2 - \alpha(R\theta_{xx} + 2\theta_x R_x) - R^3,$$

$$R\theta_t = R\theta_{xx} + 2\theta_x R_x + \alpha(R_{xx} - R\theta_x^2) - \beta R^3. \tag{12.20}$$

We can easily verify that the pair of solutions (R, θ) with $R = R_0 = 1 - q^2$ and $\theta = \theta_0 = qx + \omega t$ is our stationary solution, in agreement with the plane wave solution (see conditions (12.18)), $\omega = -\beta - (\alpha - \beta)q^2$.

---

7. Note that even though problems giving rise to a stationary bifurcation are fundamentally linked to non-equilibrium states (such as Rayleigh–Bénard convection), the amplitude equation does have a variational character. However, this is only true for any analysis done in the neighborhood of the bifurcation. Thus, in following the behavior of a warmed-up fluid in the case of Rayleigh–Bénard convection, the tidy convection rolls start to break apart as we enter the chaotic (or turbulent) regime, which cannot follow any variational principle and keep its erratic character, markedly different from monotonic behavior.

Now we look at the stability of the plane wave by introducing a small perturbation ($\rho(x,t)$ and $\psi(x,t)$) to the wave's amplitude and phase:

$$R = R_0 + \rho(x,t), \quad \theta = \theta_0 + \psi(x,t). \tag{12.21}$$

As you know, if the perturbations decay with time, the solution of a plane wave is stable; if not, it is unstable. If we carry the expression for the perturbed solution (12.21) into the system of evolution equations (12.20), and keep the linear terms in ρ and ψ for a linear analysis, we have:

$$\rho_t = -2R_0^2\rho + \rho_{xx} - R_0(\alpha\psi_{xx} + 2q\psi_x) - 2\alpha q\rho_x,$$

$$R_0\psi_t = \alpha\rho_{xx} + R_0\psi_{xx} + 2q(\rho_x - \alpha R_0\psi_x) - 2\beta R_0^2\rho\alpha. \tag{12.22}$$

Since the system is autonomous with respect to x and t (that is, the coefficients of the equation are independent of x and t), let us find solutions for ρ and ψ of the following form: $\rho = ae^{iQx+\sigma t}$ and $\psi = be^{iQx+\sigma t}$, where a and b are constants which are a priori complex, Q is the perturbation's wavevector and σ is the (generally complex) growth or attenuation rate. We carry these solutions into the system above to find a homogeneous system of algebraic equations in a and b, with a non-trivial solution when the determinant of the system vanishes. This gives us a characteristic equation of second order in σ, with solutions given by:

$$\sigma_\pm = -R_0^2 - Q^2 - 2i\alpha qQ$$

$$\pm\sqrt{R_0^4 + \alpha^2Q^4 - 2\alpha Q^2\beta R_0^2 + 4q^2Q^2 + 4iqQ\beta R_0^2}. \tag{12.23}$$

While the real part of σ_- is always negative, the real part of σ_+ can become positive, characterizing the possibility of emergence of an instability.

Also note that when the perturbation's wavevector Q is zero, σ_+ is zero whereas σ_- is negative (attenuation of the instability). The mode in which the perturbation's wavevector is zero defines a neutral mode (i.e. a mode associated to a zero eigenvalue), characteristic of a perturbation that neither grows nor decays. The mode $Q = 0$ corresponds to a homogeneous translation along the Ox-axis vis-à-vis the amplitude A. However, given that the total field of the perturbation is composed of a fast oscillation and a slow envelope A (equation (12.8)), the mode $Q = 0$ corresponds to a rigid translation of the wave along the axis; this mode reflects the arbitrary aspect of the choice of origin along Ox.

This type of mode defines a Goldstone mode, as we have seen in section 10.3 devoted to the Eckhaus instability. This invariance by translation along the Ox-axis signifies that the evolution of the system – and thus of the physical system it governs – does not depend on the observer's choice of reference point.

12.3.2. Benjamin–Feir Instability

As seen above, σ_+ is thus zero when the wavevector Q is zero, but it can become positive, making it possible for an instability to develop for small wavenumbers Q, just as in the Eckaus instability (see section 10.3). To study this scenario, let us expand σ_+ for small wavenumbers Q, obtaining:

$$\sigma_+ = -2i(\alpha - \beta)qQ - \left(1 + \alpha\beta - \frac{2q^2(\beta^2 + 1)}{R_0^2}\right)Q^2 + O(Q^3). \quad (12.24)$$

It is important to make a clear distinction between q and Q. q is the wavevector of the plane wave solution whose stability is being studied, while Q is the wavenumber of the perturbation applied on top of the plane wave solution. The first term is imaginary and does not affect the stability of the system. The second term, of second order in Q (i.e. Q^2) can become positive, causing the development of an instability. Thus, we can distinguish between two cases:

1. The plane wave solution is stable against perturbations if the coefficient of Q^2 is negative, which corresponds to the following:

$$1 + \alpha\beta - 2q^2\frac{\beta^2 + 1}{R_0^2} > 0. \quad (12.25)$$

In other words, since $R_0 = 1 - q^2$ (see conditions (12.18)), the wavevectors q which satisfy the following inequality:

$$q^2 < \frac{1 + \alpha\beta}{3 + \alpha\beta + 2\beta^2}, \quad (12.26)$$

correspond to stable modes.

2. By contrast, the plane wave solution becomes unstable if the coefficient of Q^2 becomes positive. This happens for wavevectors q of the plane waves satisfying the following inequality:

$$q^2 > \frac{1 + \alpha\beta}{3 + \alpha\beta + 2\beta^2}. \quad (12.27)$$

This is a generalization of the result found for the Eckhaus instability (see section 10.3). When constants α and β are zero (i.e. $\alpha = \beta = 0$), the instability condition corresponds to $|q| > 1/\sqrt{3}$, which is simply the Eckhaus criterion. Lastly, note that if the following inequality, called the Benjamin–Feir criterion [8] is satisfied:

$$1 + \alpha\beta < 0, \quad (12.28)$$

the entire band is unstable, including wavenumber $q = 0$: there is thus no stable wave solution. In the unstable domain of Benjamin–Feir, there exist complex dynamics, of which we will later give a brief overview.

12.4. Illustration of the Dynamics for Some Typical Cases: Numerical Analysis

We will now review some properties of the amplitude equation, found with the help of a numerical solution.

12.4.1. Plane Waves

If we take up the complex amplitude equation (12.13) and give some specific values to the real constants α and β, such as $\alpha = 1$ and $\beta = 2$, to study the dynamics (see figure 12.1), we find that the Benjamin–Feir criterion for instability (12.28) is not satisfied, since we have $1 + \alpha\beta = 3 > 0$. The criterion for stability (12.26) defines a band of wavevectors which correspond to plane waves whose wavenumber q satisfies: $-\sqrt{3/13} < q < \sqrt{3/13}$. The dynamics shown in figure 12.1 show the evolution of amplitude A, more specifically, of modulus $|A|$, and of the real component in the spatio-temporal plane $(x$–$t)$; this type of description is sometimes called a *spatio-temporal phase portrait*.

Starting with a noisy, or random state, we observe that the modulus of A becomes homogeneous over a long enough time, and everywhere in space (figure 12.1, left). The evolution of the imaginary component (figure 12.1, right)

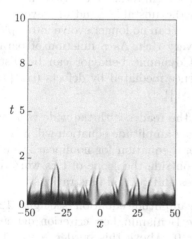
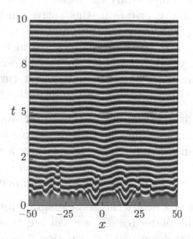

Figure 12.1 – Spatio-temporal portrait: the horizontal axis represents position, the vertical axis time, and the color represents the modulus of A (left) and the phase (right) with random initial conditions (gray corresponds to the maximum and light gray to the minimum). We have chosen $\alpha = 1$, $\beta = 2$. The system finds itself in the stable Benjamin–Feir region, and the final state is a plane wave.

of the perturbation amplitude shows that the initial disturbance (figure 12.1, bottom of right-hand graph) disappears, giving rise to a periodic modulation which shows the plane property of the wave. Note that the associated wavelength takes longer to reach a steady state than modulus $|A|$: we say that the dynamics of the amplitude, given by $|A|$, is fast compared to that of the wavelength. The dynamics are thus limited by the phase; that is why the Benjamin–Feir instability, like the Eckhaus instability, is an instability in the phase. The study of dynamics at longer timescales reveals that the wave-vector becomes homogeneous throughout space, with the bands becoming straight and parallel to each other (with an inclination angle determined by the phase velocity of the wave, q/ω).

Now, keeping the same values for parameters α and β (here $\alpha = 1$ and $\beta = 2$), we modify the initial condition with a plane wave as our perturbation, with wavevector q within the unstable Benjamin–Feir band ($|q| > \sqrt{3/13}$). What are the dynamics for this type of system? In fact, the system tends to head to a smaller wavenumber q (i.e. increases the wavelength) such that the system evolves into a stable Benjamin–Feir region ($|q| < \sqrt{3/13}$). There is an elimination of bands, and a readjustment of the wavelength toward a final state in the form of a plane wave.

12.4.2. Phase Turbulence, and Turbulence Mediated by Topological Defects

We will now study the scenario for $1 + \alpha\beta < 0$ (in terms of the Benjamin–Feir instability criterion, this corresponds to the unstable band of wavevectors q of plane waves). In this regime, the system can no longer evolve into a plane wave and so the dynamics are instead very rich. As a function of parameter values of α and β, two classes of dynamic behavior can be distinguished: (i) phase turbulence, or turbulence mediated by defects (see [95]); and (ii) spatio-temporal intermittency (see [18]).

In this section, our goal is to familiarize the reader with the wide variety of behaviors which a relatively simple-seeming amplitude equation will give rise to, since (don't forget!) it is also a general equation for nonlinear systems. An exact analysis of the dynamics falls outside the scope of this work, but we would like to make some commentaries and general illustrations.

If we now study the phase diagram for the amplitude equation (see figure 12.2) in the plane (α–β), we will see that the Benjamin–Feir criterion defines a border described by equation $1 + \alpha\beta = 0$. Above this border $\alpha > -1/\beta$, the entire band corresponding to plane waves is unstable. Below it, plane waves may exist, but they coexist with more complex solutions called coherent structures, defined in a following subsection (12.4.4). In the region $\alpha > -1/\beta$, two types of dynamics are possible (see figure 12.2).

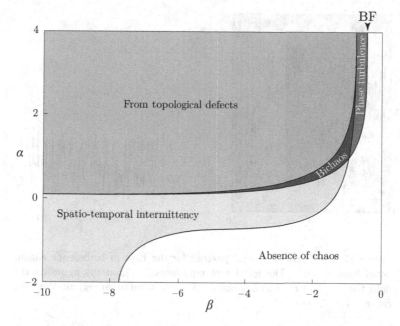

Figure 12.2 – Phase diagram for the complex amplitude equation. The curve labeled BF is the Benjamin–Feir border, below which there exist plane waves which are stable (no chaos). Other coherent structures coexist with these waves (see text). Below the BF curve, the dynamics are complex. The region labeled "bichaos" is a region where phase turbulence and turbulence coming from topological defects coexist (see text). The manifestation of one or the other turbulence depends on initial conditions.

1. *Turbulence set off by defects.* In this case, the dynamics are such that the modulus of the amplitude is zero (i.e. $|A| = 0$) at points which we call *defects*; the phase of the perturbation is thus arbitrary at those points. When coming across these points, the phase may vary by a multiple of $2\pi/L$, where L is the lateral extent of the system (see the spatio-temporal portrait of this instability and the configuration of $|A|$ in the permanent regime, figure 12.3).

2. *Phase turbulence.* In this case, modulus $|A|$ of the amplitude does not approach zero and the total phase is conserved. This type of dynamics happens at the Benjamin–Feir frontier of instability, and goes by the same name (see figure 12.2). In this region, the most important variable is the phase of the structure, since the amplitude is slaved to the phase (see the spatio-temporal portrait showing this instability, and also the configuration of the amplitude's modulus $|A|$ in the permanent regime, figure 12.4). We can show that the phase θ of the complex amplitude A

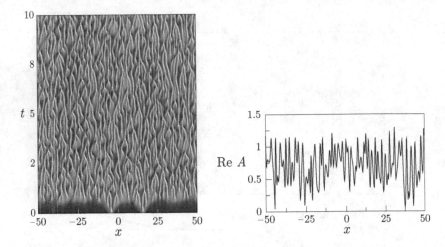

Figure 12.3 – Spatio-temporal portrait for the case of turbulence engendered from defects. The left figure represents $|A|$, and the figure on the right the real part of A as a function of x in a permanent regime. We have chosen $\alpha = 2$, $\beta = -2$.

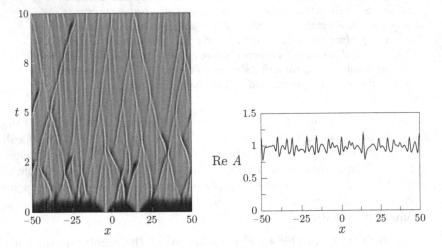

Figure 12.4 – Spatio-temporal portrait represented by $|A|$ (on the left) and the real part of A (on the right) in the case of phase turbulence. We have chosen $\alpha = 2$, $\beta = -1$.

is described by a famous equation called the Kuramoto–Sivashinsky equation (see solved problem 12.5.1)[8].

12.4.3. Spatio-temporal Intermittency

In the regime of stable plane waves, found in the region beyond the Benjamin–Feir frontier in the phase diagram (figure 12.2), most initial conditions bring the system to evolve toward a plane wave solution. However, when the initial conditions are localized and have large amplitudes, the system evolves into another type of solution, called *spatio-temporal intermittency* (see figure 12.5). As the initial conditions suggest, spatio-temporal intermittency is characterized by localized modes (the dark regions in the spatio-temporal portrait in figure 12.5), like deep *holes*, and they evolve in a complex way, carving out regions of different amplitudes. The localized modes are visualized by dark zones in the form of a half-open umbrella (figure 12.5) "interrupting" the subtle and complex movement (see the clearer spots in figure 12.5), the reason for which the term *intermittency* is attributed to the solution.

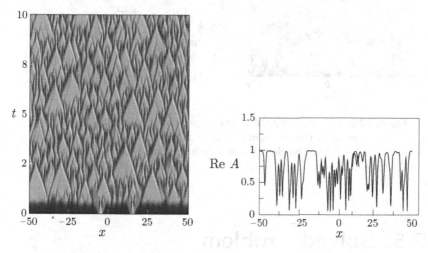

Figure 12.5 – Spatio-temporal portrait represented by $|A|$ (on the left) and the real part of A as a function of x (on the right) showing an example of spatio-temporal intermittency. We have chosen $\alpha = 0$, $\beta = -3$.

---

8. The Kuramoto–Sivashinsky equation is: $\theta_t + \nabla^4\theta + \nabla^2\theta + |\nabla\theta|^2/2 = 0$, where ∇^4 is the biharmonic operator, ∇^2 is the Laplacian operator, and ∇ is the gradient operator. The unidimensional Kuramoto–Sivashinsky equation was introduced by Kuramoto in his studies of phase turbulence in a diffusion-reaction system (the Belousov–Zhabotinsky reaction, see [57]); Sivashinsky derived it independently in his research on thermo-diffusive instabilities, see [97]. The Kuramoto–Sivashinsky equation is used for studying connections between microscopic and macroscopic scales in many physical contexts.

12.4.4. Bekki–Nozaki Holes

Besides the plane waves associated with the stable Benjamin–Feir region, a variety of alternate solutions is possible. The most well-known of these are called *Bekki–Nozaki holes* (see [7]). These solutions correspond to deep "holes" in the amplitude (see figure 12.6), and belong to a larger family of solutions called *coherent structures*. These solutions consist of fixed spatial profiles which can vary via propagation or oscillation; other coherent structures include sinks, sources, and fronts (for a review, see [106]).

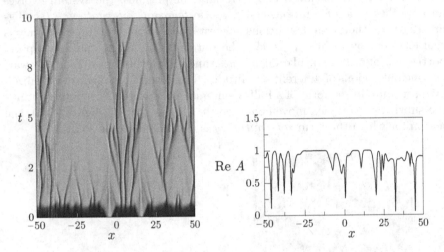

Figure 12.6 – Spatio-temporal portrait represented by $|A|$ (on the left) and the real part of A (on the right). In this case the amplitude remains fairly constant but has some deep holes, which are called Bekki–Nozaki holes. We have chosen $\alpha = 0$, $\beta = 1.5$.

12.5. Solved Problem

12.5.1. Kuramoto–Sivashinsky Equation in the Neighborhood of the Benjamin–Feir Instability

Show that in the neighborhood of the Benjamin–Feir instability, the phase obeys the Kuramoto–Sivashinsky equation.

Solution. The Benjamin–Feir condition is given by the criterion (12.26). In the neighborhood of the instability, quantity $1 + \alpha\beta$ is close to zero.

We introduce a small parameter ϵ such that:

$$-\epsilon^2 = 1 + \alpha\beta \quad \text{or} \quad \beta = -\left(\frac{1}{\alpha} + \frac{\epsilon^2}{\alpha}\right). \tag{12.29}$$

To simplify the analysis, we will limit ourselves to the case where the wavenumber is zero (i.e. $q = 0$)[9]. Remember that according to the condition on the evolution of the instabilities (equation (12.27)) and for the case where the Benjamin–Feir condition is satisfied, the entirety of the band with wavevectors q gives rise to plane waves which are unstable, including the mode corresponding to the zero valued wavevector $q = 0$. Furthermore, since the band of the perturbation wavevectors, ΔQ, corresponding to unstable modes (see equation (12.24)), are of order $\Delta Q \sim \sqrt{|1 + \alpha\beta|} \sim \epsilon$, we can deduce that the growth rate of the instability is of the order $\sigma_+ \sim \epsilon^4$ (see equation (12.24)). If we introduce the two slow variables X and T we get:

$$X = \epsilon x, \quad T = \epsilon^4 t. \tag{12.30}$$

We also know that the phase of the perfect plane wave is given by $\theta_0 = qx + \omega t$ and that $\omega = -\beta - (\alpha - \beta)q^2$ (see subsection 12.3.1). For $q = 0$, we have $\omega = -\beta$, and the phase is equal to $\theta_0 = -\beta t$.

Remember that our aim is to describe the spatio-temporal modulation of the structure; the wavevector becomes a local quantity. Denoting the total phase by θ, we make the following expansion:

$$\theta = -\beta t + \epsilon^2 \psi(X, T), \tag{12.31}$$

where the slow phase ψ depends on the slow variables X and T. We make this choice because the local wavevector is given by

$$Q(X, T) = \partial\theta/\partial x = \epsilon^{-1}\partial\theta/\partial X = \epsilon\partial\psi/\partial X \sim O(\epsilon).$$

Thus, the wavevector of the modulated structure is of order ϵ, in agreement with the fact that the wavenumber is such that $Q \simeq \epsilon$ in the unstable band. Like the phase, the amplitude of A is a function of the slow variable, and is written as:

$$R = R(X, T). \tag{12.32}$$

In terms of the new variables θ and R, and without approximation, the system (12.20) takes the following form:

$$\epsilon^4 R_T = R + \epsilon^2 \left[R_{XX} - R\psi_X^2 - \alpha\epsilon^2\left(R\psi_{XX} + 2\psi_X R_X\right)\right] - R^3$$

$$\left(\frac{1}{\alpha} + \frac{\epsilon^2}{\alpha}\right) R + \epsilon^6 R\psi_T = \epsilon^2 \left[\epsilon^2\left(R\psi_{XX} + 2\psi_X R_X\right) + \alpha(R_{XX} - \epsilon^4 R\psi_X^2)\right]$$

$$+ \left(\frac{1}{\alpha} + \frac{\epsilon^2}{\alpha}\right) R^3. \tag{12.33}$$

---

9. For other cases ($q \neq 0$), the same analysis can be done, giving us just a slightly more sophisticated result (see [47, 59]).

Now we look for a solution in powers of the small parameter ϵ. Given that the power of ϵ is even in the system of equations (12.33), we will look for a solution that is a power series of the following form:

$$R = R_0 + \epsilon^2 R_2 + \epsilon^4 R_4 + \epsilon^6 R_6 + \cdots, \tag{12.34}$$

$$\psi = \psi_0 + \epsilon^2 \psi_2 + \epsilon^4 \psi_4 + \epsilon^6 \psi_6 + \cdots. \tag{12.35}$$

We now carry this solution into system (12.33) and identify the orders of ϵ.

To order ϵ^0, the two equations of the system (12.33) give: $R_0(1 - R_0^2) = 0$, that is $R_0 = 0$ and $R_0 = 1$. The second choice is the interesting one for us.

To order ϵ^2, we find the condition $R_2(1 - 3R_0^2) = 0$, which implies $R_2 = 0$.

To order ϵ^4, we obtain, from the first equation of system (12.33), the following:

$$R_6 = -\frac{\alpha}{4}\psi_{0XXXX} - \frac{1}{2}\psi_{0X}^2 - \frac{\alpha}{2}\psi_{2XX}. \tag{12.36}$$

Carrying this into the second equation of the same system (ψ_2 cancels automatically):

$$\psi_{0T} = -\psi_{0XX} - \frac{1}{2}\left(1 + \alpha^2\right)\psi_{0XXXX} - \left(\alpha + \alpha^{-1}\right)\psi_{0X}^2. \tag{12.37}$$

This equation is known as the Kuramoto–Sivashinsky equation and was derived in many non-equilibrium contexts (see [74] for an overview). It is known for engendering spatio-temporal chaos; this state is sometimes called phase chaos because it is the phase which is the main variable in the full complex amplitude equation.

12.6. Exercises

12.1

We have seen in this chapter the amplitude equation for a Hopf bifurcation. We have restricted ourselves to the case where the modes propagate in a given direction (say the positive x direction) with a complex amplitude denoted by A. In reality, propagation can take place in both directions. The goal of this exercise is study the situation where a right wave with amplitude A coexists with a left wave with amplitude B.

1. Explain why the coupled equations of the right and left waves should have (to leading order) the following form:

$$A_t = \epsilon A - (1 + i\alpha)|A|^2 A - \beta(1 + i\gamma)|B|^2 A$$

$$B_t = \epsilon B - (1 + i\alpha)|B|^2 B - \beta(1 + i\gamma)|A|^2 B, \tag{12.38}$$

where ϵ, α and β and γ are real numbers.

Let us consider the following two possibilities: (i) propagative modes where $A \neq 0$ and $B = 0$; (ii) stationary waves with $A = B$.

2. Find the stability conditions for the propagative wave solution.

3. Find the stability conditions for the standing wave solution.

12.2

We have seen in subsection 12.5.1 that for a Hopf bifurcation, the phase ψ obeys a Kuramoto–Sivashinsky equation (equation (12.37)):

$$\psi_t = -\psi_{xx} - \psi_{xxxx} - \psi_x^2, \qquad (12.39)$$

where we have set all coefficients here to unity (which is always possible thanks to an adequate change of variables). We have also seen that when the bifurcation is stationary (see equation (10.60)), the phase obeys a different equation:

$$\psi_t = -\psi_{xx} - \psi_{xxxx} + \psi_x\psi_{xx}. \qquad (12.40)$$

Explain the form of the equations obeyed by the phase in each case on the basis of symmetries. Hint: write the mode in the linear regime as $u = Ae^{iq_c x} + c.c.$ (for the stationary bifurcation and for a Hopf bifurcation (see equation (12.8))), and decompose A in a modulus and a phase. By considering the form of the linear mode, describe which symmetries should leave the form of the evolution equation invariant.

12.3

Extend the form of the system of equations (12.38) to next order in amplitude and justify your answer.

12.4

We consider the following equation:

$$u_{tt} = u_{xx} - \sin(u), \qquad (12.41)$$

known as the sine-Gordon equation.

1. Look for a solution in the form of $u = ae^{-iqx+i\omega t} + c.c.$, and by linearizing equation (12.41), determine the dispersion relation $\omega(q)$. Does this equation exhibit an instability? Or do we have a wave? Is the medium dispersive? Justify your answer.

2. Beyond the linear regime the solution of the form $\sim e^{iqx+i\omega t}$ ceases, in general, to be valid. In the nonlinear regime the solution can be written as the product of this form times a slowly varying envelope A. The hypothesis

of a slowly varying amplitude means that in Fourier space the amplitude of the wavepacket has a maximum around a wavevector q_0 (the wavepacket looks Gaussian in shape). Write ω as a function of q around $q = q_0$ to leading order.

3. Justify that the solution of the wavepacket can be written as $A(X, T_1, T_2)e^{i(-q_0 x + \omega(q_0)t)}$, where $X = \epsilon x$ and $T_1 = \epsilon t$, $T_2 = \epsilon^2 t$, with $\epsilon \sim q - q_0$.

4. By adopting the method of multiscales, transform the operators ∂_x and ∂_t in terms of the slow variables.

5. Rewrite equation (12.41) in terms of multiscales.

6. We seek solutions in the form $u = \epsilon u_1 + \epsilon^2 u_2 + \cdots$. Show that the solution u_1 is given by:

$$u_1 = A(X, T_1, T_2)e^{i(-q_0 x + \omega(q_0)t)} + c.c. \tag{12.42}$$

7. Show that to some order in ϵ (to be specified) we must impose a solvability condition (which is equivalent to eliminating secular terms) that amounts to setting:

$$A_{T_1} = U_0 A_X. \tag{12.43}$$

Provide the expression for U_0 and provide an interpretation for this equation. What is the general solution of this equation?

8. Show that to next order in ϵ, the solvability condition yields:

$$A_{T_2} = i\alpha A_{XX} - i\beta|A|^2 A. \tag{12.44}$$

Provide the expressions of α and β (these coefficients can be real or complex). This equation is known as the nonlinear Schrödinger equation.

Chapter 13

Two Dimensional Patterns

Abstract *This chapter introduces two dimensional patterns, such as honeycombs (hexagonal patterns) and square structures. By use of symmetry, we will write down the equations describing systems in the neighborhood of the instability threshold giving rise to these patterns. The equations are universal. We will explain why most ordered patterns found in nature have a honeycomb or hexagonal shape, and discuss the stability of different structures.*

Up until now, we have looked at one dimensional structures and have seen that order can exist in the form of bands. The dynamics in the neighborhood of the instability threshold is described by a universal amplitude equation, which we first saw when describing the emergence of one dimensional spatial order in non-equilibrium systems (see chapter 10). In this chapter, we will look at spatial structures that are two dimensional, studying their dynamics and the emergence of two dimensional order. We will only look at instabilities associated with stationary bifurcations, that is, we will focus on systems evolving from one stationary state toward another, and not, for example, the case of a Hopf bifurcation, in which there is a transition from a stationary to an oscillating solution (as described in chapters 5, 6 and 12).

13.1. Two Dimensional Order

As we have seen throughout the book, an important step in the study of nonlinear phenomena is the analysis of linear stability for homogeneous and stationary solutions. In the case of one dimensional systems, we established that the form of a perturbation in the neighborhood of such a stationary and homogeneous solution is that of a plane wave: $e^{\omega t + iqx}$ where q is the wavevector and ω is the growth or attenuation rate. When the real component of ω is positive (i.e. $\Re(\omega) > 0$), any initial perturbation grows with time: the solution is thus unstable (see chapter 9). The dispersion relation $\omega(q)$ is obtained from the linear stability analysis.

In the two dimensional case, the wavevector has two components q_x and q_y in the x and y directions of the plane. If the plane (Oxy) does not show a preference for any direction, we can assume that the dispersion relation will not depend on q_x and q_y independently of each other, but rather on the modulus of wavenumber $|q|$. For example, take the nonlinear model of a 2-D version of the isotropic (i.e. with no privileged spatial direction) Swift–Hohenberg equation (10.2). We replace the second derivative with respect to x with the Laplacian ∇^2 in two dimensions (i.e. $\nabla^2 = \partial_{xx} + \partial_{yy}$) to obtain the bidimensional Swift–Hohenberg equation:

$$\frac{\partial \phi}{\partial t} = \left[\epsilon - (1 + \nabla^2)^2\right]\phi + \alpha\phi^2 - \phi^3. \tag{13.1}$$

$\phi = 0$ is a trivial solution of the bidimensional Swift–Hohenberg equation (13.1). To begin our study of the linear stability, we superpose a small perturbation ϕ_1 and look for a solution like:

$$\phi_1 = \nu e^{\omega t + iq_x x + iq_y y} \equiv \nu e^{\omega t + i\boldsymbol{q}\cdot\boldsymbol{r}}, \tag{13.2}$$

where ν is a small parameter and $\boldsymbol{r} = (x, y)$ is the position vector. Substituting this into equation (13.1), we find the following dispersion relation:

$$\omega = \epsilon - (q^2 - 1)^2, \tag{13.3}$$

where we have posited $q = |\boldsymbol{q}|$ and ignored terms nonlinear in ν.

When $\epsilon = 0$, the instability mode associated with the critical wavevector (i.e. $q \equiv q_c = 1$) is neutral: ω is zero. The other modes associated to other wavevectors (i.e. $q \neq q_c$) are stable: ω is negative. The critical condition is thus characterized by the pair $(\epsilon_c, q_c) = (0, 1)$. When $\epsilon > 0$, the plane wave (q_x, q_y) has a ring of unstable modes of thickness $|q_+ - q_-|$. If ϵ is sufficiently small, we can write $|q_+ - q_-| \equiv \Delta q \simeq \sqrt{\epsilon}/2$ (see figure 13.1). Because $\omega > 0$ for any q belonging the ring, we have an exponential amplification of the instability, and can no longer ignore the nonlinear terms. To determine the two dimensional amplitude equation, we can follow the same approach as we used in describing the emergence of unidimensional spatial order (chapter 10). We will present the results below, inviting the reader to see details of the calculation in the worked out example of a bidimensional amplitude equation for two different symmetries (in subsections 13.5.1 and 13.5.2).

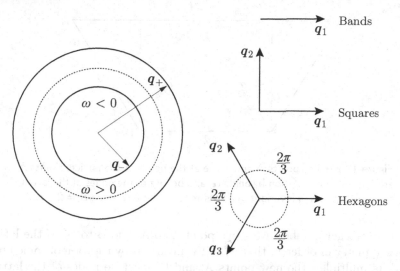

Figure 13.1 – All the modes in which the wavevector is within the ring (left) in thick lines are linearly unstable. We can represent all solutions as a linear combination of wavevectors found in the unstable band (inside the ring). The dotted line corresponds to the critical circle of radius q_c. On the right, we show three possible combinations of wavevectors giving birth to ordered structures. We can have a banded structure (really this corresponds to the case of a unidimensional solution, but we can study its stability vis-à-vis perturbations in the plane; see further ahead), and square and hexagonal structures. Other ordered structures are possible (see text).

13.1.1. Different Types of Two Dimensional Spatial Order

Unlike the unidimensional problem for which only one kind of order exists (that of a periodic solution along Ox), a few different kinds of spatial order are possible in bidimensional problems. The most common are: (i) bands (which are like the rolls in the case of Rayleigh–Bénard convection); (ii) squarely symmetrical order (also known as square lattice); (iii) hexagonally symmetrical order. We will further differentiate the different types of order through some general considerations.

The existence of order means there is an invariance of the pattern if we displace ourselves by translation by a certain distance; the minimum distance at which the pattern repeats is equal to the pattern's wavelength. The pattern's condition of invariance by translation restricts the number of possible ordered solutions. There exist a limited number of possible types of distinct lattices. The lattices each have a particular symmetry; first, let us show how we find the different lattices, designating the smallest possible period of the lattice by λ. Take two points A and B, found at the nodes of the lattice, separated by a length λ (see figure 13.2). If we now perform a rotation of angle $\alpha = 2\pi/n$,

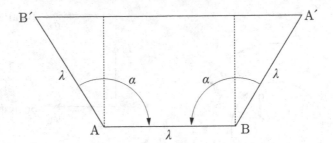

Figure 13.2 – The points A and B are at the nodes of the lattice. Points A′ and B′, found by rotation by angle α, also belong to the nodes of the lattice, since the system is invariant vis-à-vis rotation by any angle $\alpha = 2\pi/n$.

with n an integer number, we move point A to A′ and B to B′. If the lattice has a symmetry of order n, that is, if it is invariant by rotation of some angle α and its multiples, the new points A′ and B′ must be nodes of the lattice. This condition is only satisfied if the two points are some distance $p\lambda$ apart, where p is an integer. The distance between A′ and B′, called $p\lambda$, is also given by the following geometric relation:

$$p\lambda = \lambda + 2\lambda \sin(\alpha - \pi/2) = \lambda - 2\lambda \cos(\alpha). \tag{13.4}$$

This result, easily inferred from figure 13.2, imposes the following: $\cos\alpha = (1-p)/2$; and since the absolute value of the function cosine is always less than or equal to one (i.e. $|\cos(\alpha)| \leqslant 1$), the integer p can only take on the following values: 3, 2, 1 or 0. These values correspond to rotations of $\alpha = 2\pi/n$, where $n = 2, 3, 4$, and 6. The axis around which a rotation of order n leads to an equivalent configuration is called an axis of order n. One of the practical consequences of this mathematical result is that it clearly shows that it is impossible to continuously tile a flat floor with tiles whose basic unit is a five- or seven-sided figure.

On the other hand, a continuous layer in a plane tiled by elementary tiles composed of 6, 4, 3, or 2 sides is possible. An axis of order 6 corresponds to a hexagonal elementary tile, order 4 to a square symmetry, order 3 to a triangular symmetry, order 2, rectangular symmetry.

Having determined which rotation symmetries leave the ordered pattern invariant, we are in a position to present the so-called Bravais lattices. An ordered lattice is a collection of points (called nodes) in which the neighborhood of each point is the same as the neighborhood of every other point under translation. In other words, the pattern consists of a repeated array of points with an arrangement and orientation that appears exactly the same from whichever point the array is viewed. Of course these lattices must comply with the four types of rotation symmetries evoked above. One can easily guess that there must be at least four different types of lattices.

We can in fact classify five different lattice types (see figure 13.3). Figure 13.4 shows examples of periodic patterns. The ordered lattices are often called Bravais lattices[1]. In this book, we will not examine each symmetry, focusing instead on the two most common bidimensional structures[2]: the hexagonal lattice, the most abundant structure in nature, and the square lattice. To whet our appetite, we will begin by giving a simple explanation for the abundance of hexagonal structures in nature.

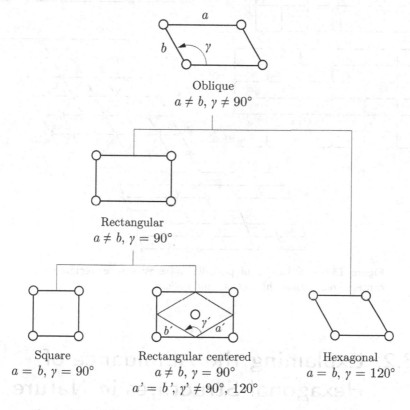

Figure 13.3 – The five types of lattices possible in two dimensions.

---

1. Named after Auguste Bravais (1811–1863), known for his work in crystallography.
2. Note that the band structure, also common, is a unidimensional structure!

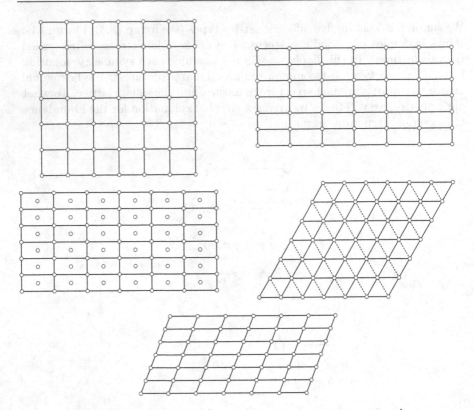

Figure 13.4 – Examples of periodic patterns: square, rectangular, centered rectangular, hexagonal, and oblique.

13.2. Explaining the Abundance of Hexagonal Structures in Nature

Hexagonal patterns, also called honeycombs, are constantly seen in our natural environment. A few examples are: (i) sedimentation figures (see figure 13.5); (ii) soil or other surfaces when exposed to rain and then to heavy evaporation (fractures caused by thermal constraints); (iii) the surface of liquid heated from below (see figure 1.6). The frequent appearance of this pattern is not arbitrary; it is linked to a natural nonlinear mechanism that we can understand from a few general considerations[3].

---

3. The following argument does not depend on any dogma or principles, like that of maximum compactness, but rather on a natural nonlinear mechanism.

Figure 13.5 — Top: basalt columns. [© forcdan #28975739 Fotolia]
Bottom: fractured clay soil. [© Sahara Nature #3863433 Fotolia]

13.2.1. Origin of Hexagonal Patterns

We have already seen a few times throughout this book, that eigenmodes in a linear regime are Fourier modes selected from a Fourier series decomposition containing all harmonics.

Though these modes do not couple in the linear regime, the same is not true in the presence of nonlinearities. Suppose that a Fourier mode emerges due to an instability. We can write it as $A_1 e^{i q_1 \cdot r}$. Once the instability is reached, the nonlinear terms must be taken into account, the first natural leading nonlinearity being quadratic. The mode[4] $A_1 e^{i q_1 \cdot r}$ will couple to other modes because of the nonlinearities.

---

4. At the instability threshold, the module with wavenumber q_1 (i.e. $|q_1|$) describes a circle of radius q_c, see figure 13.1.

For a square pattern, the basic solution in the linear regime is written $A_1 e^{i q_1 \cdot r} + A_2 e^{i q_2 \cdot r}$ with q_2 orthogonal to q_1 (the two wavevectors have the same amplitude). For a hexagonal order, the solution is of the form $A_1 e^{i q_1 \cdot r} + A_2 e^{i q_2 \cdot r} + A_3 e^{i q_3 \cdot r}$ with $q_1 + q_2 + q_3 = 0$, etc.

Generally, all modes of the form $A_n e^{i q_n \cdot r}$ (with integer n and $|q_n| = q_c$) are possible solutions. The emergence of order consists of a particular combination of these basic modes (such as indicated above for the square and hexagonal cases). In the initial quasi-linear regime, and in the most general case, the solution is a superposition of basic modes (for example, two modes for squares, three for hexagons), and of their higher harmonics with wavevectors of moduli $2q_c, 3q_c, \ldots$ In the neighborhood of the instability's threshold, only the principal mode with vector of norm q_c is unstable, and all the higher order modes are stable.

The terms which couple to the initial mode $A_1 e^{i q_1 \cdot r}$ are those in which the wavevector is almost equal to the initial mode's wavevector, q_1. The proximity of wavevectors ensures a dominant coupling, or resonance. In the nonlinear regime, the different Fourier modes appear as products of modes. Each product of two Fourier modes (in the case of quadratic nonlinearities) corresponds to the addition of their wavevectors; for example, coupling between modes "2" and "3" gives the following mode: $A_2 e^{i q_2 \cdot r} A_3 e^{i q_3 \cdot r} = A_2 A_3 e^{i (q_2 + q_3) \cdot r}$. We investigate further all the quadratic combinations which can happen in the presence of three different modes with amplitudes A_1, A_2, and A_3:

$$A_1^2, A_1^{*2}, |A_1|^2, A_1 A_2, A_1 A_2^*, A_1^* A_2, A_1^* A_2^*, A_1 A_3, A_1 A_3^*, A_1^* A_3, A_1^* A_3^*,$$
$$A_2^2, A_2^{*2}, |A_2|^2, A_2 A_3, A_2^* A_3, A_2 A_3^*, A_2^* A_3^*,$$
$$A_3^2, A_3^{*2}, |A_3|^2. \tag{13.5}$$

The quadratic combinations (equation (13.5)) are associated with the following wavevectors:

$$2q_1, -2q_1, 0, q_1 + q_2, q_1 - q_2, q_2 - q_1, -q_1 - q_2,$$
$$q_1 + q_3, q_1 - q_3, q_3 - q_1, -q_1 - q_3,$$
$$2q_2, -2q_2, 0, q_2 + q_3, q_3 - q_2, q_2 - q_3, -q_2 - q_3,$$
$$2q_3, -2q_3, 0. \tag{13.6}$$

For resonance, it is required that each combination be equal to q_1. This immediately rules out many possibilities if we remember that the vectors q_1, $q_1 + q_3$ must have the same norm (i.e. non-zero). Only the last four combinations in the second line of equation (13.6) are potential candidates; all the others either yield zero wavevectors, or do not lead to the fact that all the wavevectors have the same norm. Equating these combinations to q_1, we are thus left with the following possibilities: $q_2 + q_3 - q_1 = 0$, $q_1 + q_2 - q_3 = 0$, $q_1 + q_3 - q_2 = 0$ and $q_1 + q_2 + q_3 = 0$. The first two relations are geometrically represented in figure 13.6. Even though the two combinations seem triangular, any simple rotation by an angle of order $n = 2, 3, 4, 6$ (the only angles allowed for bidimensional order) will never bring the generated vector base to coincide with the initial base (see figure 13.6). The same conclusion obviously holds for the third equation.

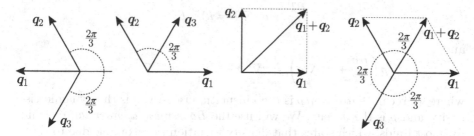

Figure 13.6 – Wavevectors satisfying $q_2 + q_3 - q_1 = 0$ and $q_1 + q_2 - q_3 = 0$ (the two schemas on the left). No rotation by an angle of order $n = 2, 3, 4, 6$ can bring the base back to the same location. These combinations are thus not allowed. The third schema illustrated shows square order. A quadratic nonlinearity amounts to adding two wavevectors (see text). However, for a square pattern, $q_1 + q_2$ cannot coincide with q_1, and therefore a quadratic nonlinearity does not occur in a square symmetry. On the right we show a hexagonal configuration where the sum of two vectors is equal to the opposite of the third. This would mean that the amplitude of a mode (say A_1) is coupled to the complex conjugate of the product of the other two.

The only possibility left is that giving rise to $q_1 + q_2 + q_3 = 0$ (arising from equating $-q_2 - q_3$ to q_1). In other words, the only nonlinearity which couples to the dominant order in the dynamics of A_1 is:

$$A_2^* A_3^*. \tag{13.7}$$

We can also verify geometrically that only the hexagonal order can satisfy a resonance if we limit ourselves to quadratic nonlinearities. For example, for a square symmetry, no combination of two wavevectors can create a quadratic resonance (see figure 13.6).

13.2.2. An Exception: Rayleigh–Bénard Convection

There are some exceptions to the general rule of hexagonal dominance in nature. Sometimes, due to the particular intrinsic symmetries of a system, quadratic terms are not allowed in the amplitude equation. Consequently, we cannot carry out the same reasoning in which the quadratic nonlinearity is dominant (subsection 13.2.1) and brings about hexagonal structures. One famous example of this kind of system is Rayleigh–Bénard convection, which we studied in chapter 9 when describing the emergence of unidimensional order.

Though not a priori obvious, given that the equations of thermoconvection contain quadratic nonlinearities such as the $\boldsymbol{v} \cdot \nabla \boldsymbol{v}$ term (see equation (9.53)), a hidden symmetry in the equations forbids the emergence of hexagons. To prove it, we can rewrite the equations of thermal convection as:

$$\frac{\partial T}{\partial t} + \boldsymbol{v} \cdot \nabla T = \kappa \nabla^2 T$$

and

$$\rho \left(\frac{\partial \boldsymbol{v}}{\partial t} + \boldsymbol{v} \cdot \nabla \boldsymbol{v} \right) = \nu \nabla^2 \boldsymbol{v} - \nabla p + \rho \boldsymbol{g}, \quad \nabla \cdot \boldsymbol{v} = 0,$$

where we recall that $\nu = \eta / \rho$ is the kinematic viscosity, η is the dynamic viscosity, and ρ is the density. We will use the *Boussinesq approximation*, valid for most fluids, which states that density variation (for example due to thermal gradient) is sufficiently small to be neglected, except where it appears in the term multiplied by $\boldsymbol{g}$, the acceleration due to gravity. Accordingly, we can say that everywhere $\rho = \rho_0$, except in front of term $\boldsymbol{g}$ where we can define $\rho = \rho_0(1 - \alpha(T - T\downarrow))$, with α the thermal expansion coefficient (see chapter 9), and $T\downarrow$ the temperature of the bottom plate. Defining $\theta = T - T_0(z)$, where $T_0(z)$ is the temperature for the purely conductive state, we can rewrite the equations above as follows:

$$\frac{\partial \theta}{\partial t} + \boldsymbol{v} \cdot \nabla \theta = \kappa \nabla^2 \theta$$

and

$$\rho_0 \left(\frac{\partial \boldsymbol{v}}{\partial t} + \boldsymbol{v} \cdot \nabla \boldsymbol{v} \right) = \nu \nabla^2 \boldsymbol{v} - \nabla p + \rho_0 g \hat{\boldsymbol{z}} \theta, \quad \nabla \cdot \boldsymbol{v} = 0,$$

where $\hat{\boldsymbol{z}}$ is the unitary vector directed along the vertical axis Oz. Remember that a Rayleigh–Bénard cell goes from $z = 0$ (bottom plate) to $z = d$ (upper plate). Note that θ (such as defined above) satisfies the boundary conditions $\theta(x, y, z = 0, t) = \theta(x, y, z = d, t) = 0$, and since the two surfaces are identical, the boundary conditions for temperature and velocity are the same on each plate. Each quantity (temperature, velocity, and pressure) found in the above equation is, in general, a function of time t and of the three spatial variables (x, y, z). We can easily see that the system of equations for thermal convection above (including the boundary conditions) is invariant vis-à-vis the following symmetry operation:

$$(v_x, v_y, v_z, \theta, p)(t, x, y, z) \longrightarrow (v_x, v_y, -v_z, \theta, p)(t, x, y, d - z), \qquad (13.8)$$

where v_x is the x component of velocity vector $\boldsymbol{v}$, etc.

This symmetry, which flips the top with the bottom, is called "up-down symmetry". In the Boussinesq approximation, the thermoconvection equations have up-down symmetry. This is the reason why hexagonal patterns cannot take place. Indeed, as seen in the previous section (and as will be seen below), hexagons are associated with a quadratic nonlinearity (for example, the amplitude equation has a term that looks like $A_t \sim A^2$) in the amplitude equation which breaks the up-down symmetry. In changing the amplitude from A to $-A$, the equation does not remain unchanged. Note finally, that in the cases in which the Boussinesq approximation stops being valid, hexagonal structures may appear.

13.3. General Form of the Amplitude Equation in Two Spatial Dimensions

13.3.1. Hexagonal Symmetry

In this section, we derive the amplitude equation from considerations of symmetry. The direct derivation of this equation is shown in the worked out problem starting with the two dimensional Swift–Hohenberg model (see solved problems 13.5.1 and 13.5.2).

Let us begin with hexagonal symmetry. It is common to introduce three unitary base vectors, for example a_1, a_2, and a_3 with an angle of $2\pi/3$ between two of them. Note that q_c is the module of the critical wavevector and that q_1, q_2, and q_3, the three vectors oriented along the base vectors a_1, a_2, and a_3 such that $|q_i| = q_c$ (see figure 13.1), satisfy the following relation:

$$q_1 + q_2 + q_3 = 0. \tag{13.9}$$

Writing A_i ($i = 1, 2, 3$) to describe the amplitude of the perturbation in direction i, we can describe the system with the physical field ϕ of interest (which can be any field representing the system under study, for example the concentration in a chemical reaction):

$$\phi(r) = \sum_{n=1}^{3} A_n e^{iq_n \cdot r} + A_n^* e^{-iq_n \cdot r}. \tag{13.10}$$

When the amplitude is the same in each direction (i.e. $A_1 = A_2 = A_3$) this is the description of a hexagonal structure; the field ϕ can then be written as:

$$\phi(r) = 2A \left(\cos(q_c x) + \cos\left(\frac{q_c}{\sqrt{3}}(x - y) \right) + \cos\left(\frac{q_c}{\sqrt{3}}(x + y) \right) \right). \tag{13.11}$$

For what follows, we will use the general case description where the values of amplitude A_i are all different[5], to keep our analysis general. To cubic order[6], the amplitude equation has the following form:

$$\dot{A}_1 = \epsilon A_1 - \left[|A_1|^2 + \nu(|A_2|^2 + |A_3|^2)\right] A_1 + \gamma A_2^* A_3^*, \qquad (13.12)$$

$$\dot{A}_2 = \epsilon A_2 - \left[|A_2|^2 + \nu(|A_3|^2 + |A_1|^2)\right] A_2 + \gamma A_3^* A_1^*, \qquad (13.13)$$

$$\dot{A}_3 = \epsilon A_3 - \left[|A_3|^2 + \nu(|A_1|^2 + |A_2|^2)\right] A_3 + \gamma A_1^* A_2^*, \qquad (13.14)$$

where ϵ is a small parameter, γ a parameter that we choose as positive[7], and ν a parameter which may be positive or negative. These equations, besides the cubic nonlinearities which we have seen in one dimension, contain a quadratic nonlinearity having the form discussed in subsection 13.2.1 arising from the condition of resonance. Invariance by rotation of an angle $2\pi/3$ imposes that each equation is obtained from the previous one via circular permutation. Below, in this section, we provide symmetry arguments supporting the form of equations (13.12)–(13.14).

The above equations are general, and contain special limits: bands (if we set all amplitudes except one to zero), and squares if we set $\gamma = 0$ and set $A_3 = 0$. We will see (subsection 13.4.4) that square structures exist only if $\nu + 1 > 0$. Since our goal will be to compare the stability of bands, hexagonal, and square patterns, we will pick $\nu + 1 > 0$. The negative sign in front of the cubic terms is there to guarantee a nonlinear saturation of the instability to cubic order[8]. When parameter ϵ is negative, the homogeneous solution of zero amplitude (i.e. $A = 0$) is linearly stable; it loses its stability when ϵ becomes positive.

In chapter 10, dedicated to the universality at the threshold for emergence of order, we found the form for the unidimensional amplitude equation by considering the symmetry of the system (see section 10.2). Here, taking A_2 and A_3 to zero (i.e. $A_2 = A_3 = 0$), the amplitude equation of A_1 (equation (13.12)) becomes identical to the amplitude equation (10.24) obtained in one dimension. Note that the quadratic terms disappear in one dimension. Invariance by translation (or more exactly, invariance vis-à-vis the choice of origin) requires that the following operation leaves the equation unchanged:

---

5. Solved problems 13.5.1 and 13.5.2 show that the derivation of the amplitude equation does not need to have any statement about the equality of the three amplitudes. This is why we prefer to work in the most general case possible, which allows us, for example, to study the stability of ordered structures by adding a different perturbation to each of the amplitudes.

6. See also solved problems 13.5.1 and 13.5.2.

7. Changing the sign of γ is akin to changing A to $-A$, which would give us identical results.

8. Remember here that the sign of the cubic term is not systematically negative, see solved problem 3.9.2.

$r \rightarrow r + r_0$, where r_0 is an arbitrary constant. This operation is equivalent to making the following operation on the amplitude:

$$A_n \longrightarrow A_n e^{i q_n \cdot r_0}. \tag{13.15}$$

This condition of invariance must be satisfied for all terms in the three equations (13.12)–(13.14). Consider, for example, the amplitude equation describing A_1 (equation (13.12)). The translation invariance (condition (13.15)) is obvious for the linear term and the cubic terms. Indeed, each of these terms is simply changed by a factor $e^{i q_1 \cdot r_0}$. In the bidimensional case, we have the newly permitted quadratic term (forbidden in the unidimensional case, see subsection 10.2.1), which transforms under translation like:

$$A_2^* A_3^* \longrightarrow A_2^* A_3^* e^{-i(q_2 + q_3) \cdot r_0} = A_2^* A_3^* e^{i q_1 \cdot r_0}, \tag{13.16}$$

where we used the relation between the three wavevectors, equation (13.9). In other words, the quandratic term transforms like a linear term and, as such, is allowed by symmetry. This applies for the other two amplitude equations (13.13) and (13.14) as well.

It is useful to specify that the quadratic term in the amplitude equation destroys the $A_n \rightarrow -A_n$ symmetry. In other words, hexagons break the "up-down" symmetry. In practice, we can find systems for which the point-of-departure model displays the up-down symmetry. This is the case, for example, for the bidimensional Swift–Hohenberg equation (13.1), when parameter α is zero (i.e. $\alpha = 0$). In this case, even if the symmetries of the amplitude equation allow for a quadratic term, the symmetry of the original equation would not allow this kind of term: thus the coefficient of the quadratic term of the amplitude equation must be zero (see solved problems 13.5.1 and 13.5.2). This also happens for Rayleigh–Bénard convection, as we have seen in the last section. Finally, note that if the bifurcation of the homogeneous state ($\phi = 0$) toward the modulated state is stationary, the coefficients ϵ, ν and γ are real (see subsection 10.2.2, where the proof used in one dimension can easily be generalized to two dimensions).

13.3.2. Square Symmetry

For square symmetry, the steady-state solution of physical field ϕ takes, to leading order, the form:

$$\phi(r) = \sum_{n=1}^{2} A_n e^{i q_n \cdot r} + A_n^* e^{-i q_n \cdot r}, \tag{13.17}$$

where A_i is the amplitude in direction i with $i = 1, 2$ being the two directions of the square structure. Note that the wavevectors q_1, q_2, perpendicular to each other, have the same modulus equal to q_c (see figure 13.1).

The amplitude equations for the square structure are obtained from the general equations (13.12)–(13.14) by setting $A_3 = 0$ (which in turn indicates $\gamma = 0$)[9]. We get:

$$\dot{A}_1 = \epsilon A_1 - \left(|A_1|^2 + \nu|A_2|^2\right) A_1, \qquad (13.18)$$

$$\dot{A}_2 = \epsilon A_2 - \left(|A_2|^2 + \nu|A_1|^2\right) A_2. \qquad (13.19)$$

In the last section, the rules of invariance via symmetry (either by rotation of an angle of $\pi/2$, or by translation) are what determine the final form of the equation. We can confirm that setting one of the two amplitudes to zero gives us back the unidimensional equation, called the band equation, or sometimes the roll equation, in reference to Bénard rolls (see subsection 9.11.2 in which we study Rayleigh–Bénard convection).

13.4. Stability of Band, Square, and Hexagonal Structures

Before going into the stability of each solution, we should specify some properties which will be useful for the analysis. Firstly, as we have seen, the general equations (13.12)–(13.14) give us squares (see subsection 13.3.2) and bands as particular limit cases. Let us begin with the simplest case, patterns in the form of bands. In this case, the associated systems of amplitude equations (equations (13.12)–(13.14)), reduce to a unidimensional equation of the form:

$$\dot{A}_1 = \epsilon A_1 - |A_1|^2 A_1, \qquad (13.20)$$

where A_1 is the only non-zero amplitude. We can easily check that the phase of the complex amplitude A_1 is constant (i.e. $\dot{\theta} = 0$) by setting $A_1(t) = R_1(t)e^{i\theta(t)}$. Owing to translational invariance, the phase is arbitrary, and can thus be taken to be equal to zero: amplitude A_1 is thus real. The amplitude equation associated with A_1 has two fixed points: $A_1 = 0$ and $A_1 = \pm\sqrt{\epsilon}$, with $\epsilon > 0$. The field associated with this solution (in the form of bands) is given by:

$$\phi = A_1(t)e^{iq_c x} + c.c., \qquad (13.21)$$

where c.c. stands for complex conjugate. Unlike the purely unidimensional case (of a unidimensional band), the space here is bidimensional and as such

---

9. It is useful to specify that the coefficients of the amplitude equation of square structures cannot be deduced from those for hexagons (for example, through setting $A_3 = 0$). Determining the coefficients must be done case by case. We recommend the same method as the one presented in solved problem 13.5.1 for hexagons, and adopting the case to a square symmetry. For now, we will use the same notations for coefficients here for simplicity.

the bands can be perturbed in two spatial directions. Remember that the bands are a special case of the hexagonal patterns. The stationary solution for bands is thus given by $R_1 = \pm\sqrt{\epsilon}$ and $R_2 = R_3 = 0$.

Rewriting the amplitude equations (13.12)–(13.14) by setting $A_n(t) = R_n(t)e^{i\theta_n(t)}$, we get:

$$\dot{R}_1 = \epsilon R_1 - \left[R_1^2 + \nu(R_2^2 + R_3^2)\right] R_1 + \gamma R_2 R_3, \qquad (13.22)$$

$$R_1\dot{\theta}_1 = -\gamma R_2 R_3 \sin(\theta_1 + \theta_2 + \theta_3). \qquad (13.23)$$

The four other equations, associated with two amplitudes and two phases, are derived from these two equations via circular permutation. By the following change of variables $\psi = (\theta_1 + \theta_2 + \theta_3)$ and using the amplitude equations associated with θ_1 (equation (13.23), as well as the analogous equations given by θ_2 and θ_3), we get the following equation:

$$\dot{\psi} = -P\sin(\psi), \qquad (13.24)$$

with quantity P such that:

$$P = \gamma \left(\frac{R_2 R_3}{R_1} + \frac{R_3 R_1}{R_2} + \frac{R_1 R_2}{R_3} \right) > 0, \qquad (13.25)$$

where γ is a parameter whose sign is arbitrarily chosen to be positive[10]. The phase equation (13.24) has the following fixed points: $\psi_0 = 0$ and $\psi_0 = \pi$. Taking $\psi = \psi_0 + \psi_1(t)$, where $\psi_1(t)$ is a perturbation, and linearizing the equation with respect to ψ_1, we get

$$\dot{\psi}_1 = -P\cos(\psi_0)\psi_1. \qquad (13.26)$$

Let us look for a solution of the type $e^{\sigma t}$ for perturbation ψ_1. Starting from the evolution equation of the perturbation ψ_1 (equation (13.26)), the growth (or attenuation) rate of the perturbation is given by $\sigma = -P\cos(\psi_0)$. When the stationary solution is the fixed point $\psi_0 = 0$, the solution is stable since σ is negative ($P > 0$, see equation (13.25)). On the other hand, when instead $\psi_0 = \pi$, σ is positive. Consequently, only the states where $\psi_0 = 0$ can be stable; we thus take that as our phase value. Note that had we chosen $\gamma < 0$, we would have had a stable solution $\psi_0 = \pi$, and choosing this value as our phase value would have led us to the same general conclusions we reach in what follows.

---

10. Recall that the sign of γ is inverted via the operation $A_n \to -A_n$ which is applied to the original set of equations (13.12)–(13.14). Thus the choice of a positive sign does not change the generality of our results.

13.4.1. Bands Stability

To perform a study of the stability of bands, we start with the solution:

$$R_1 = R_1^0, \quad R_2 = 0, \quad R_3 = 0, \tag{13.27}$$

where R_1^0 is the stationary solution $\pm\sqrt{\epsilon}$. Adding small perturbations $\delta_1(t)$, $\delta_2(t)$, and $\delta_3(t)$ to test the stability, we look for solutions of the form:

$$R_1 = R_1^0 + \delta_1(t), \quad R_2 = \delta_2(t), \quad R_3 = \delta_3(t). \tag{13.28}$$

We carry equation (13.28) into the amplitude equation for R_1 (equation (13.22)) as well as into R_2 and R_3 obtained by circular permutation, and linearize with respect to the small perturbation δ_i. We obtain the following system:

$$\dot{\delta}_1 = -2\epsilon\delta_1, \tag{13.29}$$

$$\dot{\delta}_2 = \epsilon\delta_2 - \nu\epsilon\delta_2 + \gamma\sqrt{\epsilon}\delta_3, \tag{13.30}$$

$$\dot{\delta}_3 = \epsilon\delta_3 - \nu\epsilon\delta_3 + \gamma\sqrt{\epsilon}\delta_2. \tag{13.31}$$

We impose $\epsilon > 0$ to ensure the existence of a solution in bands (because $R_1^0 = \pm\sqrt{\epsilon}$) and look for a solution of the form: $\delta_i = a_i e^{\sigma t}$. The first equation indicates that the mode δ_1 is stable. Setting the determinant of the other two equations to zero gives us the following criterion:

$$(\sigma - \epsilon + \nu\epsilon)^2 = \gamma\epsilon > 0. \tag{13.32}$$

The largest algebraic value of the eigenvalues determines the stability condition. This is given by:

$$\sigma = \epsilon(1 - \nu) + \gamma\sqrt{\epsilon}. \tag{13.33}$$

Thus, when $\nu < 1$, the bands are always unstable. In the opposite case of $\nu > 1$, the bands are unstable unless:

$$\epsilon > \epsilon_b \equiv \frac{\gamma^2}{(\nu - 1)^2}. \tag{13.34}$$

13.4.2. Some Remarks on the Stability of Bands

To study band stability, we can also start from the equations corresponding to a square symmetry, (13.18) and (13.19), instead of those determined by a hexagonal symmetry. Taking the square symmetry as the point of departure, we obtain, for the amplitude equations of R_1 and R_2, the following system:

$$\dot{R}_1 = \epsilon R_1 - \left(R_1^2 + \nu R_2^2\right) R_1, \tag{13.35}$$

$$\dot{R}_2 = \epsilon R_2 - \left(R_2^2 + \nu R_1^2\right) R_1. \tag{13.36}$$

The stationary solution is identical to that found above. A study of stability gives us the same equation for δ_1, while the one we got for δ_2 must satisfy the following: $\dot{\delta}_2 = \epsilon(1 - \nu)\delta_2$, and we obtain, for parameter σ, the following condition: $\sigma = \epsilon(1 - \nu)$. Thus, when $\nu < 1$, we obtain the same result as we got when we started with a hexagonal symmetry, namely: the bands are always unstable. On the other hand, for $\nu > 1$, we have no restriction on the stability of the bands: they are always stable.

Thus, starting from equations corresponding to square structures, we obtain that the bands are stable for $\nu > 1$ and that, whatever the value for parameter ϵ, just as for the special case of hexagonal structures, we have shown that the bands are stable if the following condition is satisfied: $\epsilon > \epsilon_b$ (condition (13.34)).

How can we reconcile these two seemingly contradictory results? In fact, the difference in conditions for $\nu > 1$ comes from the presence of a quadratic term in γ in the hexagonal case. Remember that a solution is called stable if it conserves its stability when subjected to any perturbation allowed by symmetry.

Thus, if the initial equations, such as the modified Swift–Hohenberg (equation (13.1)), do not have up-down symmetry (i.e. operator $\phi \to -\phi$), the presence of a quadratic term (proportional to γ) in the amplitude equations becomes possible, and consequently perturbations of this type are allowed. The presence of these quadratic nonlinearities pose restrictions on the domain of stability, limiting the values which parameter ϵ may take (equation (13.34)). When, on the other hand, the initial equations do have a top/bottom symmetry (i.e. $\phi \to -\phi$), the quadratic term in γ is absent and the bands are stable for any value of ϵ (as long as $\nu > 1$).

13.4.3. Stability of Hexagons

A stationary hexagonal solution corresponds to variables R_i all having the same value, thus equal to R_0 ($R_i = R_0$, $i = 1, 2, 3$). Departing from the amplitude equation for R_1 (equation (13.22)), we can obtain the stationary equation which R_0 satisfies:

$$\epsilon R_0 + \gamma R_0^2 - R_0^3(1 + 2\nu) = 0. \qquad (13.37)$$

There are three fixed points:

$$R_0 = 0, \quad R_{0\pm} = \gamma \pm \sqrt{\frac{\gamma^2 + 4\epsilon(1 + \nu)}{2(1 + \nu)}}. \qquad (13.38)$$

The solutions $R_{0\pm}$ impose that the term under the root must be positive. Given that $\nu > -1$ (as we assumed in subsection 13.3.1 to ensure the existence of square structures), parameter ϵ is restricted to variation within the following range:

$$\epsilon > -\frac{\gamma^2}{4(1+\nu)} \equiv \epsilon_e. \tag{13.39}$$

We thus see that the two solutions $R_{0\pm}$ exist for negative values of parameter ϵ (i.e. $\epsilon < 0$), before reaching the condition of linear instability of the zero value solution $R_0 = 0$, which is reached when ϵ is zero. As such, this is a subcritical bifurcation (see section 3.3; and for transcritical bifurcations, section 3.4).

Let us write the time-dependent equation obeyed by the amplitude R_0 ($R_1 = R_2 = R_3 = R_0(t)$). From amplitude equation R_1 (equation (13.22)), we find:

$$\dot{R}_0 = \epsilon R_0 + \gamma R_0^2 - (1 + 2\nu)R_0^3$$
$$= -\frac{\partial}{\partial R_0}\left(-\epsilon\frac{R_0^2}{2} - \gamma\frac{R_0^3}{3} + (1 + 2\nu)\frac{R_0^4}{4}\right) \equiv -\frac{\partial V}{\partial R_0}, \tag{13.40}$$

where

$$V = V(R_0) = -\epsilon\frac{R_0^2}{2} - \gamma\frac{R_0^3}{3} + (1 + 2\nu)\frac{R_0^4}{4}$$

is a potential (see figure 13.7), from which the diagram of stationary solutions can be deduced (see figure 13.8).

Superposing a perturbation δ_i with the stationary solution and making a linearization of the amplitude equations R_i (equation (13.22) and the associated equations for R_2 and R_3, deduced from (13.22) by circular permutation), we find:

$$\dot{\delta}_1 = (\epsilon - 3R_0^2 - 2\nu R_0^2)\delta_1 - (\gamma R_0 - 2\nu R_0^2)(\delta_2 + \delta_3). \tag{13.41}$$

The equations for δ_2 and δ_3 are deduced from equation (13.41) by circular permutation. Let us rewrite the system by calculating the difference between each pair of the equations associated with δ_i. For the first equation associated with $\delta_1 - \delta_2$ we get:

$$(\sigma - \lambda_1)(\delta_1 - \delta_2) - \lambda_2(\delta_1 - \delta_2) = 0, \tag{13.42}$$

where λ_1 and λ_2 are two parameters such that: $\lambda_1 = (\epsilon - 3R_0^2 - 2\nu R_0^2)$ and $\lambda_2 = -(\gamma R_0 - 2\nu R_0^2)$. The two other equations are obtained by circular permutation of (13.42). Setting the determinant to zero in this system of equations gives us the degenerate eigenvalue:

$$\sigma = \lambda_1 = \epsilon - 3R_0^2 - 2\nu R_0^2. \tag{13.43}$$

Using the equation for R_0 (equation (13.37)), we can rewrite this equation as

$$\sigma = \frac{2\epsilon(\nu - 1)}{1 + 2\nu} - 2\frac{R_0(2 + \nu)\gamma}{1 + 2\nu}. \tag{13.44}$$

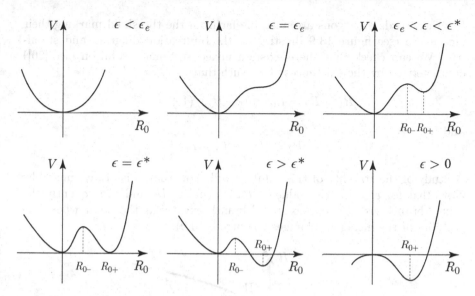

Figure 13.7 – The potential $V(R_0)$ for different values of ϵ. $\epsilon = \epsilon^*$ corresponds to bistability, and the curve $\epsilon = \epsilon^*$ corresponds also to the Maxwell plateau (see solved problem 3.9.1).

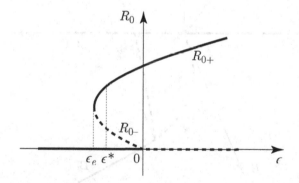

Figure 13.8 – The bifurcation diagram for hexagons. The dotted line represents unstable branches.

The sign of σ is found in the following expression:

$$\epsilon(\nu - 1)(1 + 2\nu) - R_0(2 + \nu)\gamma.$$

Remember that the solution R_{0-} is always unstable and that replacing solution R_0 by R_{0+} gives us a positive σ if the following is true:

$$\epsilon > \epsilon_H = \frac{\gamma^2(2 + \nu)}{(1 - \nu)^2}. \tag{13.45}$$

In other words, hexagons are unstable far from the threshold marking their emergence (see figure 13.9 illustrating the bifurcation diagram and stability). We can check that there exists a mixed stationary solution (see [20]) characterized by the existences of R_i such that:

$$R_1 = R_2 = R_0 = [(\epsilon - U^2)/(1 + \nu)]^{1/2}$$

and

$$R_3 = U = \gamma/(n - 1).$$

A study of the stability of this solution will show that it is always unstable. Note that for $\epsilon = \epsilon_H$, the value of R_0 is equal to U, and consequently the mixed branch crosses the hexagonal branch exactly at the point where the stability of the hexagons changes (see figure 13.9).

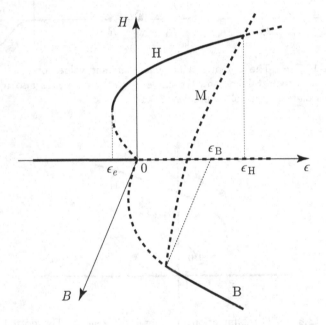

Figure 13.9 – Bifurcation diagram for hexagonal structures, band structures, and their stability. On the vertical axis we denote the amplitude of the hexagonal solution by H. Axis B refers to the amplitude of bands. The branches of hexagons, bands, and of a mixture of the two modes (see text) are designated by the symbols H, B, and M. The hexagons (H) are stable in the interval $\epsilon_e < \epsilon < \epsilon_H$, and unstable for $\epsilon > \epsilon_H$. The bands (B) are unstable for interval $\epsilon < \epsilon_B$ and stable thereafter. The mixed branch (M) is always unstable. It crosses branch H at its point of instability.

13.4.4. Stability of Square Structures

To study square structures, we begin with the system of equations (13.35) and (13.36). There exists a stationary solution,

$$R_1^0 = R_2^0 = R_0 = \pm\sqrt{\epsilon/(1+\nu)}.$$

For $\epsilon > 0$, we must have $\nu > -1$. By positing $R_i = R_0 + \delta_i$ ($i = 1, 2$) and linearizing the system composed of equations (13.35) and (13.36), we find the following system:

$$\dot{\delta}_1 = -2\epsilon\delta_1 - 2\nu R_0^2\delta_2,$$
$$\dot{\delta}_2 = -2\epsilon\delta_2 - 2\nu R_0^2\delta_1.$$

Searching for an exponential solution of the form: $\delta_i \sim e^{\sigma t}$ and setting the determinant to zero, we find the two eigenvalues:

$$\sigma_1 = \frac{2\epsilon(1-\nu)}{1+\nu}, \qquad \sigma_2 = -2\epsilon. \tag{13.46}$$

Given that $\nu > -1$, eigenvalue σ_1 is positive (the case of instability) for $\nu > 1$. Remember that ϵ has to be positive as the necessary condition for the fixed points to be non-trivial ($R_0 \neq 0$). Thus, square structures are unstable when $\nu > 1$ and stable for $-1 < \nu < 1$.

To summarize, now that hexagonal symmetry is excluded from the possible solutions because of the symmetries, we must instead compare the solution for square symmetry with that of banded structures. Taking into account the stability of each case, we reach the conclusions:

– square structures exist for $\nu > -1$ and are stable for $-1 < \nu < 1$;

– bands always exist and are stable for $\nu > 1$.

In general we should compare stability of hexagons relative to square structures. Since hexagons emerge from the weakest nonlinearity, we can expect that in general they would dominate over other structures. However, it is possible to imagine situations in which square (or rectangular, etc.) structures may compete against the hexagonal structures, or may coexist. This important point about competition between structures of different symmetries is beyond the scope of this book, the main goal being to show the rich variety of solutions in nonlinear dynamical systems[11].

---

11. It is useful to specify that the coefficients of the amplitude equation of square structures do not derive from those of hexagonal structures (for example by positing $A_3 = 0$). The coefficients must be determined case by case.

13.5. Solved Problems

13.5.1. Detailed Derivation of the Amplitude Equation for a Hexagonal Pattern

Derive the amplitude equation associated with bidimensional structures of hexagonal symmetry by using the Swift–Hohenberg model as the point of departure (equation (13.1)).

Solution. The first step in the procedure adopted in chapter 10, dedicated to the universal amplitude equation, was to introduce slow variables with which to describe the problem. In full generality (i.e. considering the ensemble of unstable modes), amplitude A_n is a function of the slow spatial variables X, Y, and of slow temporal variable T (i.e. $A_n(T, X, Y)$). The final result gives an amplitude equation with derivatives of A_n with respect to X, Y, and T. To simplify the calculation, we only look at the mode associated to wavevector q_c, not the entire unstable band Δq. Amplitude A_n is thus only a function of time, of slow temporal variable $T = \epsilon t$, and not any spatial variable. We can easily extend the solution to the inhomogeneous case using the same principle as in the unidimensional case (see chapter 10). We leave this exercise for interested readers.

The bidimensional Swift–Hohenberg equation (13.1) is written:

$$\epsilon \frac{\partial \phi}{\partial T} = \left[\epsilon - (1 + \nabla^2)^2 \right] \phi + \sqrt{\epsilon} \bar{\alpha} \phi^2 - \phi^3, \qquad (13.47)$$

where we have imposed that $\alpha = \bar{\alpha}\sqrt{\epsilon}$ and $\bar{\alpha}$ is of order one; the reason for this choice will become clear later.

Following the approach of chapter 10 in expanding field ϕ using powers of small parameter $\sqrt{\epsilon}$, we set:

$$\phi(x, T) = \epsilon^{1/2} \phi_1 + \epsilon \phi_2 + \epsilon^{3/2} \phi_3 + \cdots \qquad (13.48)$$

to find, from equation (13.47) to order $\epsilon^{1/2}$, the following equation:

$$(1 + \nabla^2)^2 \phi_1 = 0 \qquad (13.49)$$

with the solution:

$$\phi_1 = A(T) e^{\boldsymbol{q} \cdot \boldsymbol{r}} + c.c., \qquad (13.50)$$

where *c.c.* stands for complex conjugate and $|\boldsymbol{q}| = q_c = 1$.

As we have seen in section 13.1 in which we looked at bidimensional spatial order in an istotropic plane, all wavevectors $\boldsymbol{q}$ with modulus one are valid solutions for perturbation ϕ_1. Thanks to the linearity of the equation that perturbation ϕ_1 must satisfy (equation (13.49)), we obtain the general solution: $\phi_1 = \sum_{\boldsymbol{q}} A_{\boldsymbol{q}}(T) e^{i\boldsymbol{q} \cdot \boldsymbol{r}} + c.c.$

Still, all elementary Fourier modes $e^{i\boldsymbol{q}\cdot\boldsymbol{r}}$ do not necessarily represent a state with a two dimensional spatial organization. An ordered state only emerges in specifically ordered wavevectors (e.g. square or hexagonal symmetry, see subsection 13.3.1).

For the study of hexagonal order we restrict ourselves to the Fourier modes $\boldsymbol{q}_i$, with $i = 1, 2, 3$ which satisfy the following condition: $\boldsymbol{q}_1 + \boldsymbol{q}_2 + \boldsymbol{q}_3 = 0$ (see figure 13.1), giving the following solution for perturbation ϕ_1 (see equation (13.10)):

$$\phi_1(\boldsymbol{r}, T) = \sum_{n=1}^{3} A_n(T)e^{i\boldsymbol{q}_n\cdot\boldsymbol{r}} + A_n^*(T)e^{-i\boldsymbol{q}_n\cdot\boldsymbol{r}}. \tag{13.51}$$

Writing equation (13.47) to order ϵ we have:

$$(1 + \nabla^2)^2\phi_2 = 0. \tag{13.52}$$

This equation is identical to that found to order $\epsilon^{1/2}$ for ϕ_1 (equation (13.49)), so ϕ_2 has the same form as ϕ_1 (equation (13.51)) with amplitudes denoted by $B_n(T)$. To this order, field ϕ is given by: $\phi = \epsilon^{1/2}\phi_1 + \epsilon\phi_2$.

Looking at solutions ϕ_1 and ϕ_2, we see that ϕ satisfies the same equation associated with ϕ_1 (equation (13.49)) and ϕ_2 (equation (13.52)) and as such, the solution for field ϕ is the same as for ϕ_1 (equation (13.51)) when we substitute A_n by $A_n + \sqrt{\epsilon}B_n \equiv \tilde{A}_n$. This means that we do not actually need function B_n.

Now take a look at equation (13.47) to order $\epsilon^{3/2}$. We find:

$$(1 + \nabla^2)^2\phi_3 = -\frac{\partial\phi}{\partial T} + \phi_1 + \bar{a}\phi_1^2 - \phi_1^3. \tag{13.53}$$

The solution is composed of a solution for the homogeneous case plus a particular solution. Remember that only the resonant terms, sometimes also referred to as secular, of the right-hand side of equation (13.53) are important (see chapter 10). These resonant terms are solutions of equation $(1 + \nabla^2)^2\phi = 0$ and can be described by the function: $e^{i\boldsymbol{q}_n\cdot\boldsymbol{r}}$ (with $n = 1, 2, 3$). Consider the terms proportional to $e^{i\boldsymbol{q}_1\cdot\boldsymbol{r}}$ which come from the right-hand side of equation (13.53). If we keep only the linear terms $-(\partial\phi/\partial T) + \phi_1$, we obtain terms of the form $[-(\partial A_1/\partial T) + A_1]e^{i\boldsymbol{q}_1\cdot\boldsymbol{r}}$. Cubic term ϕ_1^3 gives us contributions of the form $A_1A_1^*A_1e^{i\boldsymbol{q}_1\cdot\boldsymbol{r}}$ as well as terms like $A_2A_2^*A_1e^{i\boldsymbol{q}_1\cdot\boldsymbol{r}}$ and $A_3A_3^*A_1e^{i\boldsymbol{q}_1\cdot\boldsymbol{r}}$ (remember that the condition for hexagonal symmetry is $\boldsymbol{q}_1 + \boldsymbol{q}_2 + \boldsymbol{q}_3 = 0$). The quadratic term produces contributions of the form $A_2^*A_3^*e^{i\boldsymbol{q}_1\cdot\boldsymbol{r}}$.

The solutions to equation (13.53) must be devoid of resonant terms on the right-hand side of the equation (see chapter 10). This means the coefficient of

$e^{i\boldsymbol{q}_1 \cdot \boldsymbol{r}}$ on the right-hand side of (13.53) must be zero. This condition results in a nonlinear equation, which is the evolution equation we were looking for:

$$\frac{\partial A_1}{\partial T} = A_1 + 2\bar{\alpha}A_2^* A_3^* - 3|A_1|^2 - 6(|A_2|^2 + |A_3|^2)A_1. \qquad (13.54)$$

The amplitudes A_2 and A_3, found by eliminating the terms resonant with $e^{i\boldsymbol{q}_2 \cdot \boldsymbol{r}}$ and $e^{i\boldsymbol{q}_3 \cdot \boldsymbol{r}}$, obey similar equations, found by circular permutation of equation (13.54).

Thus, we find that the quadratic term $A_2^* A_3^*$ of equation (13.54) is induced by the α term of the original equation (13.53). This is markedly different from the one dimensional case (seen in chapter 9) where the quadratic term was unimportant in determining the amplitude equation's form. Finally, we have assumed that α is of order $\sqrt{\epsilon}$ so as to ensure that both cubic and quadratic terms appear to the same order. The next problem shows that the amplitude equation can be obtained without this assumption.

13.5.2. Derivation of the Amplitude Equation Based on Harmonics Expansion

As an exercise, we can derive the amplitude equation without assuming α is small. The idea is to expand the general solution in a Fourier series and regroup the resonant terms. Then, by adiabatic elimination, we drop the higher order harmonics.

The expansion into a Fourier series for a hexagonal base gives us the following first few terms:

$$\begin{aligned}
\phi(\boldsymbol{r}, t) = \sum_{n=1}^{3} & A_n(t)e^{i\boldsymbol{q}_n \cdot \boldsymbol{r}} + C_n(t)e^{2i\boldsymbol{q}_n \cdot \boldsymbol{r}} \\
& + D(t)e^{i(\boldsymbol{q}_1 + \boldsymbol{q}_2) \cdot \boldsymbol{r}} \\
& + E(t)e^{i(\boldsymbol{q}_1 - \boldsymbol{q}_2) \cdot \boldsymbol{r}} + F(t)e^{i(\boldsymbol{q}_2 + \boldsymbol{q}_3) \cdot \boldsymbol{r}} \\
& + G(t)e^{i(\boldsymbol{q}_2 - \boldsymbol{q}_3) \cdot \boldsymbol{r}} \\
& + H(t)e^{i(\boldsymbol{q}_1 + \boldsymbol{q}_3) \cdot \boldsymbol{r}} + J(t)e^{i(\boldsymbol{q}_1 + \boldsymbol{q}_3) \cdot \boldsymbol{r}} + c.c. \qquad (13.55)
\end{aligned}$$

The second harmonic, coming from the terms $\sum_{n=1}^{3} A_n(T)e^{i\boldsymbol{q}_n \cdot \boldsymbol{r}} + C_n e^{i\boldsymbol{q}_n \cdot \boldsymbol{r}}$, gives the double of wavevector $\boldsymbol{q}_n$ (term C_n), but also terms whose wavevectors are $\boldsymbol{q}_1 \pm \boldsymbol{q}_2$ (plus circular permutations), represented here by D, E, F, G, H, J. Now, substituting this expression into equation (13.1), we can identify and group together, term by term, the same Fourier modes. Consider, for example, the case of mode A_1. We find:

$$\frac{\partial A_1}{\partial t} = \epsilon A_1 + 2\alpha A_2^* A_3^* + \text{cubic terms in } A_n + etc., \qquad (13.56)$$

with *etc.* indicating the existence of higher order terms, containing the other amplitudes $C, D \ldots$ Writing similar equations for those amplitudes and using adiabatic approximation (that is, by setting the time derivative of higher harmonics equal to zero), we can express the amplitudes of the higher order harmonics as a function of the different A_n. These are thus "injected" into equation (13.56), with the aim of extracting the equation associated with amplitude A_1 (using circular permutation to deduce the equations associated with A_2 and A_3).

In this approach, it is not a power expansion of a small parameter which is important, but the order of the harmonic. The two methods (power expansion of a small parameter, and expansion of harmonics) coincide. We took care to have a small parameter attached to ϕ^2 in order to make the expansion coherent from the start (letting the quadratic and cubic terms appear at the same level), thereby also making the derivation of the amplitude equation more concise.

13.6. Exercises

13.1

We consider the following nonlinear model equation in two dimensions:

$$\frac{\partial \phi}{\partial t} = \left[\epsilon - \left(1 + \frac{\partial^2}{\partial x^2} + \frac{\partial^2}{\partial y^2} \right)^2 \right] \phi - \phi^3. \qquad (13.57)$$

1. Show that in the linear regime we have the following dispersion relation $\omega = \epsilon - (1 - q^2)^2$, where $q = \sqrt{q_x^2 + q_y^2}$, with q_x and q_y the wavevectors in the two spatial directions.

2. Show that the critical wavevector for instability is $q = q_c = 1$. In the $(q_x$–$q_y)$-plane the critical wavevector represents a circle of radius unity. What is the critical value of ϵ for which there is an instability?

3. Let us first assume that at the bifurcation there is a band structure along the Oy direction (i.e. the bands are periodic in the Ox direction). Justify that at the bifurcation $q_x = q_c$ and $q_y = 0$.

4. Let us analyze the linear stability of the band structure in both the Ox and Oy directions. For that purpose, we first need to determine the nonlinear amplitude equation in the vicinity of the bifurcation. We set $q_x = 1 + \Delta q_x$ and $q_y = 0 + \Delta q_y = \Delta q_y$, where Δq_x and Δq_y are small deviations with respect to the critical wavevector. Show that $q \simeq 1 + \Delta q_x + \Delta q_y^2/2$.

5. Expand the dispersion relation in the vicinity of the bifurcation and show that:

$$\omega \simeq \epsilon - a(\Delta q_x)^2 - b(\Delta q_x)(\Delta q_y)^2 - c(\Delta q_y)^4. \qquad (13.58)$$

Provide the expressions of the coefficients a b and c. Justify why we have $\Delta q_x \sim \epsilon^{1/2}$ and $\Delta q_y \sim \epsilon^{1/4}$. How does ω behave with ϵ?

6. The preceding question hints that we can introduce, apart from rapid variable x, slow variables:

$$X = \epsilon^{1/2}x, \quad Y = \epsilon^{1/4}y, \quad T = \epsilon t. \qquad (13.59)$$

Write the expression of the operators entering equation (13.57) in terms of the new variables.

7. Following the same spirit as in chapter 10, we seek solutions in the form of $\phi = \epsilon^{1/2}\phi_1 + \epsilon\phi_2 + \cdots$. Show that to order $\epsilon^{1/2}$, we have:

$$\phi_1 = A(X, Y, T)e^{ix} + c.c. \qquad (13.60)$$

Find the solution to order ϵ. Show that to order $\epsilon^{3/2}$ we must impose a solvability condition (by eliminating resonant or secular terms), yielding:

$$\partial_T A = \alpha A + \beta\left(\partial_X - \frac{i}{2}\partial_Y^2\right)^2 A - \gamma|A|^2 A. \qquad (13.61)$$

Provide the expressions of α, β and γ. Show that with the help of an adequate change of variables, the equation can take the following canonical form:

$$\partial_T A = A + \left(\partial_X - \frac{i}{2}\partial_Y^2\right)^2 A - |A|^2 A. \qquad (13.62)$$

Note that, for simplicity, we have kept the same notations as before the change of variables.

13.2

Equation (13.62) can be written in a variational form. Provide the corresponding functional.

13.3

Let us study the linear stability of the periodic of equation (13.62).

1. Show that equation (13.62) admits spatially periodic solutions $A = a_K e^{iKX}$ and provide the expression of a_K as a function of K. Specify the condition on K under which these solutions exist.

2. We set $A = \rho(X,Y,T)e^{i\psi(X,Y,T)}$. Provide the equations obeyed by ρ and ψ. Express the solution of question 1 in terms of amplitude and phase, denoted by $\rho_0(X)$ and $\psi_0(X)$. Set $\rho = \rho_0(X) + \rho_1(X,Y,T)$ and $\psi = \psi_0(X) + \psi_1(X,Y,T)$, with ρ_1 and ψ_1 small perturbations. Linearize the equations in ρ_1 and ψ_1. We seek solutions in the form of $(\rho_1,\psi_1) = (a,b)e^{i\boldsymbol{Q}\cdot\boldsymbol{R}+\Omega T}$ with $\boldsymbol{R} = (X,Y)$, and a and b real constants. We do not focus on the negative eigenvalue (meaning stability), but only on the one which may become positive. Show from the system obeyed by (a,b) that the condition of non-zero solution provides us with a condition yielding:

$$\Omega = -\left(1-K^2\right) - \frac{1}{2}\left(U_+ + U_-\right) + \left[\left(1-K^2\right) + \frac{1}{4}\left(U_+ - U_-\right)^2\right]^{1/2},$$
(13.63)

with $U_\pm = \left(K^2 \pm Q_X + Q_Y/2\right)^2 - K^2$.

3. We set $Q_x = 0$, and study the possibility for transverse instability (in the direction perpendicular to the bands). Expand (13.63) for small Q_y and show that we have:

$$\Omega = -KQ_y^2 - \frac{Q_Y^4}{4}.$$
(13.64)

Recall that K (deviation from critical value) can take both positive and negative values. The general solution (of field ϕ) reads $\phi \sim a_K e^{i(x+K\sqrt{\epsilon}x)}$ and $K\sqrt{\epsilon}$ refers to deviation from critical wavevector $q_c = 1$. The solution (13.64) indicates that for $K < 0$ the bands are unstable; $K < 0$ means that the band on the left of the neutral curve is unstable (we suggest the reader refers to section 10.3 in order to link K to the neutral curve). The present instability is referred to as "zig-zag" instability since the perturbation corresponds to modulation in the perpendicular direction of the band structure. Recall that we have already seen the bands undergo the Eckhaus instability. We have just seen that in addition they are unstable against zig-zag modes for $K < 0$. We can thus conclude (see figure 10.4) that only the band of wavevectors on the right side of the Eckhaus stable band are stable.

Chapter 14
Wavelength Selection

Abstract *In this chapter we explore the dilemma of selection: how to choose
one stable nonlinear solution from among the many other possible stable
solutions. More specifically, we will discuss wavelength selection for patterns,
and then summarize different mechanisms, such as boundary effects, the effect
of the invasion of a localized pulse or of a front, and the effect of defects.
We will see that defects provide a robust wavelength selection mechanism.*

14.1. Introduction

We have come across many scenarios in which two or more solutions co-exist.
The most simple example is that of a subcritical bifurcation, where a stable
solution coexists with a metastable solution. We briefly discussed (in subsection 3.3.1) that if, because of initial conditions, the system is in a metastable
state, it is generally difficult for it to evolve to the stable state because the
fluctuations in a macroscopic system are too weak (for a discussion on the
effects of fluctuations in Rayleigh–Bénard convection, see [45]). In this scenario it is difficult to know a priori which state the system will choose to
settle on when we cannot possess perfect control on the system's history
(initial conditions, etc.).

Another scenario, which we encountered in chapter 10, describes a continuous family of periodic and steady-state solutions (for the amplitude equation) which differ by their wavenumbers denoted by q_0 (see subsection 10.3.2).
Here, any value of q_0, such that $|q_0| < 1$ is a possible solution. In other words,
the system admits an infinite number of solutions, since by continuous family
we mean that q_0 can take any arbitrary value in the interval $[0, 1]$. This is
generally the case, not just for an amplitude equation. Thus we may ask:
by what criterion can we distinguish between the solutions? In chapter 10

© Springer Science+Business Media B.V. 2017
C. Misbah, *Complex Dynamics and Morphogenesis*,
DOI 10.1007/978-94-024-1020-4_14

we made a linear study of the stability of the periodic solutions' engendering patterns, characterized by wavevector q_0. As an example, we looked at the Eckhaus (or phase) instability. Not all periodic solutions show this instability: only those which have a wavevector q_0 such that $|q_0| > 1/\sqrt{3}$; the other solutions remain stable. In other words, the study of stability tells us that the initial band of wavevectors, $-1 < q_0 < 1$, is partially and simply reduced to a smaller interval defined by $-1/\sqrt{3} < q_0 < 1/\sqrt{3}$, i.e. only solutions whose wavevectors lie within this new band are stable. Still confronted with an infinite number of stable solutions, the question naturally arises as to whether the system may pick up a unique solution, and if so, by which mechanism? This is a problem of wavelength (or wavevector) selection. To avoid any ambiguity, we specify anew that q_0 represents the difference with respect to the critical wavevector and, as we already discussed in section 10.3, the total wavevector of the structure is given by $q = 1 + q_0\sqrt{\epsilon}$ (where $q_c = 1$ is the critical wavevector), such that the stable band (defined by $-1/\sqrt{3} < q_0 < 1/\sqrt{3}$) corresponds, for the total wavevector q, to the condition:

$$\Delta q \equiv |q - 1| < \sqrt{3\epsilon}. \tag{14.1}$$

In other words, the stable band is of order $\sqrt{\epsilon}$, and (since ϵ is small in our assumptions) is thus wider than ϵ, which measures the distance from the threshold of order emergence.

14.2. Different Mechanisms for Wavelength Selection

14.2.1. Selection via Walls or Boundaries

Real systems are finite, confined to some region of space. Up until now, we have assumed that spatial order is extended far, or even infinitely, along the Ox-axis. The question arises: how does the finite size of a system, its boundary conditions, affect the solutions? Let us take Rayleigh–Bénard convection as an example, where amplitude A is proportional to the hydrodynamic velocity of the convection rolls. If the system is confined in the Ox direction by two rigid plates, amplitude A must go to zero at each plate (the so-called no-slip hydrodynamic condition).

Consider the amplitude equation:

$$A_t = \epsilon A - |A|^2 A + A_{xx}, \tag{14.2}$$

and to simplify, assume the system is unilaterally confined by a rigid wall located at $x = 0$. The amplitude equation is thus subject to the following boundary condition: $A(x = 0) = 0$. The stationary solution satisfying this boundary condition is:

$$A = \sqrt{\epsilon}e^{i\psi}\tanh\left(x\sqrt{\frac{\epsilon}{2}}\right),\qquad(14.3)$$

where ψ is an arbitrary constant. This solution is shown in figure 14.1. The amplitude is zero at the boundary, and the transition zone between a zero and a positive constant amplitude has a length of order $\epsilon^{-1/2}$. Note that the presence of phase ψ indicates that the pattern (band, convection roll, etc.) has the possibility of slipping with respect to the wall. We can think of the wall as a defect in the system. Further on, we will see that defects in patterns play a fundamental role in selection procedures, particularly wavelength selection, of these structures.

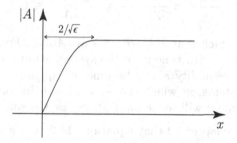

Figure 14.1 – Solution A as a function of x in the presence of a wall at $x - 0$, and the length of the transition zone (of order $\epsilon^{-1/2}$) between 0 and a constant amplitude.

The stability of the solution for a system bounded by a wall has been amply discussed in the literature [24]. It is found that the presence of a wall does not directly lead to a unique selection of a wavelength, but considerably reduces the breadth of the Eckhaus band, such that the condition for stability becomes:

$$\Delta q \equiv |q - 1| < a\epsilon,\qquad(14.4)$$

where a is a number which can be determined by detailing the conditions (dependent on whether there are one or two walls, as well as the nature of the walls). In other words, the allowed band which behaves like $\epsilon^{1/2}$ (inequality (14.1)) in the absence of any wall is reduced to order ϵ in the presence of a wall (remember that in the case of an amplitude equation, ϵ is a dimensionless number which is small with respect to one). In addition, the existence of walls defines a discrete ensemble of wavevectors (see exercises of chapter 9). We still have a finite (albeit smaller) band of possible solutions, and not a unique solution.

14.2.2. Selection by Variational Principle

An interesting and useful property of amplitude equation (14.3), as you may
have observed throughout the book (especially in chapter 10), is the fact
that it derives from a potential, or that it can be written as the derivative
of physical quantity which we call the potential, or a Lyapunov function
(or functional). More specifically, we can write

$$A_t = -\frac{\delta}{\delta A^*} V[A] = -\frac{\delta}{\delta A^*} \int dx \left(-\epsilon \frac{|A|^2}{2} + \frac{|A|^4}{4} + \frac{|A_x|^2}{2} \right), \qquad (14.5)$$

where $\delta/\delta A^*$ is the functional derivative with respect to A^* (the complex
conjugate of A), with the property

$$\frac{\partial V[A]}{\partial t} = \frac{\delta V}{\delta A^*} \frac{\partial A^*}{\partial t} = -\left| \frac{\delta V}{\delta A^*} \right|^2 \leqslant 0, \qquad (14.6)$$

where we used the fact that $\partial A^*/\partial t = -\delta V[A]/\delta A$. Functional $V[A]$ decreases monotonically with time until the system attains a stationary state,
in which case the inequality above become an equality. This is not unlike
thermodynamic systems, in which the second law helps determine which of
many states the system will settle on. Can we use the same approach here?

We already saw in chapter 10 that equation (14.2) has a continuous family
of periodic solutions:

$$A_0(X) = \sqrt{\epsilon - q_0^2} e^{iq_0 X}, \qquad |q_0| < \epsilon, \qquad (14.7)$$

where q_0 is any real number satisfying $|q_0| < \epsilon$. Since the solution is stationary,
it corresponds to an extremum of $V[A]$, as seen in (14.5). That is, we have
$\delta/\delta A^* V[A] = 0$. This is, a priori, a local extremum. Still, from all possible
extrema we must choose the absolute one, since that would be the one which
corresponds to the absolute minimum (say the minimal energy by analogy
with systems at global thermodynamical equilibrium). Using (14.7) and the
definition of the potential (14.6), we can write:

$$V(q_0) = -L \frac{(\epsilon - q_0^2)^2}{4}, \qquad (14.8)$$

where L is the size of the system (which can be arbitrarily large). Thus, $V(q_0)$
has an absolute minimum for $q_0 = 0$. This means that V is an extremum for
$q = q_c$, the critical wavevector at the bifurcation (remember that q_0 is the
difference with respect to the critical wavevector q_c).

We must insist on the fact that, generally, we can only count on the existence of a potential (i.e. that the dynamical equation takes the form of a
variational character, as seen above) when we are very close to the instability

threshold. Once we move away from the threshold, higher order terms must be taken into account and the amplitude equation loses its variational nature (see exercises at the end of this chapter). Furthermore, even if a potential exists, we will see that systems do not necessarily choose the state which corresponds to the extremum of this potential. In fact, the system can be found at a local extremum without the ability to transfer to other configurations because of a too-high potential barrier. The numerical solution to the amplitude equation (14.2) reveals that we can get any state allowed by the Eckhaus band if we choose the right initial conditions. Once the system reaches a stationary state, it will remain there. Unlike systems in thermodynamical equilibrium, where in general, thermal fluctuations can help the system to find an absolute extremum, these fluctuations are generally too small to allow the systems studied here (which are macroscopic) to explore other states. In other words, this principle of selecting a minimal potential would only make sense if fluctuations were important enough. In the opposite limit, the existence of a potential does not guarantee an absolute selection of wavelength.

14.2.3. Selection by a Propagating Front

Another possible selection mechanism is that of a propagating front. This mechanism, put forward by Dee and Langer in 1983 [26] (see also [106]) does not depend on the extrema of a potential (if there is a potential) in its selection of a wavelength. Consider: in chapter 11 we saw that the numerical simulation of equation (14.2) described a propagating front starting from an initial condition localized, denoted by $A_0(x)$, and invading the system, leaving in its wake an amplitude $A = \pm\sqrt{\epsilon}$ which serves as the envelope of a periodic structure whose wavevector is q_c (the total wavevector is $q = q_c + q_0\sqrt{\epsilon}$). We saw in chapter 11 that the front picks an invasion velocity (equal to $v^* = 2\sqrt{\epsilon}$ for equation (14.2)) corresponding to a marginal stability. Since this book is an introductory text, we will skip the more complicated mathematics and simply present the result (curious readers may follow a trajectory to unearth these results in exercise 11.2).

Marginal stability is based on the following idea. Consider a perturbation in the system of the form $e^{(\omega+iqv)t}$ where q may be the wavevector, ω the growth rate, and v the propagation velocity. Unlike the ordinary notion of wavevector (which is real), this theory allows q to be, generally, a complex number. The principle of marginal stability gives us these three results:

$$v^* = i\frac{d}{dq}\omega(q)\Big|_{q=q^*}, \quad v^* = \frac{\Re(\omega)}{\Im(q)}\Big|_{q=q^*}, \quad q_s = \Re(q^*) + v^{*-1}\Im(\omega(q^*)),$$

(14.9)

where q^* is an abstract wavevector (see exercise 11.2) which may be complex, q_s is the wavevector of the periodic stationary structure, and ω is the dispersion relation in the linear regime. From these three conditions we can determine the three unknowns, q^*, v^* and q_s. Equation (14.2) gives us:

$$\omega = \epsilon - q^2, \tag{14.10}$$

where q and ω are treated as being a priori complex numbers.

Now, the first two conditions of (14.9) tell us $q^* = i\sqrt{\epsilon}$ and $v^* = 2\sqrt{\epsilon}$. Note that q^* is a pure imaginary number and that the value v^* coincides with the value derived in chapter 11 through a different set of considerations. The third condition of (14.9) tells us $q_s = 0$. In other words, the system will select wavevector q_c, the critical wavevector (remember the total field is $Ae^{iq_c x}$ and the q_s above refers to that of the envelope A). This result is identical to the one we found in the last section, where the choice had been based on a variational principle, but as far as we know this is a coincidence. To see this, consider the Swift–Hohenberg equation (discussed in chapter 10):

$$\frac{\partial \phi}{\partial t} = \left[\epsilon - \left(1 + \frac{\partial^2}{\partial x^2} \right)^2 \right] \phi + \alpha \phi^2 - \phi^3, \tag{14.11}$$

where ϕ is a scalar function of two variables x and t describing the dynamic evolution of the nonlinear system (for example, a component of the convection velocity). The linear dispersion relation is given by:

$$\omega = \epsilon - (1 - q^2)^2. \tag{14.12}$$

Using (14.9) we find the propagation velocity of the front, selecting the wavevector:

$$v^* = \frac{4}{3\sqrt{3}} \left(\sqrt{1 + 6\epsilon} + 2 \right) \left(\sqrt{1 + 6\epsilon} - 1 \right)^{1/2} \tag{14.13}$$

$$q_s = \frac{3 \left(\sqrt{1 + 6\epsilon} + 3 \right)^{3/2}}{8 \left(\sqrt{1 + 6\epsilon} + 2 \right)}. \tag{14.14}$$

We can check that for $\epsilon \to 0$ we have $q_s \simeq 1$ and $v^* \simeq 4\sqrt{\epsilon}$. These results agree with those of the amplitude equation. Remember that here, in equation (14.11), the wavevector contains both the fast and slow spatial parts, which is why $q_0 = q_s = 1$ (the total wavevector), whereas the amplitude equation (which gives $q_s = 0$) concerns only the wavevector of the envelope. Also remember that the linear part of the amplitude equation, which comes directly from equation (14.11), is given by $A_t = \epsilon A + 4A_{xx}$[1], whereas the canonical form (14.2) is simply given by $A_t = \epsilon A + A_{xx}$. If we transform

---

1. See (10.22), where we single out ϵ by looking at physical variables; see also subsection 10.2.3.

the equation that comes from (14.11) in a canonical form (by appropriate rescaling as we have seen many times in this book), we find $v^* \simeq 2\sqrt{\epsilon}$. Equation (14.11) can be written as a variational equation (see exercise 14.2), and the absolute minimum of the potential (for small enough ϵ) [85]:

$$q_m = 1 - \frac{\epsilon^4}{1024}. \tag{14.15}$$

In the limit of small ϵ, q_s (equation (14.14)) is given by:

$$q_s = 1 + \frac{\epsilon}{8}. \tag{14.16}$$

In conclusion, the wavevector corresponding to the optimum of the potential is different from that selected by the propagation of a front. The question arises as to which of the two mechanisms is the more robust. Numerical integration of (14.11) with non-localized initial conditions (a white noise or a vaguely sinusoidal form) shows that in the end we find a wavevector which depends on the initial conditions. All wavevectors which lie within the Eckhaus stable band are possible choices. The system can become trapped in a local extremum (because all stationary solutions are extrema for variational dynamics). One might conjecture that if we add some noise to equation (14.11), the system might settle on the solution corresponding to the extremum of a potential. However, the numerical simulation [44] shows that it goes nowhere. In fact, the system selects the wavevector which corresponds to the most linearly unstable mode, that is, the one corresponding to the maximum of ω, given by (14.12). Thus, it seems that the variational principle does not give a real selection criterion.

On the other hand, if in our initial conditions we begin with a fairly localized perturbation, it will invade the system with velocity v^* and leave behind it a periodic structure with a well-defined wavevector, corresponding to the marginal stability. This selection principle has been studied at length and seems to be robust [106]. We can think of the propagation of the front as a defect which propagates throughout the system. As we will see, the presence of defects in systems generally plays a robust role in the selection of structure.

14.2.4. Selection by a Ramp

One scenario in which a wavelength must be selected arises when the control parameter (such as ϵ in equation (14.12)) is a slowly varying function over a spatial dimension, i.e. ϵ is not a constant but a slowly varying function of x, $\epsilon(x)$.

Suppose that $\epsilon(x)$ is chosen such that $\epsilon < 0$ in one region of space, and slowly changes to be $\epsilon > 0$ at some point x. We know that the solution with zero amplitude is stable when $\epsilon < 0$ and is unstable otherwise. Since $\epsilon(x)$ depends on x, we will have zero amplitude in the region of x where $\epsilon < 0$, which is connected to non-zero amplitude in the region of x with $\epsilon > 0$. It has been shown that for any initial condition, the periodic structure always chooses the same wavevector! This phenomenon was predicted theoretically and checked numerically [56, 86], and later also seen experimentally [24].

In the case of Rayleigh–Bénard convection, we can imagine a slowly varying gap (slowly varying distance between two plates). The selection can be described as follows. In classical mechanics [58] one knows that that if one of the mechanical parameters defines the properties of the system itself (e.g. stiffness of a spring) or of the exterior field in which it evolves, and is slowly varying (in mechanics we speak of a slow variation in time, in contrast to our examples of patterns where we speak of slow variation in space), then there exists an adiabatic invariant I (which depends, in mechanics, on physical quantities, for example frequency native to the harmonic oscillator), which remains constant as the parameter varies. Here, the variable analogous to time in the mechanical scenario is the spatial variable x. Adiabatic invariance relates the wavelength of the structure to control parameter $\epsilon(x)$ like $I(q, \epsilon) = C$ where C is a constant (expressing invariance). If we impose when $\epsilon(x) = 0$ (the instability threshold) that $q = q_c$ (equal to 1 in equation (14.12)) we can determine constant C, which gives us a unique relation between q and ϵ, and thus proving that we can achieve a wavelength selection via a ramp. In section 14.4 (exercises), we explicitly analyze this issue with a concrete example.

14.2.5. Selection through Defects

Selection by a Solitary Wave

Even if defects are rarely present in unidimensional structures, there are many situations in which they play important roles for the selection of a wavelength. A defect might be, for example, an arch which is wider than the rest (see figure 14.2). We could see this arch as an occasional imperfection of structure – a defect. If the system is not of potential or variational nature (which is more or less the rule and not the exception for non-equilibrium systems) and if the defect is not symmetrical (with respect to the inversion $x \to -x$), then it will drift sideways, along the front, as shown schematically in figure 14.2 (see the exercises section for an exposition of the connection between having a non-variational character and the drift of the pattern). This defect is also known as a solitary wave. It was first seen in a problem where the dynamics of an interface separated a liquid crystal phase from its isotropic phase [96] (growth of liquid crystal at the expense of its isotropic phase), and thereafter was seen in a great many systems. As the defect makes its way (i.e. it moves sideways), the system is made to pick a single wavevector [21]. This is one of the first examples of selection by defects.

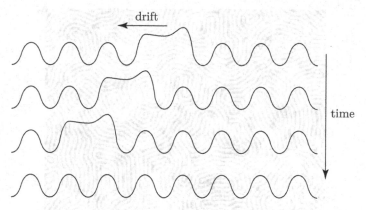

Figure 14.2 – An asymmetric arch along the front leaves behind a well-determined wavevector.

Selection by Dislocations

By contrast, real structures of two dimensions are rarely exempt of defects. The numerical solution [34] of the bidimensional version of equation (14.11) (obtained by replacing $\partial^2/\partial x^2$ by $\partial^2/\partial x^2 + \partial^2/\partial y^2$) can lead to a structure which seems disordered and has many defects (figure 14.3). This situation is complex and not easily lent to a simple qualitative or quantitative discussion. However, there are other cases where we can distinguish a periodic structure which is ordered enough, despite the presence of many isolated defects. The type of defects typically observed are called dislocations[2]; figure 14.4 shows an example.

An important result of the defect comes from the fact that a stationary state is only possible if the wavevector has a well-defined value, which is of course simply the wavevector being selected. If the dislocation in figure 14.4 climbs to the top, once it disappears (having reached the end of the domain), the number of bands will have diminished, which means the wavevector will have gotten larger. If on the other hand the dislocation moves toward the bottom, the number of bands increases and the wavelength is diminished. The velocity with which a dislocation advances depends on the wavevector of the structure, denoted q_0. We can show (see exercises) that if $q > q_0$ the dislocation will rise, and that it descends in the opposite case. The velocity of the dislocation becomes zero for a particular wavevector q_0. Thus the successive movement of dislocations (which are often present in great numbers) will finish by leading the system to a state where $q = q_0$.

---

2. This name is traditionally used for crystalline material.

Figure 14.3 – Solution for the bidimensional version of equation (14.11). [From [23] M. Cross & H. Greenside. *Pattern Formation and Dynamics in Nonequilibrium Systems*, 2009, © Cambridge University Press]

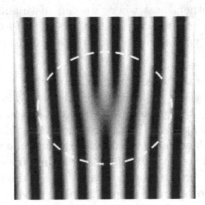

Figure 14.4 – A structure with a dislocation. [From [23] M. Cross & H. Greenside. *Pattern Formation and Dynamics in Nonequilibrium Systems*, 2009, © Cambridge University Press]

Selection by Grain Boundaries

Another type of defect is called grain boundaries, with two periodic structures of two wavevectors and different relative orientations (see figure 14.5).
At the intersection between the two structures we find a defect called
"grain boundary". The solution is stationary only if the two wavevectors are
equal to each other and to a particular value as determined by the stationary character of the grain boundary. The situation is akin to that of dislocations seen above.

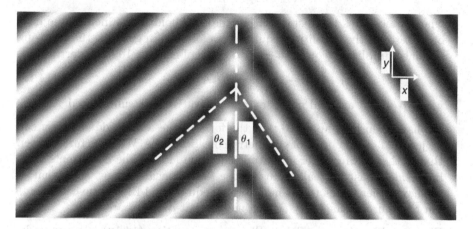

Figure 14.5 – A structure showing a grain boundary. [From [23] M. Cross &
H. Greenside. *Pattern Formation and Dynamics in Nonequilibrium Systems*,
2009, © Cambridge University Press]

14.3. Conclusion

The problem of selection of a solution in the case of non-equilibrium systems
is complicated. We have briefly reviewed the essential points of the problem
without being exhaustive. We saw that even for systems that have a variational nature (such as in equilibrium thermodynamics), the choice is not
necessarily determined by an optimal wavelength. The most efficient mechanisms for wavelength selection seem to be selection by defects and by propagating fronts, which come from very localized perturbations. When many
mechanisms are at play, it is not clear which will prevail. The goal of this
chapter was to give a general idea of the current understandings in this domain, without getting overly technical. Interested readers who are willing to
deepen their knowledge about the fascinating question of wavelength selection
can consult more specialized works [23] (and references therein).

14.4. Exercises

14.1

1. We have seen several times that in one dimension, and close to the bifurcation point, systems are described by a universal equation called the amplitude equation, which reads (for a steady bifurcation):

$$\epsilon A + A_{xx} - |A|^2 A = 0, \qquad (14.17)$$

where ϵ measures the distance from the threshold. If we include the next order term in the amplitude equation, we obtain (in general) the following modified amplitude equation (see exercises of chapter 10 for a derivation):

$$\epsilon A + A_{xx} - |A|^2 A = i \left(\alpha A_x |A|^2 + \beta A_x^* A^2 \right), \qquad (14.18)$$

where α and β are real numbers, and derivatives are subscripted.

We have seen in chapter 10 that the form of the amplitude equation can be justified on the basis of symmetry considerations. We have further seen that in the leading order amplitude equation (14.18) the coefficients (left-hand side) must be real. By using the same considerations, show that in the next order terms (right-hand side of equation (14.18)), the coefficients must be purely imaginary (which means that α and β are real).

2. First consider the case $\beta = 0$. Show that equation (14.18) has a periodic solution in the form $A = \rho e^{iqx}$ and determine ρ as a function of q and the coefficients entering equation (14.18).

3. Equation (14.18) can be derived from a potential:

$$V = \int \left[\epsilon |A|^2 - |A_x|^2 - \frac{1}{2}|A|^4 - \frac{i\alpha}{4} \left(A^* A_x |A|^2 - A A_x^* |A|^2 \right) \right]. \qquad (14.19)$$

Show that equation (14.18) does indeed derive from this potential.

4. Explain why the solution $A = \rho e^{iqx}$ renders the potential V optimal (minimum or maximum).

5. Show that the optimal wavenumber is given by $q = \alpha\epsilon/4$. Compare this with the case studied in this chapter for the leading order amplitude equation.

6. When $\beta \neq 0$, argue that equation (14.18) cannot be written as a functional derivative of a potential. As discussed in this book, the potential character of the equation ceases to be valid when higher order contributions are taken into account.

7. Show that if ϵ is a constant, then equation (14.18) has a periodic solution in the form:

$$A = \rho(\epsilon, q) e^{iqx}. \qquad (14.20)$$

Determine ρ as a function of q and ϵ (and the coefficients entering equation (14.18)).

8. We consider now the problem of wavenumber selection by a ramp. That is, we consider that ϵ depends on x (in a slow manner) such that for $x \to -\infty$, $\epsilon < 0$ (and thus $A = 0$ is a stable solution), while for $x \to +\infty$, $\epsilon > 0$ ($A = 0$ is unstable and we have a pattern having a form proportional to $Ae^{iq_c x}$, with q_c the critical wavenumber). The passage from negative to positive values is supposed to be smooth, for example $\epsilon(x) = \tanh(x\nu)$ where ν is a small parameter. We shall not consider a specific expression of ϵ in what follows, but rather focus on the general result. Argue that the wavenumber cannot now be constant (independent of x) throughout the system.

9. The wavenumber q is now a function of x, and we can talk about a local wavenumber. We define a phase $\psi(x)$ such that the local wavenumber q is defined by $q = \psi_x$. In the adiabatic approximation (if the ramp is slow enough) we are tempted to expect that the solution given by equation (14.20) constitutes a reliable starting point. Show that equation (14.18) is no longer solved by:

$$A = \rho(\epsilon(x), q(x))e^{i\psi(x)}, \qquad (14.21)$$

where the amplitude ρ depends on x only, via ϵ and q. Write down the condition under which the above solution would become a solution of equation (14.18).

10. Argue that the condition found just above cannot be satisfied in general.

11. For a perfectly periodic solution $\psi = qx$ (with q a constant). For a slowly varying environment this ceases to be valid. If ϵ depends on x, then (14.21) is no longer a solution of equation (14.18). In order to express the ramp nature (slow variation) we introduce a small parameter ν and state that ϵ is a function of $X = \nu x$ only, and we expect that the local wavenumber also depends on X only (since a variation of q with space is directly implied by that of ϵ). Since, as stated above, (14.21) is no longer a solution of equation (14.18) (except when $\nu = 0$), we must seek an alternative method. For that purpose, we expand the solution in the power series of ν:

$$A = f + \nu g + \cdots . \qquad (14.22)$$

We also introduce the multiscale spirit where we have the fast variable x and the new slow variable $X = \epsilon x$. Write in the multiscale spirit how the operators $\partial/\partial x$ and $\partial^2/\partial x^2$ should be transformed. As we have seen in this book, once the operators are adequately transformed, the two variables x and X must be treated as if they were independent.

12. Show that to order ν we have:

$$\epsilon(X)f + f_{xx} - |f|^2 f - i\left(\alpha f_x |f|^2 + \beta f_x^* f^2\right) = 0. \qquad (14.23)$$

Provide a periodic solution to this equation and determine the amplitude.

13. Show that to order ν, equation (14.18) yields:

$$\epsilon g + g_{xx} - f^2 g^* - 2|f|^2 g$$
$$- i \left[\alpha(f_x f^* g + f_x f g^* + |f|^2 g_x) + \beta(2 f_x^* f g + |f|^2 g^*) \right]$$
$$= -i \left[2\rho q_X + \rho_X (2q - (\alpha + \beta)\rho^2) \right] e^{i\psi(X)}, \quad (14.24)$$

where we have used on the right-hand side the solution of the zeroth order solution (recall that ρ and q depend on X only via ϵ).

14. Consider equation (14.23) and take the derivative with respect to x and set $g = f_x$. By comparing the result to the expression on the left-hand side of equation (14.24), what conclusion do you draw?

15. The mode f_x is called the Goldstone mode which we have encountered in this book (see solvability condition, subsection 6.1.5). By following the same spirit, show that the solvability condition yields:

$$2\rho q_X + \rho_X \left[2q - (\alpha + \beta)\rho^2 \right] = 0. \quad (14.25)$$

16. We anticipate that $q \sim \epsilon$. By keeping leading terms in ϵ, show that:

$$\epsilon_X \left[2q - (\alpha + \beta)\epsilon \right] + 2q_X \epsilon = 0. \quad (14.26)$$

17. Show that:

$$\Phi = 2q\epsilon - (\alpha + \beta)\frac{\epsilon^2}{2} \quad (14.27)$$

is a constant of motion[3].

18. When $\epsilon = 0$ we know that the wavenumber is equal to the critical one q_c, which corresponds to $q = 0$ (recall that here q corresponds to the wavenumber of the envelope and the total field is a product of the complex envelope A and the pattern oscillation $e^{iq_c x}$). Show that the selected wavenumber by the ramp is given by:

$$q = \frac{(\alpha + \beta)\epsilon}{4}. \quad (14.28)$$

Compare this result with that obtained from a variational principle. Do you understand why we considered the higher order amplitude equation?

---

3. This constant of motion is akin to the adiabatic invariant in mechanics [58] where a particle moves in an external field which is slowly varying in time. The slow variation in time corresponds here to the slow evolution in space ensured by ϵ (the control parameter).

Chapter 15

Conclusion

Abstract *Our final chapter will summarize the content covered in this book, tying the spirit and ideas together which led to the elaboration of the subject matter, and end with an overview of several topics we did not cover but which would constitute an interesting extension of this presentation.*

We designed this volume so as to give a qualitative and quantitative introduction to the field of nonlinear science through the use of simple, concrete, examples. Considering familiar systems like springs, pendulums, etc., we built an intuitive image from which the language and basic concepts used in the field could more easily be defined. By this method, we first introduced the main bifurcations, which we then divided into the seven categories of elementary catastrophes. These seven types of catastrophe describe the ensemble of possible scenarios that a system with no more than four independent control parameters can experience[1]. We then highlighted the universal nature of the classification of catastrophes and bifurcations by showing how the classifications depend exclusively on the number of dimensionless parameters, and not on degrees of freedom or any other specificity of the system under consideration. We went on to briefly examine the phenomenon of chaos, comparing it to randomness; we arrived at a statistical description of deterministic chaos. This concluded the first part of the book, dedicated to the purely temporal aspect of the dynamics of a system.

From chapter 9 onwards, we started looking at morphogenesis (or the emergence of forms and patterns), first through some examples (the Turing system and Rayleigh–Bénard convection), from which we then extracted the universal properties of such phenomena. In particular, we showed that the dynamics near the instability threshold were described by a universal amplitude

---

1. Note the adimensional nature of these parameters; the systems in question can have a higher number of dimensional parameters.

© Springer Science+Business Media B.V. 2017
C. Misbah, *Complex Dynamics and Morphogenesis*,
DOI 10.1007/978-94-024-1020-4_15

equation; solving this, we could then analyze the nature and stability of the solutions. We then looked at the Eckhaus or phase instability, which can be classified as a secondary instability in contrast to the primary instability which describes the passage from homogeneous to ordered state. Then we examined the invasion of an unstable solution by a stable solution, comparing it to the more classic problem of invasion of a metastable solution by a stable solution. We saw that in the latter case, there is a unique invasion velocity that the system will settle on, whereas in the former we are confronted with an infinite number of possible solutions. We then went through the several arguments by which a system can choose a velocity from among this infinitude.

Review of the Important Problems We Came Across

First we looked at the case of a stationary bifurcation, with its relatively simple dynamics. Then we looked at the problem of spatial and temporal dynamics in the case of the Hopf bifurcation, which introduces a temporal oscillation. Unlike the stationary bifurcation, the dynamics of a Hopf bifurcation cannot be described by a potential and loses its variational character[2]. This is an essential point about non-equilibrium systems: the loss of this variational character is what allows for such rich dynamics from order to spatio-temporal chaos. Spatio-temporal chaos is complex in and of itself: it can refer to a phase chaos, a chaos brought about by defects, etc. Our treatment of the unidimensional problem of spatio-temporal chaos allowed us a glimpse of the spatial and temporal complexity found in non-equilibrium systems.

The examples described above are limited to one dimensional structures. We presented some bidimensional problems in a later chapter, introducing "structures" or "patterns" seen in two dimensions. Here, we restricted ourselves to a study of stationary bifurcations so as to introduce the essential concepts clearly.

Finally, we addressed the question of wavelength selection, which is still a topic of ongoing debate in the literature. We listed various scenarios for the selection of wavelengths, showing that the most robust method for a firm selection seems to be via the dynamics of defects in the system.

Overall, this book is a sort of open sesame to the nonlinear sciences, and several chapters are accessible to most first year students, teacher-researchers, and researchers. Nonetheless, the nonlinear sciences are so rich that several other, more subtle, questions deserve to be looked at in closer detail. We leave you with such a (non-exhaustive) list.

---

2. The variational character in the case of a stationary bifurcation remains true only insofar as we remain close to the instability threshold. When we leave this neighborhood, this variational character is also broken and we find more complex behaviors; see the short discussion in section 15.1.

15.1. Secondary Instabilities

In this book, we only saw one secondary instability, the Eckhaus instability (similar to the Benjamin–Feir instability that we saw in our study of the Hopf bifurcation). The Eckhaus instability is actually the only secondary instability which can occur in systems subject to stationary bifurcations when looking in the neighborhood of the primary instability's threshold. If we go a short distance from this threshold, instabilities associated with other Fourier modes besides the principal harmonic can be activated. If, for example, the second harmonic becomes unstable in turn, other secondary instabilities are observed. The most well known of such instabilities gives rise to sudden parity symmetry breaking: unidimensional spatial patterns suddenly lose their mirror symmetry. For example, a convective Bénard roll suddenly loses its circular symmetry. This instability was first seen experimentally in 1988 in a study investigating the dynamics of a growth front between two phases[3] (see [96] for the experiment and [21,50,61,74] for theoretical analyses). Later experiments using different systems showed the same phenomenon (see [14,31,88]). One consequence of this symmetry breaking is that we see a lateral drift of the structure at a constant velocity (for example, the Bénard rolls travel horizontally).

Other secondary instabilities may also emerge, such as the vacillation-respiration mode in which consecutive arches (in other words, a front modulated between two phases) oscillate in time, in opposite phases (or the length of two successive bands in a chemical pattern oscillate with opposite phases).

If we think in terms of symmetry, ten generic secondary instabilities can be named (see [22]), the three most common examples being: the Eckhaus instability, the instability associated with parity breaking, and the oscillation-respiration mode. As we go further away from the primary instability, other harmonics become unstable and the dynamics become increasingly complex, until we encounter dynamics like spatio-temporal chaos.

---

3. The front, originally flat, grows in the Oz direction. It develops a primary instability, for a critical growth velocity, characterized by a one dimensional spatial modulation, along the Ox-axis, and with $x \to -x$ symmetry. Beyond a second threshold in the growth velocity, the front tilts and loses the $x \to -x$ symmetry; this is a symmetry breaking bifurcation. The structure then drifts with a constant velocity along the Ox-axis, in the direction orthogonal to the growth.

15.2. Two Dimensional Structure for a Hopf Bifurcation

In this book we only looked at the bidimenisional patterns emerging from stationary bifurcations. If we also look at Hopf bifurcations, the dynamics become very complex. In particular, we would see the Benjamin–Feir instability again (seen before in one dimension) and spatio-temporal chaos. Furthermore, many more typical two dimensional patterns emerge, such as spirals, or spatio-temporal chaos mediated by spirals, wave sources and sinks (see [2] for a review).

15.3. Systems Not Reducible to an Amplitude Equation

Throughout this book we have insisted that, close to instability thresholds, the dynamics of a system can be reduced to a single universal amplitude equation with real or complex coefficients. Remember that this amplitude equation describes the envelope of the main Fourier mode's fast oscillation (see chapter 10). This reduction of the dynamics into one equation is only possible when only the main harmonic is unstable and thus all the higher order harmonics are stable. In fact, several systems do not fall under this classification even when we are very close to the instability threshold, such as systems in which the critical wavevector is very small (or even zero). The type of dispersion relation that emerges in such a scenario looks like:

$$\omega = \epsilon q^2 - q^4, \tag{15.1}$$

where ω is the growth rate of the mode of the instability and q is the wavevector. The maximum of ω, defining the most unstable mode, is obtained for wavevector $q = \sqrt{\epsilon/2} \sim \sqrt{\epsilon}$, and the band of wavevectors which corresponds to the unstable modes, that is, modes whose growth rate ω is positive, are of the order $\Delta q \sim \sqrt{\epsilon}$ (more exactly, this band goes from $q = 0$ to $q = \sqrt{\epsilon}$).

For all wavevectors q which are small enough (say, as way of example, around the order of $\sqrt{\epsilon}/100$), we have an instability (i.e. $\omega > 0$) not only in the principal Fourier mode, but also for all its multiples (i.e. nq, with $n < 100$). In this situation the number of unstable modes can be large even near the instability's threshold (since we are assuming ϵ to be small), and we can no longer reduce our description to any single amplitude equation. Nevertheless, we can take advantage of the small size of these unstable modes' wavevectors to reduce the equations of the model into another kind of simple equation, no longer an amplitude equation. This limit is sometimes called the hydrodynamic limit (with no direct relation to fluid flow). For example, we can make an expansion using ϵ in which the variables describing the dynamics of a system, such as the velocity in hydrodynamics, or chemical concentration, slowly vary. Note that here we are talking about the entire quantity describing dynamics and not only of the fast oscillation's envelope, e^{iqx}, introduced in our derivation of the amplitude equation; see chapter 10. In general, this type of expansion allows us to reduce the original set of equations which are strongly nonlinear and which might even be non-local (the dynamics at a point depending on faraway points) into weakly nonlinear partial differential equations (thus, local equations). A typical example of this is the Kuramoto–Sivashinsky equation, which we briefly mentioned in subsection 12.5.1, which can be written as:

$$u_t = -u_{xx} - u_{xxxx} + u_x^2, \tag{15.2}$$

where $u(x,t)$ is a function which may represent a chemical concentration, a front between two phases in a growth problem, etc. (see [74] for an overview of the Kuramoto–Sivashinsky equation). Unlike the amplitude equation, which is the same for all systems, the generic equations of these systems, studied at a large distance from the primary instability's threshold, are diverse. A list of the evolution equations can be organized relatively systematically using symmetries and conservation laws of a system (see [25]). This field of research is one of the most active (for a recent review article, see [73]).

Lastly, certain problems describing nonlinear systems are completely irreducible to any local equation (like the Kuramoto–Sivashinsky equation (15.2)). The non-local dynamics are important even in the so-called hydrodynamic limit (i.e. even when the wavevectors of the active modes are small, see [51]).

15.4. Coarsening in Non-equilibrium Systems

In our study of morphogenesis, we studied the emergence of both ordered periodic structures, characterized by regular spatial patterns (see chapter 9), as well as the emergence of some non-ordered structures, such as spatio-temporal chaos (see chapter 12). It must be noted that in the non-ordered structures that we studied, there does exist a spatial modulation in the solutions, which though complex does maintain a characteristic wavelength over time (see, for example, figure 12.3).

This example is fundamentally different from another category of systems which do not have any characteristic wavelength over time. These systems are instead characterized by a pattern which initially has a certain wavelength (or characteristic lengthscale) and then undergoes a coarsening process in which the characteristic lengthscale of the pattern grows with time (i.e. $\lambda \sim t^\alpha$, where α is called the coarsening exponent).

Though the concept of coarsening is relatively classic for systems in global equilibrium (think spinodal decomposition[4] [13]), open systems (or non-equilibrium systems) pose a formidable challenge and constitute one of the current main fields of research. Among the first promising results are the establishment of criteria which allow us to know if a nonlinear equation will or will not exhibit coarsening, without needing to solve the full equation on a computer (see [83], or, for a review, [73]). However we do not yet have any universal classification, nor is there one on the horizon.

15.5. Complex Forms in both Inert and Living Material

Morphogenesis is a domain of studies which is extremely rich and varied. We have chosen to limit our study in this book to the emergence of periodic structures over a spatial dimension (for example, the chemical patterns seen in chapter 9, like the waves in sand) without going into more general emergent phenomena. This first step, both simple in its physical incarnation as well as logically helpful before talking about forms in a larger sense, is a good introduction to the emergence of more or less complex forms in nature. The most famous example is that of snowflakes (see figure 15.1). Conceptually, the snowflake problem is similar to the Saffman–Taylor finger[5] (see figure 15.1) and to the most important diffusion-limited aggregation model, in which the growth is dominated by diffusion[6] (e.g. the *diffusion-limited aggregation* (DLA) model, see figure 15.1, introduced by Witten and Sander in 1981 [109]).

---

4. Take a binary alloy in a solid state, and suppose that it is made of two chemical substances A and B which are miscible. Most miscible alloys cease to be miscible below a certain critical temperature T_c. If we abruptly lower the temperature below T_c (i.e. a fast transition, or quench, to use the appropriate word), the miscible alloy becomes unstable and we see a separation of phases in the form of irregular patches of high concentration of either A or B: this is spinodal decomposition. Over time, the patches grow and join to form bigger patches until we are left with two macroscopic phases, one rich in A and the other rich in B; the size of the patches evolve following a temporal law $t^{1/3}$.

5. This structure, which resembles a finger, as indicated by the name, emerges from an instability also called the Saffman–Taylor instability, which occurs at the interface of two immiscible fluids of different viscosity.

6. This model describes the branched growth of a cluster of particles which are initially diffused randomly, and join the cluster when they come into contact with it.

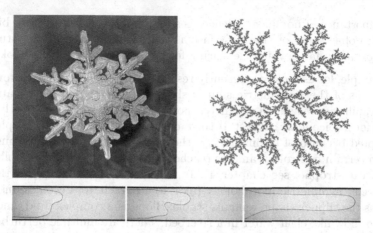

Figure 15.1 − Above, left: a snowflake [© Yaroslav Gnatuk #507798 Fotolia]; above, right: a structured branch found from diffusion-limited aggregation [All rights reserved]. The three bottom images represent the evolution of a Saffman−Taylor finger [With the permission of Marc Rabaud]. This happens by pushing a viscous fluid, such as oil, through a less viscous fluid, such as air, in a rectangular cell which has a certain length and width. The thickness in the perpendicular direction must be small with respect to the other two directions (we call this a Hele−Shaw cell). After an initial transitory stage (left and center figures), the finger attains a stationary state (right figure) and occupies about half of the canal.

These types of branched structures are also found in more complex fields, such as growth of bacteria colonies (see figure 15.2, and [5]), or in angiogenesis (the formation of blood vessels), to name a couple of examples. The snow crystal, Saffman−Taylor finger, and DLA models are now understood both theoretically and experimentally, but to include them in this book would have called for an introduction to many other advanced concepts.

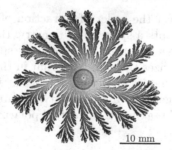

10 mm

Figure 15.2 − Bacteria (known as *Bacillus subtilis*) organize themselves, multiplying in the form of branches which sometimes resemble DLA structures. [All rights reserved]

The importance of nonlinear sciences is seen in all scientific domains; biology seems to offer some of the most fascinating opportunities for the study of morphogenesis, far beyond the Turing forms we have seen in this book.

For example, let us look at a recently resolved problem (2009) which revealed the origins of the asymmetric slipper shape of red cells when circulating in blood capillaries (even in a symmetric flow) as opposed to their perfectly symmetric form (biconcave) at rest. It turns out that this asymmetric morphology is adopted because of an instability: the symmetric form becomes unstable when a certain control parameter reaches a critical value (see [49]; following a *cusp* catastrophe, see chapter 4). It seems that when in a flow, the red blood cells' symmetric morphology (akin to a parachute) becomes unstable, whereas the slipper shape remains stable. Blood is a complex fluid, and thus does not flow like clear water in a riverbed. There are multiple perturbations which emerge in the flow through blood vessels. These small perturbations are responsible for initiating the bifurcation from parachute toward the blood cells' slipper shape when flowing through veins. Furthermore, the hemoglobin responsible for the transportation of oxygen remains at rest with respect to the red cell membrane when in the form of a parachute, whereas the slipper shaped (asymmetric) red globules stimulate an internal circulation of the hemoglobin, oxygenizing the cells and tissues. The model thus developed by researchers has allowed us to discover how hematites adopt their asymmetric, rather than symmetric, form in small blood vessels, and to identify the benefit this brings.

Let us cite another example in biology in which the importance of morphogenesis is found at the heart of the immune system. White cells (leucocytes) use a polymerization process engendering a network made of a protein in the cytoplasm (actin). Simplified laboratory experiments (see [10, 63, 110]) and theoretical analyses (see [48]) have recently shown that a layer of polymerized actin formed on the surface of artificial spherical beads[7] will undergo a spontaneous symmetry breaking.

Initially, the thickness (of the order of a fraction of a micron) of the coat of actin on the bead adopts a spherical symmetry. Beyond a certain critical thickness, the actin layer spontaneously loses its spherical symmetry; this symmetry breaking polarizes the cell. Note that in this case the polarization is initiated by the presence of a perturbation[8], but is an intrinsic property resulting from the emergence of an instability. Whatever the case, after this spontaneous symmetry breaking, the acquired polarization of the cell follows

---

7. The bead is plunged into an aqueous solution with actin molecules. This marble is covered by an enzyme which allows the actin molecules in the solution to polymerize in branched form.

8. In other biological systems, certain cells spontaneously break symmetry without exterior agents.

a process of polymerization which seems to indicate the direction of movement for the cell.

We would also like to very briefly point to another example in biology, that of cellular division (or mitosis). When we have cellular division, certain macromolecules – called microtubules – of the cytoplasm organize themselves in the form of stars (see figure 15.3).

These few examples show the ubiquity of instabilities and symmetry breaking in the living world; the nonlinear sciences allow a great deal of development in our understanding of these phenomena.

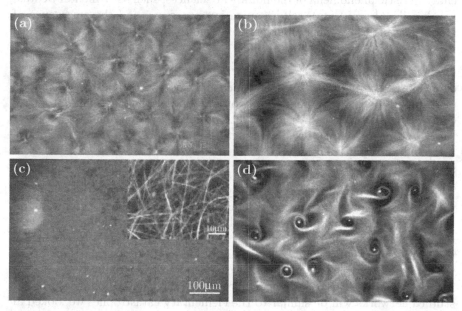

Figure 15.3 – In cellular division, macromolecules organize themselves into the shape of stars. The same type of organization is reproduced in simpler systems (a solution of microtubules and molecular motors). Depending on experimental parameters, different patterns can emerge: (a) vortices; (b) irregular stars; (c) microtubules forming filament structures; and (d) another pattern of vortices. [From [78] F.J. Nedelec, T. Surrey, A.C. Maggs & S. Leibler. Self-organization of microtubules and motors, *Nature*, **389**(22): 305–308, Sept 1997, reprinted with permission from Macmillan Publishers Ltd]

15.6. Toward a Science of Complex Systems

Besides biological sciences, nonlinear behavior is also seen in social and economic sciences, and more generally in all systems of networks, in the broad sense of the term[9], such as networks composed of nervous cells in our brain, or networks composed of financial agents in the economic market. These networks constitute a myriad of possible subjects for studying the ensemble of characteristic phenomena of the nonlinear sciences, such as elementary catastrophes (local accidents), self-organization processes, and the emergence of global properties or coherent structures, due to the attainment of a critical value by a control parameter (see [6]).

One of the notable characteristics of these systems is the presence of many elements in competition, some of which may be simple and others chaotic or disordered. This type of system is often found at the emerging borders of competitive behavior, indicating an alternation between dynamics that show the evolution of simple, ordered structure, and chaos and disorder. Several temporal and spatial scales appear in these systems, creating a hierarchy of structures.

The main challenge in studying these systems lies in the difficulty of reducing them to a simple set of equations. Hence why, generally, these systems are often first modeled using numerical simulations which take up a few interaction rules between elements in the network, rules for feedback, etc. Nevertheless, the work of describing complex systems can be made at least partially easier with the help of the nonlinear sciences discussed throughout this work. It is highly likely that the basic principles underlying catastrophes seen in the studied networks will be similar to the elementary catastrophes we looked at here, though the global macroscopic appearance may prove to be much more complex. Thus, the conceptual framework of instabilities, catastrophes, and elementary bifurcations forming part of the base of the nonlinear sciences can be useful in describing complex systems. The complexity and diversity of global appearances, as occur in nature (as happens in many natural phenomena), come from the statistical nature (such as intricate correlations) of the local catastrophes. A fundamental theory formalizing these ideas in an accessible framework remains to be constructed.

---

9. There is no consensus about the meaning of the name "complex systems". Here we have adopted a relatively simple definition which describes a system as complex when it is difficult or impossible to describe the system in the form of simple equations, which is the case for systems which have complex rules of interaction between the entities which make up the network (proteins in a cell; financial agents; etc.). Of course, it is possible that one day we will find a simple mathematical expression for these systems.

Chapter 16

Solutions to Exercises

Several exercises were inspired by a number of books and articles. Without being exhaustive, I have benefitted from the following books and articles: [6, 9, 16, 23, 24, 29, 32, 42, 58, 75, 77, 80, 86, 87, 91, 99, 101].

2.1

1. The potential energy takes the form

$$E_p = \frac{k}{2} \left\{ [(\ell_c + \theta x)^2 + x^2]^{1/2} - \ell_0 \right\}^2. \tag{16.1}$$

There are 3 fixed points given by

$$x_0 = -\frac{\ell_c \theta}{1 + \theta^2}, \quad x_0^\pm = -\ell_c \theta \pm \sqrt{\ell_c^2 \theta^2 - (1 + \theta^2)(\ell_c^2 - \ell_0^2)}. \tag{16.2}$$

The last two fixed points exist only for $\ell_c < \ell_c^* = (1 + \theta^2)^{1/2}\ell_0$, while the first fixed point always exists.

By analyzing the function $E_p(x)$ it is easy to see that for $\ell_c > \ell_c^*$, the only existing fixed point x_0 is stable (it is a minimum of E_p), whereas for $\ell_c < \ell_c^*$ it becomes unstable, while the two fixed points $x_0^\pm$ are stable.

2. The system undergoes a pitchfork bifurcation: a single solution becomes unstable at the critical point $\ell_c = \ell_c^*$, splitting into two stable branches. Close to the bifurcation point we can write

$$x_0^\pm = -\ell_c \theta \pm \sqrt{2\ell_c^*}\sqrt{\ell_c^* - \ell_c}. \tag{16.3}$$

These branches behave as $\sqrt{\ell_c^* - \ell_c}$. This is typical for a pitchfork bifurcation.

© Springer Science+Business Media B.V. 2017
C. Misbah, *Complex Dynamics and Morphogenesis*,
DOI 10.1007/978-94-024-1020-4_16

3. The energy close to the critical point x_0 reads (apart from an unimportant additive constant)

$$E_p(x) = \frac{\ell_c - \ell_c^*}{\ell_0}(x - x_0)^2 + \frac{1}{4}\frac{(\theta^2 + 1)^2}{\ell_0^2}(x - x_0)^4. \qquad (16.4)$$

Note, this is the potential form of a pitchfork bifurcation.

2.2

1. There are three fixed points

$$x_0 = 0, \quad x_0^\pm = \pm\sqrt{-\epsilon}. \qquad (16.5)$$

x_0 always exists, while $x_0^\pm$ exists only if $\epsilon < 0$.

2. The equation reads

$$\dot{x} = -V', \quad V = -\frac{\epsilon}{2}x^2 - \frac{1}{4}x^4, \qquad (16.6)$$

where prime designates a derivative with respect to x. Figure 16.1 shows the behavior of V for $\epsilon > 0$ and $\epsilon < 0$. From this figure we can deduce the stability (a maximum means instability while a minimum means stability); the solution x_0 is stable for $\epsilon < 0$ and unstable for $\epsilon > 0$. The solution $x_0^\pm$, when it exists, is always unstable.

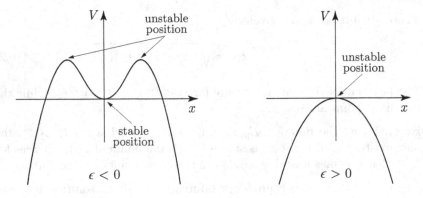

Figure 16.1 – The qualitative behavior of the potential.

3. Figure 16.2 shows the bifurcation diagram.

4. The linearized equation reads $\dot{x}_1 = (\epsilon + 3x_0^2)x_1$, and if we look for a perturbation of the form $x_1 \sim e^{\omega t}$, we find $\omega = (\epsilon + 3x_0^2)$; a positive (negative) ω signifies instability (stability). For $x_0 = 0$, we have $\omega = \epsilon$ and the solution is stable for $\epsilon < 0$ and unstable otherwise. The solutions $x_0^\pm$ give $\omega = -2\epsilon$, which are always unstable since they exist only for $\epsilon < 0$.

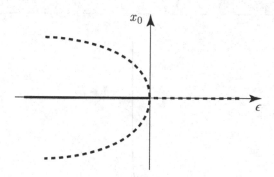

Figure 16.2 – The qualitative behavior of the bifurcation diagram.

5. For $\epsilon > 0$ there is no stable solution. In order to obtain stable solutions, there are several possibilities: for example, change the sign of the cubic term, or keep the same sign and introduce the next order quintic term with a negative sign.

6. We set $y = 1/x^2$, and the equation becomes linear $\dot{y} = -2(\epsilon y + 1)$. From here, we can easily find a solution (particular solution plus the solution of the homogeneous equation). We find (after going back to x)

$$x = \text{sign}(x(0))\sqrt{\frac{\epsilon}{(1 + \epsilon/x(0)^2)e^{-2\epsilon t} - 1}}. \tag{16.7}$$

The argument must be positive and $x(0)$ is the initial value.

For $\epsilon < 0$ one sees that at $t \to \infty$, $x \to 0$, which is in agreement with our previous result; in this case the solution always exists.

For $\epsilon > 0$ the solution exists as long as $t < t_c = \frac{1}{2\epsilon}\ln(1 + \epsilon/x(0)^2)$. As $t \to t_c$ the solution diverges. We have seen above that for $\epsilon > 0$ there is no stable solution. A ball set at the maximum of the potential will fall indefinitely (figure 16.3 shows the solution for $\epsilon > 0$) without bound. The interesting fact here is that the infinite position is obtained at finite time t_c. This is usually referred to as *finite time singularity*.

2.3

1. If ℓ_0 is the length at rest, then the potential energy reads $E = k(\ell' - \ell_0)^2/2$. It is easy to find a relation between ℓ', ℓ, r and ϕ, so that the energy takes the form

$$E = \frac{k}{2}\left[\sqrt{r^2 + (r + \ell)^2 - 2r(r + \ell)\cos(\phi)} - \ell_0\right]^2. \tag{16.8}$$

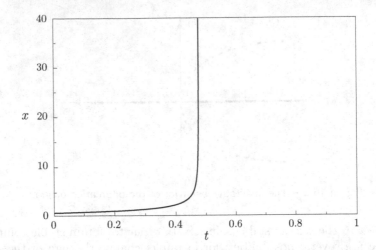

Figure 16.3 – The behavior of $x(t)$ showing a finite time singularity. We have chosen $\epsilon = 0.1$ and $x(0) = 1$. We see a finite time singularity at $t = t_c \simeq 0.476$.

2. The fixed points are

$$\phi = 0, \pi, \quad \phi_\pm = \pm \arccos \left[1 - \frac{\ell_0^2 - \ell^2}{2r(\ell + r)} \right]. \tag{16.9}$$

The first two solutions always exist, whereas the two others are subject to the condition $\ell_0 > \ell$. It can be checked by plotting $E(\phi)$ that for $\ell_0 < \ell$ the fixed point $\phi = 0$ is stable, and becomes unstable for $\ell_0 > \ell$, where a pair of two solutions $\phi_\pm$ emerge and are stable (E has a minimum at these two points). The fixed point $\phi = \pi$ is always unstable.

3. We have a pitchfork bifurcation, since one branch always exists and becomes unstable at a point where two new branches emerge. The solution $\phi_\pm$ behaves as $\pm\sqrt{\ell_0 - \ell}$ close to the bifurcation point.

4. The bifurcation point corresponds to $\ell = \ell_0$, where $\phi = 0$. Expansion of E yields

$$E \simeq \frac{k}{2} \frac{(\ell_0 - \ell)r(\ell_0 + r)}{\ell_0} \phi^2 + k \frac{r^2(\ell_0 + r)^2}{4\ell_0^2} \phi^4. \tag{16.10}$$

Note that terms like $O((\ell - \ell_0)^2)$ and $O((\ell - \ell_0)\phi^2)$ are left out of the solution. This is a typical form for the energy of a pitchfork bifurcation – we reach the same conclusions regarding the fixed points and their stability. The approximate expression of E provides the same qualitative picture as the full energy (i.e. without expansion).

2.4

1. Newton's law states that the mass m times the acceleration $r\ddot{\phi}$ is equal to the sum of the total forces. As usual, the frictional force is $\mu\dot{\phi}$. The contribution from gravity is exactly the same as with a pendulum: $mg\sin(\phi)$. The rotation of the hoop implies a centrifugal force directed along ρ with magnitude $m\rho\omega^2$. We have $\rho = r\sin(\phi)$. In addition, the centrifugal force must be projected along the circle (as with gravitational force), since the motion takes place along the circle. The projected force is thus given by $mr\omega^2\sin(\phi)\cos(\phi)$.

2. After inspecting different terms one finds that inertia is negligible if $m^2gr/\mu^2 \ll 1$.

3. We can write the equation (in the absence of inertia) as

$$\mu\dot{\phi} = mg\sin(\phi)\left[\frac{r\omega^2}{g}\cos(\phi) - 1\right]. \tag{16.11}$$

   The fixed points are $\qquad \phi = 0, \quad \phi = \pi, \tag{16.12}$

   and $\qquad\qquad \phi = \pm\arccos\left(\frac{g}{r\omega^2}\right), \quad \frac{g}{r\omega^2} < 1. \tag{16.13}$

   The first two fixed points always exist, while the two others are subject to the condition $g/r\omega^2 < 1$.

4. We can write

$$\mu\dot{\phi} = mg\sin(\phi)[\nu\cos(\phi) - 1]. \tag{16.14}$$

   If $\nu < 1$, it is clear that in the vicinity of $\phi = 0$ we have $\dot{\phi} \sim -\phi$ so that $\phi = 0$ is stable, while in the vicinity of $\phi = \pi$ we have $\dot{\phi} \sim \phi$, signifying an unstable fixed point. When $\nu > 1$, but still in the vicinity of $\phi = 0$, we have $\dot{\phi} \sim \phi$ meaning that ϕ has changed from stable to unstable fixed point, and at the same time we have two new branches $\phi = \pm\arccos(1/\nu)$. It is easy to show by a Taylor expansion around these two points that these are stable branches. We thus have a pitchfork bifurcation.

5. Rotation of the hoop creates a centrifugal force. The more the mass rises towards $\phi = \pi/2$, the greater the centrifugal force. Thus rotation will tend to raise the mass towards the position $\phi = \pi/2$. However, if ν is small, the weight will dominate and the ball slide towards the bottom. There is a critical value ($\nu = 1$) at which the centrifugal force and the weight are balanced, at which point the mass starts progressively rising. The mass, initially at the bottom (for example), has the choice to climb along the right or left arch of the hoop depending on initial conditions.

6.
$$\frac{\mu}{mg}\dot{\phi} = -\frac{\partial V}{\partial\phi}, \quad V = \frac{\nu}{4}\cos(2\phi) - \cos(\phi). \tag{16.15}$$

   Analyzing this potential, we can easily deduce the stability. Figure 16.4 shows some typical forms of the potential.

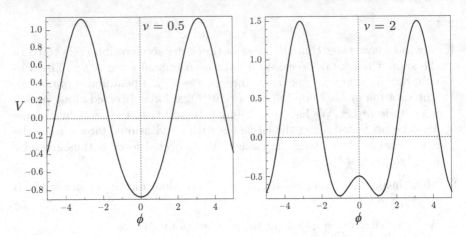

Figure 16.4 − The behavior of the potential for two different values of ν.

7. In the vicinity of the bifurcation, the energy reads (apart from an additive constant)

$$V = \frac{(1-\nu)}{2}\phi^2 + \frac{4\nu - 1}{24}\phi^4. \qquad (16.16)$$

This is the classic expression for a pitchfork bifurcation. For $\nu > 1$, the trivial solution is unstable and the quartic term always has a positive coefficient, ensuring a supercritical bifurcation.

8. Using the above results one easily finds that the equation becomes, to leading order

$$\frac{\mu}{mg}\dot\phi = (\nu - 1)\phi - \frac{4\nu - 1}{6}\phi^3. \qquad (16.17)$$

This is the form of a pitchfork bifurcation.

3.1

For the first case we have two fixed points

$$x_\pm = \frac{\epsilon \pm \sqrt{\epsilon^2 - 4\nu}}{2}. \qquad (16.18)$$

For $\nu < 0$ these fixed points always exist, whereas for $\nu > 0$ they are subject to the condition $\epsilon^2 > 4\nu$. In order to determine their stability we can rewrite the equation in a potential form; the potential is $V(x) = \epsilon x^2/2 - x^3/3 - \nu x$. For $\nu < 0$ there is neither change in stability nor in number of solutions. We cannot truly call it a bifurcation. On the other hand, we have two saddle node bifurcations for $\nu > 0$. We can intuitively understand the bifurcation diagram by first setting $\nu = 0$ (where we have a transcritical bifurcation) and guess graphically as to how the presence of ν modifies the bifurcation diagram. The results are summarized in figure 16.5.

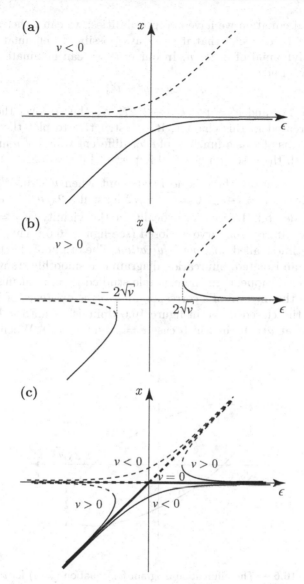

Figure 16.5 − The bifurcation diagram corresponding to equation (3.70) for different cases. (a) $\nu < 0$ where the solution always exists. (b) $\nu > 0$ where the solutions exist if $\epsilon^2 > 4\nu$. We have saddle-node bifurcations. (c) We show the case $\nu = 0$ where we have a transcritical bifurcation, from which we can deduce the two possible diagrams for $\nu \neq 0$. For $\nu \neq 0$, the branches cannot cross the ϵ-axis, since $x = 0$ is not a fixed point. Supposing ν is small, at large ϵ we should recover the branches corresponding to $\nu = 0$. Finally, for $\epsilon = 0$ we have $x^2 + \nu = 0$ which is possible if $\nu < 0$. In this case the branches cross the x-axis at $\epsilon = 0$. For $\nu > 0$ there is no crossing. With this information the bifurcation diagram can easily be drawn.

For the second equation we have two possibilities: we can either analyze the equation as it is, or recall that it is always possible to eliminate the $n - 1$ term in a polynomial of order n. In our case we can eliminate x^2. Setting $x = y - 1/3$ we find

$$\dot{y} = \epsilon' y - y^3 + \nu', \tag{16.19}$$

with $\epsilon' = \epsilon + 1/3$ and $\nu' = \nu - \epsilon/3 - 2/27$. We thus recover the imperfect bifurcation treated in this chapter. It is instructive to plot the bifurcation diagram in terms of x as a function of ϵ for different values of ν and compare the results with the case of a classical imperfect bifurcation.

For the third equation, there is no fixed point when $\nu < 0$. When $\nu > 0$, two fixed points $x_{\pm} = (-\epsilon \pm \sqrt{4\nu - 3\epsilon^2})/2$ exist if $-2\sqrt{\nu/3} < \epsilon < 2\sqrt{\nu/3}$. We have saddle-node bifurcations locally in the vicinity of $\epsilon = \pm 2\sqrt{\nu/3}$. The complete bifurcation curve is closed (see figure 16.6). This type of diagram is sometimes called an *isola bifurcation*. The origin of this name comes from the way an isolated, bifurcation diagram can smoothly transform into a closed curve or disappear. In the example studied here, reducing ν continuously reduces the interval of ϵ corresponding to the existence of fixed points accordingly; the closed curve in figure 16.6 diminishes until it becomes an isolated point at $\nu = 0$ (in which case $\epsilon = 0$ and $x = 0$). When $\nu < 0$ this point disappears.

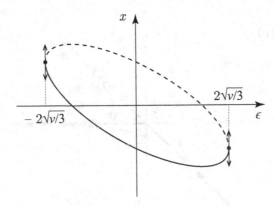

Figure 16.6 – The bifurcation diagram for equation (3.72) for $\nu > 0$.

3.2

This form corresponds to an imperfect bifurcation, as seen in this chapter. The turning point is a point where the bifurcation curve in the plane (x, ϵ) has a vertical tangent. Let us first determine that point. We have $\epsilon x_0 - x_0^3 + \nu = 0$ (we suppose $\nu > 0$). Taking the derivative with respect to ϵ we obtain $dx_0/d\epsilon = 1/(3x_0^2 - \epsilon)$, meaning that the turning point is given by $x_0 = \pm\sqrt{\epsilon/3}$. Since $\nu > 0$ it can easily be checked that the relevant solution is $x_0 = -\sqrt{\epsilon/3}$. Plugging this into the fixed point equation we find $\epsilon = 2(\nu/3)^{3/2} \equiv \epsilon^*$. Let us set

$$x = x_0 + A, \quad \epsilon = \epsilon^* + \mu, \tag{16.20}$$

where A and μ are small deviations from the turning point. The evolution equation for x becomes

$$\dot{A} = (\epsilon^* + \mu)(x_0 + A) - (x_0 + A)^3 + \nu. \tag{16.21}$$

Expanding and using the fact that $\epsilon^* x_0 - x_0^3 + \nu = 0$ and that $\epsilon^* - 3x_0^2 = 0$, one finds (neglecting higher order terms in A^3 and μA)

$$\dot{A} = \mu x_0 - 3x_0 A^2, \tag{16.22}$$

which has the form of a saddle-node bifurcation. The fixed point solution of this equation tells us that $A \sim \sqrt{\mu}$: in other words $\mu A \sim \mu^{3/2}$ is small in comparison to μ for small μ, which justifies neglecting μA above. The same applies to the term A^3.

3.3

1. There is a trivial fixed point $x = 0$. Its stability is evident for if x_1 designates the deviation from $x = 0$, we easily find $\dot{x}_1 = rx_1$ whose solution $x_1 \sim e^{rt}$ grows with time, indicating an instability.

2. The other fixed points obey

$$r\left(1 - \frac{x}{k}\right) = \frac{x}{1 + x^2}. \tag{16.23}$$

   This leads to a cubic equation, the solution of which is not always of practical interest. We can thus turn to graphical arguments.

3. Figure 16.7 shows the two curves representing the two sides of equation (16.23). By varying r for example (at fixed k) we see that there is either one or three intersections. When r is too large there is again only a single intersection.

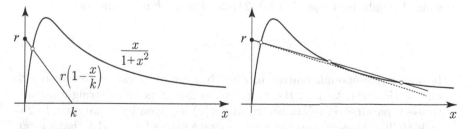

Figure 16.7 – Graphical resolution of equation (16.23). One sees that there are one or three intersections depending on k (for a given r).

4. One can plot the potential to study the stability. However, since $x = 0$ has already been determined unstable, the stability of the other fixed points can be easily deduced: the next fixed point (going from small to large x) will be stable, the next unstable, the next stable, and so on. In other words the stable and unstable fixed points alternate. For the potential this means (assuming a continuous function, as is here the case) that one goes from one minimum to the next necessarily passing through a maximum, and so on. Figure 16.8 shows the bifurcation diagram as a function of r for a given k. This bifurcation looks like a subcritical bifurcation with only half its branches ($x > 0$), but where the curve does not cross the x-axis, reminiscent of an imperfect bifurcation.

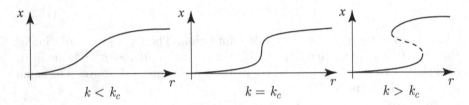

Figure 16.8 – The bifurcation diagram in the plane (x, r).

The nature of the problem loses its ambiguity if we look at it from the point of catatrophes: the critical condition for passing from one to three solutions corresponds graphically to the tangential point between the straight line and the curve $x/(x^2 + 1)$ in figure 16.7. Differentiating each term in equation (16.23) with respect to x and setting them equal we have

$$\frac{r}{k} = \frac{x^2 - 1}{(1 + x^2)^2}. \tag{16.24}$$

Inserting this into equation (16.23) yields an expression of r as a function of x

$$r = \frac{2x^3}{(1 + x^2)^2}. \tag{16.25}$$

which brought into equation (16.24) determines k as a function x

$$k = \frac{2x^3}{x - 1}. \tag{16.26}$$

It is a simple enough matter to directly express a relation between r/k and k. However, to plot the catastrophe line it is more straightforward to use a parametric relation $r(x)$ and $k(x)$ as given by equations (16.25) and (16.26). Thus by varying x we generate values for r and k that we can plot. This is represented in figure 16.9. We recognize the cusp catastrophe.

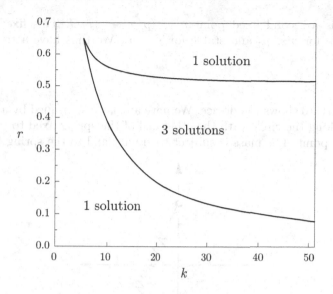

Figure 16.9 – The catastrophe portrait.

3.4

The first case corresponds to a transcritical bifurcation. For $\epsilon < 1$, $x > 0$ ($x \to \infty$ for $\epsilon \to 0^+$) is an unstable branch and $x = 0$ is a stable branch. For $\epsilon > 1$, $x < 0$ is a stable branch and $x = 0$ is an unstable branch. $\epsilon = 1$ corresponds to the critical value for a transcritical bifurcation.

The second case corresponds to a transcritical bifurcation where $x = 0$ is unstable for $\epsilon < 1$, while the branch $x \neq 0$ is stable. $x = 0$ becomes stable for $\epsilon > 1$ the same as $x \neq 0$ becomes unstable.

The third case also represents a transcritical bifurcation occurring at $\epsilon = 1$ where $x = 0$ (which exists for any ϵ) becomes unstable while the branch $x = \ln(\epsilon)$, which exists for $0 < \epsilon < \infty$, is unstable for $\epsilon < 1$ and becomes stable for $\epsilon > 1$.

3.5

1. By using the adiabatic approximation ($\dot{N} \simeq 0$) one obtains $N = p/(Gn + f)$, so that n obeys

$$\dot{n} = \frac{Gpn}{Gn + f} - kn. \qquad (16.27)$$

2. There is always a fixed point $n_0 = 0$, which is stable for $p < p_c = fk/G$ and becomes unstable for $p > p_c$.

3. There is a second fixed point $n = (p/p_c - 1)f/G$. This fixed point is unstable for $p < p_c$ and stable for $p > p_c$. We thus have a transcritical bifurcation.

3.6

1. Figure 16.10 shows the device. We have a mass constrained by a spring to move along the circle with the other end of the spring fixed to the circle's highest point. The mass is subject to gravity and to the spring force.

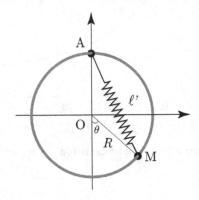

Figure 16.10 – A schematic view of the mechanical device.

We have the following relation between vectors: $\mathbf{AO} + \mathbf{OM} = \mathbf{AM}$. Taking the scalar product on both sides, we find (after simple trigonometric manipulations) $\ell' = 2R\cos(\theta/2)$, where ℓ' is the spring length at a given position and R is the radius of the circle. The spring force must be projected along the trajectory (the tangent to the circle), giving $k[2R\cos(\theta/2) - \ell]\sin(\theta/2)$ where ℓ is the spring length at rest. The force of gravity projected along the tangent is $-mg\sin(\theta)$. The system under consideration corresponds to an overdamped motion (inertia is negligible), so the equation of motion reads

$$\mu\dot{\theta} = 2Rk\left(\cos(\theta/2) - \frac{\ell}{2R}\right)\sin(\theta/2) - mg\sin(\theta), \qquad (16.28)$$

which we can rewrite as

$$\mu\dot{\theta} = c(\cos(\theta/2) - a)\sin(\theta/2) - b\sin(\theta), \qquad (16.29)$$

where $a = \ell/(2R)$, $b = mg$, $c = 2kR$.

2. One trivial fixed point, $\theta_0 = 0$, exists as well as two other fixed points, $\theta_\pm = \pm 2\arccos[k\ell/(2(kR - mg))]$, which exist only when the argument is smaller than one and positive (no bifurcation for $c < 2b$). θ_0 is stable for $k < mg/(R - \ell/2)$ (and unstable otherwise), whereas $\theta_\pm$ are stable for $k > mg/(R - \ell/2)$.

3. We have a pitchfork bifurcation in which a solution existing for all parameters loses its stability in favor of a branch of two stable solutions.

3.7

Dividing the equation by ν and setting $A' = A/(\nu^{1/3})$, $\epsilon' = \epsilon/(\nu^{2/3})$ we get $A'^3 - \epsilon'A + 1 = 0$ for the amplitude equation. The linear term can be neglected for $\epsilon' \ll 1$. This means that for the original equation the linear term can be neglected for $\epsilon \ll \nu^{2/3}$.

3.8

Searching for double roots of the equation we get $\epsilon = 5A^4 - 3A^2$ and $4A^5 - 2A^2 + \nu = 0$. The roots of the last equation correspond to the bifurcations of equation (3.81). A basic analysis suggests that for small absolute values of ν, there are 3 bifurcation points, consistent with figure 16.11. However for large absolute values of ν there is only one bifurcation point.

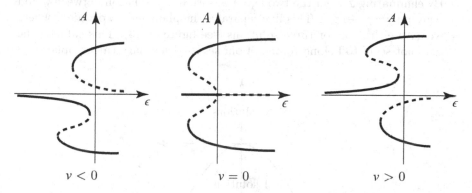

$$\nu < 0 \qquad\qquad \nu = 0 \qquad\qquad \nu > 0$$

Figure 16.11 — Different scenarios of imperfect subcritical bifurcation.

4.1

1. For $v = 0$ we have a supercritical pitchfork bifurcation. We have seen in this chapter that this bifurcation is not structurally stable. The presence of v changes the bifurcation diagram into a qualitatively new diagram. For that purpose it is easy to find a change of variable (already seen in this chapter) in order to eliminate the x^2 term, but by introducing a constant term in the equation, which is nothing but the form of an imperfect bifurcation we have seen in this chapter, and earlier.

2. When $v \neq 0$ the bifurcation diagram changes qualitatively (see figure 16.12). The solution $x = 0$ always exists and is stable for $u < 0$ and unstable for $u > 0$. There are two other solutions $x_\pm = (v \pm \sqrt{v^2 + 4u})/2$ which exist if $u > -\frac{v^2}{4}$.

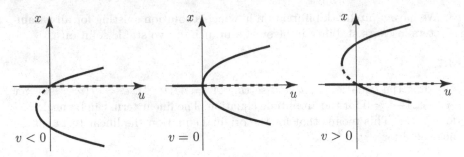

Figure 16.12 – Different scenarios of imperfect subcritical bifurcation.

3. Setting $V' = 0$ and $V'' = 0$ (where V is the potential), we obtain

$$u + vx - x^2 = 0, \quad u + 2vx - 3x^2 = 0. \tag{16.30}$$

By eliminating x from the two equations we find a relation between u and v given by $u = -v^2/4$. This line separates the plane into two regions where we have either one or three solutions (see figure 16.13). This catastrophe does not seem to belong to one of our seven elementary catastrophes.

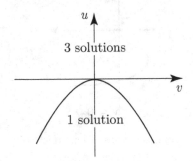

Figure 16.13 – The catastrophe line $u = -v^2/4$ separates the region with three solutions from that with a single solution.

However, if one makes the following change of variable $x = y + v/3$ one obtains

$$\dot{y} = u'y - y^3 + v', \tag{16.31}$$

where $u' = u + v^2/3$ and $v' = uv/3 + 2v^3/27$. In terms of u' and v' we know that the catastrophe line is given by the relation $u' \sim v'^{2/3}$. Using the relation between u', v' and u and v one can check that the catastrophe line is given by $u = -v^2/4$. Remember the convention used in catastrophe theory in which one eliminates from a polynomial of order n the $(n-1)$th power (this is always possible with the adequate change of variables).

4.2

1. The equation can be written as

$$PV^3 - (Pb + RT)V^2 + aV - ab = 0. \tag{16.32}$$

The solution is triply degenerate if the first and second derivatives are zero. These conditions yield

$$3PV^2 - 2V(Pb + RT) + a = 0, \quad 6PV - 2(RT + Pb) = 0. \tag{16.33}$$

Taking the difference between the first equation, and the second equation times V, one obtains $3PV^2 = a$. Then, taking the difference between twice the first equation, and the second equation times V, one obtains $V(RT + Pb) = a$. Then, in equation (16.32), replace P by $a/(3V^2)$ in the first term and $Pb + RT$ by a/V in the second to find $V = 3b \equiv V_c$. Substitute this into $3PV^2 = a$ to find $P = a/(27b^2) \equiv P_c$, and into $V(RT + Pb) = a$ to find $T = 8a/(27bR) \equiv T_c$.

2. Figure 16.14 shows the qualitative behavior of $P(V)$ for different values of T.

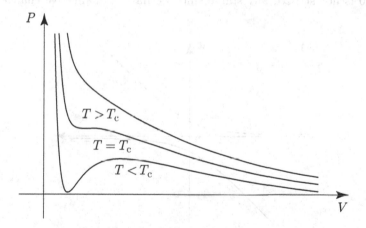

Figure 16.14 — Different behaviors of $P(V)$.

3. A substitution into equation (16.32) and simple algebraic manipulations yield $F(\epsilon, p, \phi) = 0$, where F has the form

$$F = \phi^3 + \frac{1}{3}(8\epsilon + p)\phi + \frac{2}{3}(4\epsilon - p). \tag{16.34}$$

A double solution means that not only F but the derivative of F vanishes as well

$$3\phi^2 + \frac{1}{3}(8\epsilon + p) = 0. \tag{16.35}$$

Multiply the equation given by $F = 0$ by ϕ and the above equation by 3 and take the difference to obtain $\phi = -(4\epsilon - p)/(8\epsilon + p)$, which upon substitution into equation (16.35) yields

$$81(4\epsilon - p)^2 = -(8\epsilon + p)^3. \tag{16.36}$$

If we redefine the control parameters as $u = 4\epsilon - p$ and $\nu = 8\epsilon + p$ we have $81u^2 = -\nu^3$ which is the cusp catastrophe we've already encountered in this chapter. The cusp is located at $u = \nu = 0$, or $\epsilon = p = 0$. Along this line the first and second derivatives of the potential (associated with the fixed point equation) vanish. At the special point $(\epsilon, p) = (0, 0)$ the third derivative is also zero. Thus this point corresponds to the strongest singularity, and is thus the least structurally stable.

4.3

If $\nu = 0$ we have a transcritical bifurcation.

1. There exist two solutions $x_\pm = (\epsilon \pm \sqrt{\epsilon^2 + 4\nu})/2$. For $\nu > 0$ the solution always exists, while for $\nu < 0$ it is constrained to $\epsilon < -2\sqrt{-\nu}$ and $\epsilon > 2\sqrt{-\nu}$. Figure 16.15 shows the two cases $\nu > 0$ and $\nu < 0$. The bifurcation for $\nu = 0$ is not structurally stable since we have a qualitative change when $\nu \neq 0$.

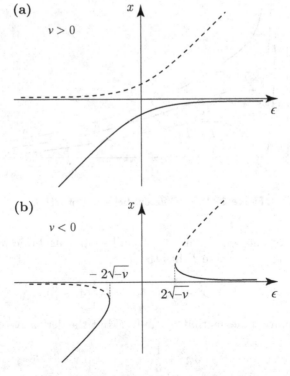

Figure 16.15 – The bifurcation diagram.

2. In order to determine the catastrophe line, we have to find the relation between ϵ and ν for which the first and second derivatives of the potential are zero. This is expressed as $x^2 - \epsilon x - \nu = 0$ and $2x - \epsilon = 0$, which yields $\nu = -\epsilon^2$, from which we trace the catastrophe line, shown in figure 16.16. Above the curve we have two solutions, below there are none. A vanishing solution is reminiscent of the saddle node bifurcation, belonging to the fold catastrophe. In fact the present catastrophe is indeed equivalent to a fold catastrophe. To see this, we set $x = y + \epsilon/2$ and find $\dot{y} = y^2 - \epsilon'$, with $\epsilon' = \epsilon^2/4 + \nu$. Recall once more that the convention in the catastrophe theory is to first eliminate from the potential the power of order $n - 1$ when the highest power is n.

3. Figure 16.16 shows the potential in each region of parameter space.

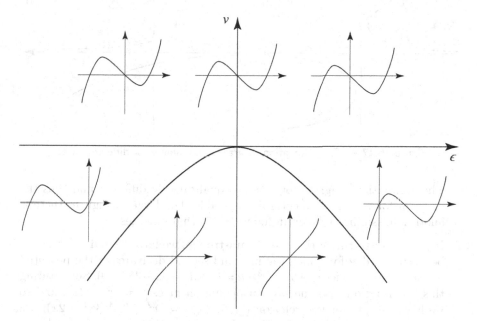

Figure 16.16 – The catastrophe line.

4.4

If $\nu = 0$, there is a trivial fixed point $x_0 = 0$. Two other fixed points $x_\pm = 1 \pm \sqrt{1 + 4\epsilon}$ exist if $\epsilon > -1/4$. Figure 16.17 shows the bifurcation diagram and stability.

1. For $\nu \neq 0$ we can first venture a graphical guess. Begin by looking at $\nu > 0$. If ν is arbitrarily small we can expect the bifurcation diagram of figure 16.17 to be quantitatively close, but since $\nu \neq 0$ the branches cannot cross the horizontal axis. Figure 16.17 shows the typical behavior.

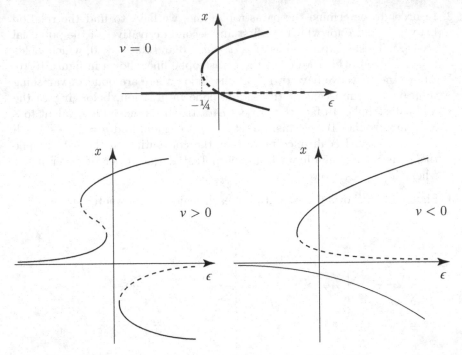

Figure 16.17 – The bifurcation diagram and stability for different cases.

The bifurcation diagram for $\nu > 0$ is qualitatively different and thus the diagram for $\nu = 0$ is structurally unstable. For large enough ν, there is another possibility (shown in figure 16.17) that emerges.

2. In order to determine the line of catastrophe, we have to find the relation between ϵ and ν for which the first and second derivates of the potential are null. This implies $\epsilon x + x^2 - x^3 + \nu = 0$ and $\epsilon + 2x - 3x^2 = 0$. Substituting this last relation into the previous equation gives $\nu = x^2 - 2x^3$, and so we have the parametric relation $(\nu(x), \epsilon(x)) = (x^2 - 2x^3, 3x^2 - 2x)$. We could attempt to eliminate x from the two equations to obtain a relation between ϵ and ν, but alternatively, we could consider x as a parameter from which we determine values for ν and ϵ. Figure 16.18 shows the line of catastrophe; we recognize a cusp catastrophe. Note that when ν is small enough, by varying ϵ from left to right, one passes from a region with three solutions, to a region with a single solution, and then again to three solutions. For large enough ν, we pass from three solutions to a single solution. The alternation of regions (from three to a single solution) is different from that of the cusp catastrophe discussed in figure 3.28, since we have not eliminated the second power in the fixed point equation, contrary to catastrophy theory convention.

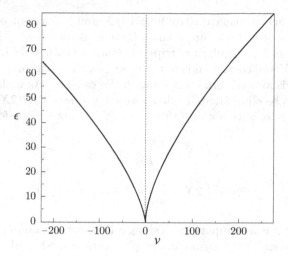

Figure 16.18 − The line of catastrophe showing a cusp catastrophe.

Remember that one can always remove the x^2 term via an adequate change of variable so that the equation becomes $\dot{y} = \epsilon' y - y^3 + \nu'$ where ϵ' and ν' are related to ϵ and ν by $\epsilon' = \epsilon + 1/3$ and $\nu' = \nu + \epsilon/3 + 2/27$. We return thus to the classical cusp catastrophe treated in this chapter. We can use the derived relation for the catastrophe line in terms of ϵ' and ν' and convert it into ϵ and ν in order to determine the form shown in figure 16.18.

3. The potential can easily be plotted in each region.

4.5

1. The point M belongs to a circle described by $(x + 1/2)^2 + y^2 = 1$, so that $x = -1/2 \pm \sqrt{1 - y^2}$. Considering x to be positive, we have $x = -1/2 + \sqrt{1 - y^2}$. Thus the point M on the circle has coordinates $(-1/2 + \sqrt{1 - y^2}, y)$.

2. The source has coordinates $(-d, y)$ so the distance between the source and M is given by $d - 1/2 + \sqrt{1 - y^2}$. The distance between point M and a target with coordinates (X, Y) is $[(X + 1/2 - \sqrt{1 - y^2})^2 + (Y - y)^2]^{1/2}$. The total length is the sum of these two lengths, yielding

$$L = d - \frac{1}{2} + \sqrt{1 - y^2} + \left[(X + \frac{1}{2} - \sqrt{1 - y^2})^2 + (Y - y)^2 \right]^{1/2}. \quad (16.37)$$

3. Using a Taylor expansion we find

$$L = - \left(2X + \frac{1}{4} \right) y^4 + 2Yy^3 + 2Xy^2 - 2Yy + (d + 1 - X). \quad (16.38)$$

We have clearly a cusp catastrophe, with X and Y our control parameters. For $X = Y = 0$ we have (up to an additive constant) $L = -(\frac{1}{4})y^4$, which is the highest order singularity (triply degenerate with $L' = L'' = L''' = 0$). Thus $X = Y = 0$ corresponds to the most singular location of the catastrophe, which is also the brightest point of the caustics. As we have seen, the y^3 term can be eliminated. To do so we set $y = w - Y/[2(2X + 1/4)]$, and carrying it into L we obtain (after expanding to order one for Y and X)

$$L = -\left(2X + \frac{1}{4}\right)w^4 + 2Xy^2 - 2Yy$$

$$= -\left(2X + \frac{1}{4}\right)[w^4 - 8Xw^2 + 8Yw], \qquad (16.39)$$

The prefactor is not important. Taking a look at the quantity $w^4 - 8Xw^2 + 8Yw$ we recognize the cusp catastrophe, with $\nu = 8Y$ and $v = -8X$ (see equation (4.11)), whose catastrophe line obeys the equation obtained from setting expression (4.13) to zero. The equation reads (in terms of X and Y)

$$Y = \pm\frac{8}{\sqrt{27}}X^{3/2}. \qquad (16.40)$$

4. The catastrophe line is plotted in figure 16.19.

5. The cusp is located at $(X, Y) = (0,0)$, which is the most singular point of the catastrophe and coincides with the brightest spot of the caustics.

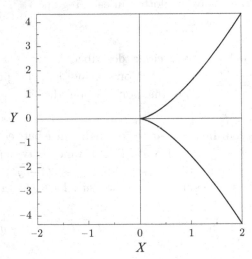

Figure 16.19 – The catastrophe line corresponding to caustics.

5.1

1. There is a fixed point at $(x, y) = (0, 0)$. Linear stability analysis (by looking for perturbations in the form $e^{\omega t}$) yields a second order equation for ω with solutions $\omega_\pm = \mu \pm i$. If $\mu > 0$ the perturbation grows exponentially. The eigenvalue has a non-zero imaginary component. This is a Hopf bifurcation, occurring at $\mu = 0$.

2. In general the nature of a bifurcation can only be handled by a careful non-linear analysis. However, here, due to the relative simplicity of the system, we can reach a conclusion by other means. To do so we set $x = r \cos(\theta)$ and $y = r \sin(\theta)$ and substitute these into the system, finding

$$\dot{r} = \mu r + r y^2. \tag{16.41}$$

   For $\mu > 0$ the trivial solution is linearly unstable (in agreement with above). The nonlinear term $r y^2$ is positive, enhancing the linear growth: thus, there is no nonlinear saturation. The bifurcation is subcritical.

5.2

The unique fixed point is $x_0 = a$, the linear stability of which yields (by seeking a solution in the form $\sim e^{\omega t}$) $\omega^2 + \mu(a^2 - 1)\omega + 1 = 0$. We have a Hopf bifurcation at $|a| = 1$ (unstable if $-1 < a < 1$ and stable otherwise), provided that $\mu > 0$. At the bifurcation point the imaginary component of ω is equal to $\pm i$. The real component vanishes for $|a| = 1$, is negative for $|a| > 1$ and positive for $|a| < 1$. In conclusion, we have a Hopf bifurcation if $\mu > 0$ and $-1 < a < 1$.

5.3

1. It is a simple matter to check the fixed points.

2. The fixed point $(0, 0)$ is stable (more precisely x is neutral and y stable). The fixed point $(1, 0)$ has a neutral eigenvalue, and another eigenvalue $\omega = 1 - a$, meaning that there is stability for $a > 1$ and instability $a < 1$. Finally, the last fixed point has eigenvalues $2\omega_\pm = a - 2a^2 \pm a\sqrt{4a^2 - 3}$. For $a = a_c = 1/2$ the two eigenvalues are purely imaginary, and for $a < a_c$ the real part is positive while the imaginary part is non-zero, indicating a Hopf bifurcation.

5.4

1. Suppose we have a general system that reads

$$\dot{x} = ax + by + f(x, y) \tag{16.42}$$
$$\dot{y} = a'x + b'y + g(x, y) \tag{16.43}$$

   where f and g are nonlinear functions (starting at least with quadratic terms). $(x, y) = (0, 0)$ is a fixed point that undergoes a Hopf bifurcation.

If eigenmodes are sought (as we have always done) for $e^{\sigma t}$, the characteristic equation imposes that at a Hopf bifurcation the following conditions must be fulfilled: (i) $b' = -a$ and (ii) $|a| < \sqrt{-a'b}$ for $a'b > 0$ (to ensure a purely imaginary eigenvalue at the bifurcation threshold). We shall assume $a' < 0$ and $b > 0$ (you can check that the other combination yields the same conclusion). Setting $y = y'\sqrt{-a'/b}$ one obtains

$$\dot{x} = ax + \beta y' + f_1(x, y') \tag{16.44}$$
$$\dot{y}' = -\beta x - ay' + g_1(x, y') \tag{16.45}$$

where f_1 and g_1 are proportional to f and g, and $\beta = \sqrt{-a'b}$. Taking the sum and difference of the above two equations we obtain (by setting $X = x + y'$ and $Y = x - y'$)

$$\dot{X} = -\lambda_1 Y + f_2(X, Y) \tag{16.46}$$
$$\dot{Y} = \lambda_2 X + g_2(X, Y) \tag{16.47}$$

with $\lambda_1 = \beta - a$ and $\lambda_2 = a + \beta$, and f_2 and g_2 linear combinations of f_1 and g_1 (we can express x and y' in terms of X and Y). Finally, by setting $Z = Y\sqrt{\lambda_1/\lambda_2}$ we get

$$\dot{Z} = \omega X + f_3(X, Z) \tag{16.48}$$
$$\dot{X} = -\omega Z + g_3(X, Z) \tag{16.49}$$

where $\omega = \sqrt{\lambda_1 \lambda_2}$, and f_3 and g_3 are proportional to f_2 and g_2 respectively. This completes our proof.

2. $a = 1/8$.

3. The bifurcation is subcritical.

4. The values of a are (i) $-1/8$ (supercritical), (ii) $3/8$ (subcritical), (iii) $-1/2$ (supercritical).

5.5

By multiplying the first equation by y and the second by x and taking the sum one easily finds (after setting $x = r\cos(\theta)$ and $y = r\sin(\theta)$) $\dot{r} = 1 - r^2$. Taking the difference one finds $\dot{\theta} = 1$. The solution of this last equation is $\theta = t + \theta_0$ (θ_0 is an integration constant). The equation for r has a stable fixed point at $r = 1$. We can solve the equation for r explicitly to find $r(t) = (ae^{2t} - 1)/(ae^{2t} + 1)$, with a, an integration constant. One sees that $r \to 1$ as $t \to \infty$.

5.6

1. Multiply the first equation by x and the second by y and take the sum and difference to obtain $\dot{r} = rf(r)$ and $\dot{\theta} = 1$, where we have set $x = r\cos(\theta)$

and $y = r\sin(\theta)$. The equation for θ has the solution $\theta = t + \theta_0$, while the fixed point for r yields $r = 0$ and $\sin(1/(r^2 - 1)) = 0$ with solution $r_j = (1 + 1/(j\pi))^{1/2}$ with $j \neq 0$ an integer. We have thus an infinite number of limit cycles parametrized by j.

2. The linear stability analysis of the r equation yields (by setting $r = r_j = \delta r_j$; δr_j is a small perturbation)

$$\dot{\delta r_j} = -2(j\pi)^2(-1)^j \left(1 + \frac{1}{j\pi}\right)\delta r_j, \tag{16.50}$$

where we have used the equations satisfied by the fixed point r_j. If j is even, δr_j decays exponentially with time (stable limit cycle). Conversely if j is odd, δr_j increases exponentially with time (unstable limit cycle).

5.7

1. The equation for ψ becomes

$$\frac{d\psi}{dt} = -\frac{1}{2} + \frac{h}{\sqrt{\Delta}}\cos(2\psi). \tag{16.51}$$

2. The fixed points are

$$\psi_0^{\pm} = \pm\frac{1}{2}\arccos\left(\sqrt{\Delta}/2h\right), \tag{16.52}$$

if $\frac{\sqrt{\Delta}}{2h} < 1$ $\left(h > h_c \equiv \sqrt{\Delta}/2\right)$. For $h < h_c$ there is no fixed point.

3. The "+" solution is stable and the "−" solution is unstable.

4. At $h = h_c$, $\psi_0^{\pm} = 0$, so we can expand equation (16.51) close to bifurcation point for small ψ and obtain

$$\frac{d\psi}{dt} = \epsilon - \frac{h}{h_c}\psi^2, \tag{16.53}$$

with $\epsilon = (h - h_c)/2h_c$. This is the classical form of a saddle-node bifurcation.

5. Substituting the suggested solution into equation (16.51), we find indeed that this is a solution with $A = 1$, $B = (h - h_c)/\sqrt{h_c^2 - h^2}$ and $C = \sqrt{1 - h^2/h_c^2}$. This solution exists if $h < h_c$, a domain in which no fixed point solution exists. This solution corresponds to tumbling. The fixed point solution is called, in the context of RBCs, a tank-treading solution, where the angle is fixed to a given value while the membrane of the RBC (which is fluid) undergoes a motion like a tank-tread.

6. The full system (5.62) has two sets of fixed points. The first set is $R \equiv R_0 = h_c$ and $\psi = \psi_0^{\pm}$. As stated above this set exists only for $h > h_c$ and

the set (R_0, ψ_0^+) is stable, while the set (R_0, ψ_0^-) is unstable. The other set of fixed points given by $(R, \psi) = (h, 0)$ always exists. Linear stability analysis (by looking for perturbations of the form $\sim e^{\omega t}$) of this solution yields

$$\omega = \pm i\sqrt{1 - \frac{h^2}{h_c^2}}. \tag{16.54}$$

For $h > h_c$, ω is real (and has two eigenvalues, one positive and one negative) so that the fixed point is unstable. In this regime the only stable fixed point is the tank-treading solution discussed above. For $h < h_c$, ω is purely imaginary and we have oscillation around $\psi = 0$. For $h < h_c$, ω has no real component and this is not a Hopf bifurcation. Indeed, a Hopf bifurcation occurs when the real component of the eigenvalue passes from negative to positive upon fulfillment of some conditions on parameters. Here, the real component is always zero (meaning that the fixed point is neutral). Therefore we do not have a limit cycle, but rather the system will select an amplitude of oscillation (or closed orbit) which depends on initial conditions (see below for further details).

7. A straightforward substitution yields the result for ρ, which can then be rewritten as

$$\frac{\dot{\rho}}{(1 - h^2/h_c^2) + \rho^2} = dt. \tag{16.55}$$

Integrating both sides directly yields the desired results. For $h > h_c$ we have real exponentials while for $h < h_c$ we have imaginary exponentials (or trigonometric functions).

8. Setting $f = 1/\zeta$ yields $\dot{f} = \rho f + 4h/\Delta$. The solution is given by the homogeneous equation plus a particular solution; the homogeneous equation is solved (after using the expression of ρ for $h > h_c$) by $f = A/(e^{\omega t} - C_1 e^{-\omega t})$, where A is an integration constant. Using the method of variation for the constant (in which A is taken to depend on time) provides us with the expression $A = (4h/\Delta\omega)[e^{\omega t} + C_1 e^{-\omega t} + C_2]$ with C_2 an integration constant. Plugging the expression of $A(t)$ into the homogeneous solution yields the desired result. ξ is obtained directly by using the definition of ρ, that is $\xi = \rho\zeta + h$. Exactly the same procedure is used for $h < h_c$.

9. The expression for $\mathcal{R}$ is obtained straightforwardly by using $\mathcal{R}^2 = \zeta^2 + \xi^2$, and that for ψ by using $\psi = (1/2)\arctan(\zeta/\xi)$. We easily find the expression for the case where $h > h_c$. When $t \to \infty$ we obtain (for $h > h_c$) that $\mathcal{R}$ tends to $\Delta/2$ and ψ tends to

$$\psi_0^+ = \frac{1}{2}\arccos\left(\sqrt{\frac{\Delta}{2h}}\right), \tag{16.56}$$

which corresponds to the fixed point discussed above. This is the tank-treading solution.

For $h < h_c$ the situation is a bit more subtle since we must make sure, in order to invert the relation $\tan(2\psi) = \zeta/\xi$, that ζ/ξ is bijective. A simple inspection shows that this is not the case over the whole interval. Naively, we may propose the inclination angle to be described by

$$\psi(t) = \frac{1}{2} \arctan\left(\omega \frac{\cos\omega t}{\Gamma + \sin\omega t}\right), \tag{16.57}$$

for all times, but in this case ψ oscillates in the interval $[-\pi/4, \pi/4]$ and cannot describe a tumbling (TB) solution. The expression (16.57) must be refined. In passing, we note that $(-\mathcal{R}, \psi \pm \pi/2)$ are solutions of the system of equations as well as $(\mathcal{R}, \psi)$. A rigorous mathematical analysis shows that the angle can be defined by

$$\psi(t) = \begin{cases} \dfrac{1}{2} \arctan\left(\omega \dfrac{\cos\omega t}{\Gamma + \sin\omega t)}\right) & \text{if } \xi > 0, \\[3mm] -\dfrac{\pi}{2} + \dfrac{1}{2} \arctan\left(\omega \dfrac{\cos\omega t}{\Gamma + \sin\omega t)}\right) & \text{if } \xi < 0,\ \zeta < 0, \\[3mm] +\dfrac{\pi}{2} + \dfrac{1}{2} \arctan\left(\omega \dfrac{\cos\omega t}{\Gamma + \sin\omega t)}\right) & \text{if } \xi < 0,\ \zeta > 0. \end{cases} \tag{16.58}$$

In the above expression the inclination angle is defined as the principal value of the real arctangent function which is the unique solution to

$$\cos(2\psi(t)) = \frac{\xi(t)}{\sqrt{\xi^2(t) + \zeta^2(t)}}, \quad \sin(2\psi(t)) = \frac{\zeta(t)}{\sqrt{\xi^2(t) + \zeta^2(t)}}, \tag{16.59}$$

satisfying $-\pi \leq 2\psi(t) \leq +\pi$. The inclination angle jumps discontinuously from $-\pi/2$ to $+\pi/2$ at $\xi < 0$ and $\zeta = 0$. The inclination angle can be written, for convenience, in the form

$$\psi(t) = \frac{\pi}{4} \frac{\Gamma}{|\Gamma|} \frac{\cos\omega t}{|\cos\omega t|} \left[1 - \frac{\Gamma}{|\Gamma|} \frac{\Gamma + \sin\omega t}{|\Gamma + \sin\omega t|}\right]$$
$$+ \frac{1}{2} \arctan\left(\omega \frac{\cos\omega t}{\Gamma + \sin\omega t}\right). \tag{16.60}$$

The variation of the inclination angle is shown in figure 16.20. Clearly, there are three cases depending on the parameter Γ which is given by (see (5.70))

$$\Gamma = \pm \frac{4h^2}{\Delta} \frac{\frac{\Delta}{4h} - \xi(0)}{\sqrt{\omega^2\zeta^2(0) + (\xi(0) - h)^2}}. \tag{16.61}$$

For $|\Gamma| > 1$ the function $t \to \chi(t) = \Gamma + \sin\omega t$ never vanishes and $\xi(t)$ is positive for all times, and so the orientation angle satisfies (16.57) for

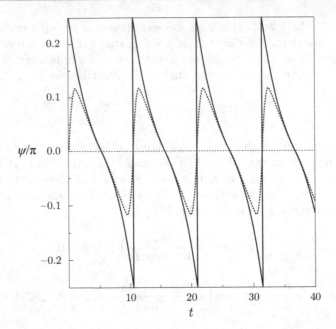

Figure 16.20 – The dynamics of the inclination angle ψ for vacillating-breathing and tumbling modes. Parameters are $\Delta = 4, h = 0.8$ and then $\Gamma_c = 0.8$. $\Gamma = -1.2$ for broken line and $\Gamma = -1$ for continuous line.

all times. Consequently, the RBC exhibits VB mode where the inclination angle oscillates between the minimal angle

$$\psi_{\text{VB}-} = -\frac{1}{2} \arctan \left(\frac{\omega}{\sqrt{\Gamma^2 - 1}} \right)$$

and the maximal angle

$$\psi_{\text{VB}+} = +\frac{1}{2} \arctan \left(\frac{\omega}{\sqrt{\Gamma^2 - 1}} \right).$$

Both minimal and maximal angles satisfy $0 < |\psi_{\text{VB}\pm}| < \pi/4$. Recall that the RBC undergoes periodic shape deformation. This is a signature of the VB regime (see the dotted line in figure 16.20).

For $|\Gamma| = 1$ the function χ is nonnegative and vanishes at finite times. In this case the inclination angle ψ spans the whole range $[-\pi/4, \pi/4]$ (see figure 16.20, full line). Hence, the inclination angle ψ reaches $\pm \pi/4$.

For $\Gamma_c \leq |\Gamma| < 1$ the function ξ changes sign and ψ oscillates between $\pm \pi/2$. This is a TB regime, and the solution is given by (16.60). An example of this motion can be seen in figure 16.20.

From the above discussion one deduces that the border separating VB and TB regimes is obtained analytically from the equation $\Gamma^2 = 1$, which is equivalent to

$$\xi(0)\left[2h - \left(1 + \frac{4h^2}{\Delta}\right)\xi(0)\right] = \zeta^2(0). \tag{16.62}$$

In passing, we point out that if initially the RBC configuration satisfies (16.62) then the relation

$$\xi(t)\left[2h - \left(1 + \frac{4h^2}{\Delta}\right)\xi(t)\right] = \zeta^2(t) \tag{16.63}$$

holds for all times. The orientation angle of the border mode can be expressed as

$$\psi(t) = \frac{1}{2}\arctan\left(\omega\frac{\cos(\omega t + 3\pi/2)}{1 + \sin(\omega t + 3\pi/2)}\right), \tag{16.64}$$

or, equivalently,

$$\psi(t) = -\frac{1}{2}\arctan\left(\omega\tan\left(\frac{1}{2}\omega(t - t_1)\right)\right), \tag{16.65}$$

where t_1 is a constant.

6.1

1. $(x_0, y_0) = (0, 0)$ is a fixed point. Seeking solutions in the form of $x(t) = x_0 + x_1(t)$ and $y(t) = y_0 + y_1(t)$ and linearizing the equations we obtain

$$\dot{x}_1 = \mu x_1 - y_1, \quad \dot{y}_1 = x_1 + \mu y_1. \tag{16.66}$$

Looking for solutions of the form $x_1 = ae^{\omega t}$ and $y_1 = be^{\omega t}$ and imposing a null determinant (to avoid the trivial solution $a = b = 0$) one obtains $\omega = \mu \pm i$. When $\mu > 0$ the real component is positive meaning that the solution (x_0, y_0) is unstable, while it is stable for $\mu < 0$. The instability threshold is given by $\mu = 0$ where the imaginary part is non-zero. This is characteristic of a Hopf bifurcation.

2. In the vicinity of the bifurcation μ is small and will play the role of the distance from the instability threshold. We introduce a slow time scale $T = \mu$ (dictated by the form of the real part of ω), and substitute the following transformation of the derivative (as we have seen in this chapter)

$$\frac{\partial}{\partial t} \to \frac{\partial}{\partial t} + \mu\frac{\partial}{\partial T}. \tag{16.67}$$

We then expand the solutions x and y in power series of μ (see this chapter)

$$x = \mu^{1/2}x_1 + \mu x_2 + \cdots, \quad y = \mu^{1/2}y_1 + \mu y_2 + \cdots \tag{16.68}$$

We then bring equations (16.67) and (16.68) into the original set of equations and deduce the resulting equations order by order. Please keep in mind that in a multiscale analysis x and y have to be considered as depending on two independent variables t and T. To order $\mu^{1/2}$ one obtains

$$\frac{\partial x_1}{\partial t} = -y_1, \quad \frac{\partial y_1}{\partial t} = x_1. \tag{16.69}$$

The solution of which is given by

$$x_1 = A(T)e^{it} + c.c., \quad y_1 = -iA(T)e^{it} + c.c. \tag{16.70}$$

where $A(T)$ is an integration factor depending on T since the above system corresponds to partial differential equations.

To order μ we obtain exactly the same system for x_2 and y_2 as found above for (x_1, y_1). The solution has the same form with $B(T)$ as an integration factor, which can be taken to be zero since we can absorb the solution at this order into that of (x_1, y_1) via a proper redefinition of A (more precisely $A + \sqrt{\mu}B \to A$).

To order $\mu^{3/2}$ one obtains

$$\frac{\partial x_3}{\partial t} + y_3 = -\frac{\partial x_1}{\partial T} + x_1 + x_1 y_1^2$$
$$\frac{\partial y_3}{\partial t} - x_3 = -\frac{\partial y_1}{\partial T} + y_1 + y_1^3 \tag{16.71}$$

By using the solutions of x_1 and y_1 found to order $\mu^{1/2}$ and collecting resonant terms proportional to e^{it} (we omit terms like e^{3it} which are not resonant), one obtains

$$\frac{\partial x_3}{\partial t} + y_3 = \left(-\frac{\partial A}{\partial T} + A + A^2 A^*\right) e^{it} + c.c.$$
$$\frac{\partial y_3}{\partial t} - x_3 = \left(i\frac{\partial A}{\partial T} - A - 3A^2 A^*\right) e^{it} + c.c. \tag{16.72}$$

The solution is composed of a homogeneous solution plus a particular solution. The particular solution reads $x_3 = Ce^{it} + c.c.$ and $y_3 = De^{it} + c.c.$, where C and D are integration factors (which depend on T in general). However, remember that the determinant of the resulting system in C and D is exactly zero (as we have seen at the beginning of this exercise) imposing a condition (often called a solvability condition). This condition is the same as the amplitude equation we seek and takes the form

$$\frac{\partial A}{\partial T} = A + 2|A|^2 A. \tag{16.73}$$

Note that here the cubic term is purely real, which is not usually the case for a Hopf bifurcation (see exercises below). Can this be expected?

The coefficient of the linear term is also real – the information describing the Hopf bifurcation is contained in the prefactor e^{it}. If we redefine $A = \tilde{A}e^{-it}$ one obtains that $\tilde{A}$ obeys

$$\frac{\partial \tilde{A}}{\partial T} = (1+i)\tilde{A} + 2|\tilde{A}|^2 \tilde{A}. \tag{16.74}$$

3. We have a subcritical bifurcation.

6.2

Let us consider the first system:

1. This system has exactly the same dispersion relation, so the conclusions are the same.

2. To derive the amplitude equation, follow exactly the same procedure as before to find

$$\frac{\partial A}{\partial T} = A - \frac{1}{2}|A|^2 A. \tag{16.75}$$

3. We have a supercritical bifurcation.

Consider the second system:

1. This system has exactly the same dispersion relation, so that the conclusions are the same.

2. The derivation of the amplitude equation follows exactly the same steps as before and one finds

$$\frac{\partial A}{\partial T} = A + \frac{3}{2}|A|^2 A. \tag{16.76}$$

3. We have a supercritical bifurcation.

Consider now the third system:

1. This system has the same dispersion relation as before.

2. The only important difference from the two previous cases is that x_2 and y_2 obey different equations than x_1 and y_1:

$$\frac{\partial x_2}{\partial t} = y_2 - x_1^2, \quad \frac{\partial y_2}{\partial t} = -x_2 + 2x_1^2. \tag{16.77}$$

The solution is composed of a homogenous solution plus a particular solution. Since the homogeneous system is identical to that obeyed by (x_1, y_1) the solution is formally identical and can be absorbed into (x_1, y_1). The particular solution takes the form $x_2 = F(T)e^{2it} + G(T) + c.c.$ and $y_2 = \tilde{F}(T)e^{2it} + \tilde{G} + c.c.$ Plugging this back into the system and using the expressions of (x_1, y_1) as in the previous exercises, one obtains

$$F = \frac{1}{3}(2i - 2)A^2, \quad \tilde{F} = 2A^2, \quad G = -\frac{(1 + 4i)}{3}|A|^2, \quad \tilde{G} = |A|^2. \quad (16.78)$$

To order $\mu^{3/2}$ one obtains

$$\frac{\partial x_3}{\partial t} - y_3 = -\frac{\partial x_1}{\partial T} + x_1 - 2x_1 x_2$$

$$\frac{\partial y_3}{\partial t} + x_3 = -\frac{\partial y_1}{\partial T} + y_1 + 4x_1 x_2 \quad (16.79)$$

Using the solutions for x_1 and y_1 found to order $\mu^{1/2}$ and the solutions for x_2 and y_2 found to order μ and collecting resonant terms (i.e. we omit terms like e^{3it}), we obtain a system with exactly the same form seen in the previous two exercises. Imposing the solvability condition one obtains the amplitude equation

$$\frac{\partial A}{\partial T} = A - \frac{12 + 44i}{3}|A|^2 A. \quad (16.80)$$

3. The bifurcation is supercritical.

6.3

1. r and θ obey

$$\dot{r} = (\epsilon - r^2)r + a\cos(\theta), \quad \dot{\theta} = -\frac{a\sin(\theta)}{r}. \quad (16.81)$$

2. Suppose $a > 0$. The fixed points are given by $\sin(\theta_0) = 0$, with solution $\theta_0 = 0, \pi$ (modulo 2π). When $\theta_0 = 0$ the fixed point equation for r obeys $(\epsilon - r_0^2)r_0 + a = 0$. This is typical for an imperfect bifurcation. Since $r_0 \geqslant 0$ we can ignore the negative branch. When $\theta = \pi$ the fixed point equation for r obeys $(\epsilon - r_0^2)r_0 - a = 0$, and again we choose $r_0 \geqslant 0$. The stability of r_0 is identical to that of an imperfect bifurcation, already studied in chapter 3. The stability of θ_0 (calling the perturbation θ_1) obeys

$$\dot{\theta}_1 = -a\theta_1 \frac{\cos(\theta_0)}{r_0}. \quad (16.82)$$

The solution $\theta_0 = 0$ is stable while $\theta_0 = \pi$ is unstable. For $a < 0$, the results are inverted. Note that due to the absence of non-diagonal terms in the stability problem, studying the diagonal terms was sufficient to make conclusions regarding stability.

3. Figure 16.21 shows the branches and their stability.

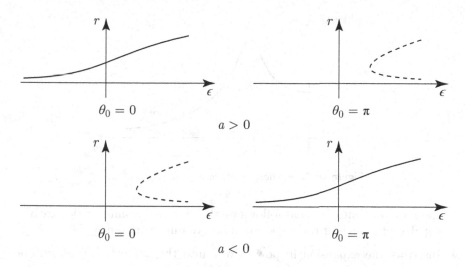

Figure 16.21 – The bifurcation diagram. We see here the branch with a turning point is always unstable, unlike the classical imperfect bifurcation. This is due to phase dynamics.

7.1

1. In this limit the equation becomes

$$\ddot{x} + ax = \Gamma \cos(\omega t). \tag{16.83}$$

The solution is composed of the homogenous plus a particular solution

$$x(t) = A \cos(\sqrt{a}t) + B \sin(\sqrt{a}t) + \frac{\Gamma \cos(\omega t)}{(a - \omega^2)}, \tag{16.84}$$

provided that $\omega^2 \neq a$ to avoid divergence. A and B are integration factors.

2. When $a = \omega^2$ it is easy to see that $A \cos(\sqrt{a}t) + B \sin(\sqrt{a}t)$ is a null eigenmode (a mode with zero eigenvalue) of the linear operator $L = d^2/dt^2 + \omega^2$ on the left-hand side of equation (16.83). Thus $L[tC \sin(\omega t)] = 2C\omega \cos(\omega t)$ and by identification one obtains $C = \Gamma/2\omega$. The full solution reads

$$x(t) = A \cos(\omega t) + B \sin(\omega t) + \frac{\Gamma t \sin(\omega t)}{2\omega}. \tag{16.85}$$

The solution grows in time without bound and the term proportional to t is called a secular term.

3. By looking for a homogeneous plus a particular solution one can easily extract the full expression. The particular solution has a maximum amplitude at $\omega^2 = a$ (resonance condition). Figure 16.22 shows the behavior of the amplitude as a function of ω. Friction modifies resonance, rendering

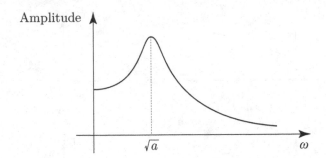

Amplitude

$\sqrt{a}$ ω

Figure 16.22 – The amplitude as a function of ω.

the response finite. In what follows we shall see that nonlinearities are also capable of rendering the response finite even in the absence of friction.

4. Inserting the expansion in powers of ϵ into the nonlinear equation, one easily finds the equations for x_0, x_1 and x_2:

$$\ddot{x}_0 + \Omega^2 x_0 = \Gamma \cos(t)$$
$$\ddot{x}_1 + \Omega^2 x_1 = x_0^3$$
$$\ddot{x}_2 + \Omega^2 x_2 = 3x_0^2 x_1 \tag{16.86}$$

The full solution for x_0 is given by

$$x_0(t) = A\cos(\Omega t) + B\sin(\Omega t) + \frac{\Gamma \cos(t)}{(\Omega^2 - 1)}, \tag{16.87}$$

where $\Omega \neq \pm 1$. If this condition is not fulfilled the equation becomes degenerate and the solution is no longer valid. The appropriate solution contains a secular term, as explained above. If we restrict ourselves to 2π-periodic solutions, and provided that Ω is not an integer, then only the last term in equation (16.87) fulfills this requirement, and the appropriate solution is

$$x_0(t) = \frac{\Gamma \cos(t)}{(\Omega^2 - 1)}, \tag{16.88}$$

5. The equation obeyed by x_1 becomes (upon using the expression for x_0)

$$\ddot{x}_1 + \Omega^2 x_1 = \frac{\Gamma^3 [3\cos(t) + \cos(3t)]}{4(\Omega^2 - 1)^3}. \tag{16.89}$$

Looking for solutions in the form of a superposition of $\cos(t)$ and $\cos(3t)$, and by identification with the right-hand side of the above equation one easily finds

$$x_1(t) = \frac{3\Gamma^3 \cos(t)}{4(\Omega^2 - 1)^4} + \frac{\Gamma^3 \cos(3t)}{4(\Omega^2 - 1)^3(\Omega^2 - 9)}. \tag{16.90}$$

Close to resonance (for example if Ω is an odd number, say 1 or 3) the solution is not valid and will contain a secular term proportional to time. After a long enough time the expansion in powers of ϵ is no longer valid for even though ϵ is small, the product ϵt becomes too large. In general this is referred to as a non-uniform expansion, since the expansion is valid only for a short time.

6. $\beta = (a/\omega^2 - 1)/\epsilon$, $b = -\epsilon\omega^2$ and $\gamma = \Gamma/(\omega^2\epsilon)$. Since γ and β are of order one, this means that $a/\omega^2 - 1$ is of order ϵ (it measures the distance from resonance), and Γ is assumed to scale like ϵ. In other words, we are choosing the vicinity of resonance, the forcing term and the nonlinear term to be of the same order.

7. Adopting an expansion of x in power series of ϵ, one easily finds that x_0 satisfies

$$\ddot{x}_0 + x_0 = 0, \tag{16.91}$$

where from now on $\dot{x}$ designates the derivative with respect to τ. Because of the choice of the scaling in ϵ the forcing term does not enter this equation, unlike the study presented above. The solution is given by

$$x_0(\tau) = a_0 \cos(\tau) + b_0 \sin(\tau). \tag{16.92}$$

8. The next order yields

$$\ddot{x}_1 + x_1 = \gamma \cos(\tau) - \beta x_0 + x_0^3$$
$$= \left[\gamma - \beta a_0 + \frac{3}{4}a_0(a_0^2 + b_0^2)\right]\cos(\tau) + b_0\left[-\beta + \frac{3}{4}(a_0^2 + b_0^2)\right]\sin(\tau)$$
$$+ \frac{1}{4}a_0(a_0^2 - 3b_0^2)\cos(3\tau) + \frac{1}{4}b_0(3a_0^2 - b_0^2)\sin(3\tau). \tag{16.93}$$

9. Eliminating the resonant terms (proportional to $\cos(\tau)$ and $\sin(\tau)$) guarantees the absence of secular terms. This is achieved by setting the coefficients of $\cos(\tau)$ and $\sin(\tau)$ to zero, which yields

$$\gamma = a_0\left[\beta - \frac{3}{4}(a_0^2 + b_0^2)\right], \quad b_0\left[\beta - \frac{3}{4}a_0(a_0^2 + b_0^2)\right] = 0. \tag{16.94}$$

This gives

$$b_0 = 0, \quad a_0\left(\beta - \frac{3}{4}a_0^2\right) = \gamma. \tag{16.95}$$

This means that $x_0 = a_0 \cos(\tau)$ and $x_1 = a_1 \cos(\tau) + b_1 \sin(\tau) - (a_0^3/32)\cos(3\tau)$, where a_1 and b_1 are integration factors (coming from the solution to the homogeneous equation). Their expressions in terms of the model parameters can be determined by eliminating the secular terms in the next order of the expansion. This task will not be dealt with here.

Equation (16.95) allows one to determine a_0 as a function of the model parameters (β and γ). This equation reminds us of the imperfect bifurcation discussed at length in chapter 3: here β plays the role of the distance from threshold (often denoted as ϵ in chapter 3) and γ the imperfection term. By referring to that chapter it is a simple matter to show that that this equation has three real roots when $\beta > (9\gamma/4)^{2/3} \equiv \beta^*$, two if $\beta = \beta^*$ (with a doubly degenerate solution) and one if $\beta < \beta^*$. We have pitchfork bifurcation in the special case when $\gamma = 0$. Figure 16.23 shows a_0 as a function of β. The forcing term has created an imperfection.

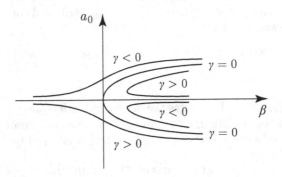

Figure 16.23 – The amplitude a_0 as a function of the closeness to the resonance β.

10. Using equation (16.95) and the relation between the old and the new parameters listed above, one easily finds $\omega^2 = a + \frac{3}{4}ba_0^2 - \Gamma/a_0$ provided $a_0 \neq 0$. If $\Gamma = 0$ then $a_0 = 0$ is a root as well as $\pm\sqrt{4(\omega^2 - a)/(3b)}$ (provided that $(\omega^2 - a)/b > 0$).

Of course here again a_0 obeys a cubic equation as in the case of an imperfect bifurcation. However, by plotting $|a_0|$ as a function of ω we reveal a form resembling resonance. Figure 16.24 shows $|a_0|$ as a function of ω for $\Gamma a_0 > 0$, $\Gamma a_0 < 0$ and $\Gamma = 0$. When $\Gamma = 0$ we have a pitchfork bifurcation with a non-zero branch for $\omega < \sqrt{a}$ (when $b < 0$), and since we are plotting $|a_0|$, the figure shows only the positive branch. We see that the resonance peak at $\omega = \sqrt{a}$ is made finite thanks to the nonlinearity.

It must be noted that it was not necessary to introduce the scaling in ϵ so that all terms on the right-hand side of our new starting equation $\ddot{x} + x = \epsilon[\gamma\cos(\tau) - \kappa\dot{x} - \beta x + x^3]$ be of the same order. We adopted this choice in order to facilitate the presentation of the effect of nonlinearity. To test your understanding you can try to go back to our original Duffing equation and repeat the analysis without a priori imposing the above scaling.

11. $\kappa = k/(\omega\epsilon)$. Since we take κ to be of order one, $k \sim \epsilon$ (small friction). Following exactly the same procedure as for the frictionless case one arrives

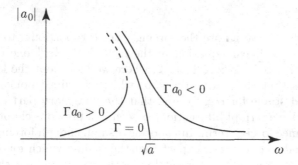

Figure 16.24 – The amplitude $|a_0|$ as a function of ω.

to the relations (7.74). Adding the squares of these relations one easily finds the expression

$$r_0^2 \left[\kappa^2 + \left(\beta - \frac{3}{4} r_0^2 \right)^2 \right] = \gamma^2, \qquad (16.96)$$

where $r_0 = (a_0^2 + b_0^2)^{1/2}$. Using the relations between parameters one finds that equation (16.96) gives

$$k^2 \omega^2 + \left[a - \omega^2 + \frac{3}{4} b r_0^2 \right]^2 = \frac{\Gamma^2}{r_0^2}, \qquad (16.97)$$

which is a cubic equation for r_0^2 and can thus have one or three real roots. Figure 16.25 shows the behavior of of r_0 as a function of ω for large and small forcing. By setting $r_0^2 = X$ one can easily write the above equation as a third order polynomial containing X, X^2 and X^3. With a simple change of variables (seen several times in chapter 4) one can eliminate the X^2 term bringing us back to a traditional form of the imperfect bifurcation. This may help in plotting figure 16.25. The dotted line represents the unstable branch. We see that friction introduces hysteresis.

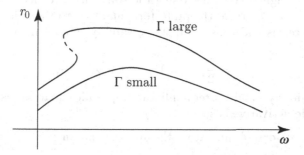

Figure 16.25 – The amplitude r_0 as a function of the driving frequency ω for $b < 0$.

7.2

First notice that if we ignore the forcing term (proportional to γ) and seek a solution for the linear equation of the form $x = Ae^{\omega t}$ one obtains $\omega = (\epsilon \pm \sqrt{\epsilon^2 - 4(1+\beta\epsilon)})/2 \simeq (\epsilon \pm 2i)/2$, where we have kept the leading order in ϵ. One sees that the solution $x = 0$ (which is a fixed point) is unstable for $\epsilon > 0$ (and stable for $\epsilon < 0$), and that the imaginary part of ω is finite. Therefore we have a Hopf bifurcation, as seen in a previous chapter, but with the new presence forcing term. In choosing a small enough forcing term (that is, proportional to ϵ) one can expect that the scaling which follows from the above dispersion relation is correct by continuity. It follows that the slow time scale can be defined as $T = \epsilon t$, and the differential operator has to be written as $\partial_t + \epsilon\partial_T$ (in multiscale fashion).

We have written the nonlinear van der Pol term as $(1-x^2)\dot{x}$ anticipating that the friction term changes sign for $x \sim 1$ (unlike in chapter 6 where we wrote the van der Pol term as $(\epsilon - x^2)\dot{x}$, the friction changing sign for $x \sim \sqrt{\epsilon}$). Thus, the expansion of the solution starts at first order (and not order $\sqrt{\epsilon}$ as in chapter 6). We therefore seek a solution of the form $x = x_0 + \epsilon x_1 + \epsilon^2 x_2 + \cdots$. Carrying this into the nonlinear equation we obtain

$$\partial_{tt}x_0 + x_0 = 0. \tag{16.98}$$

the solution of which reads

$$x_0 = A(T)e^{it} + c.c., \tag{16.99}$$

A is an integration factor depending on T (the familiar multiscale result). To next order one gets

$$\partial_{tt}x_1 + x_1 = \frac{\gamma}{2}(e^{it} + e^{-it}) - \beta x_0 + (1 - x_0^2)\partial_t x_0 - 2\partial_{tT}x_0. \tag{16.100}$$

Since the left-hand side is solved by e^{it} we must find the resonant terms appearing on the right-hand side. Using the solution x_0, collecting terms proportional to e^{it} and setting the prefactor equal to zero (in order to suppress the resonant terms and thus fulfill the solvability condition), we straightforwardly obtain

$$A_T = \frac{1}{2}(1 + \beta i)A - \frac{1}{2}|A|^2 A - i\frac{\gamma}{4}, \tag{16.101}$$

which is our dearly sought-after nonlinear amplitude equation (as throughout this book, the derivative is subscripted).

Setting $A = re^{i\theta}$ one obtains from the above equation

$$r_T = \frac{1}{2}(1 - r^2)r - \frac{\gamma}{4}\sin(\theta)$$

$$r\theta_T = r\frac{\beta}{2} - \frac{\gamma}{4}\cos(\theta). \tag{16.102}$$

The fixed points (denoted as r_0 and θ_0) obey

$$(1 - r_0^2)r_0 = \frac{\gamma}{2}\sin(\theta_0)$$

$$r_0\beta = \frac{\gamma}{2}\cos(\theta_0). \tag{16.103}$$

Adding the squares of the two equations yields

$$r_0^2(1 - r_0^2)^2 + r_0^2\beta^2 = \left(\frac{\gamma}{2}\right)^2. \tag{16.104}$$

For each forcing γ one can find a pair r_0 and β, so fixed points exist everywhere on the plane (r_0, β). Note that a fixed point corresponding to the pair (r_0, θ_0) means that the full solution x_0 is given by $x_0(t) = r_0 e^{i(t+\theta_0)} + c.c.$ The amplitude oscillates in time with pulsation unity. It must be kept in mind that the system oscillates with the period of the forcing since the amplitude is fixed by γ (check that if $\gamma = 0$ then the amplitude is that of the the pure van der Pol oscillator, while for $\gamma \neq 0$ we have an amplitude and a phase which are fixed by γ, the forcing). The solution is not a limit cycle but a forced oscillation, the amplitude of which is fixed by that of the forcing.

The linear stability analysis can be performed by setting $r = r_0 + r_1(T)$ and $\theta = \theta_0 + \theta_1(T)$. Plugging this into system (16.102), linearizing in r_1 and θ_1, and looking for solutions in the form $r_1 = ae^{\sigma t}$ and $\theta_1 = be^{\sigma t}$ (where a and b are constant amplitudes) we obtain a linear and homogeneous system of equations for a and b. Imposing a zero determinant (in order to avoid the trivial solution $a = b = 0$) we obtain the characteristic equation

$$\sigma^2 - S\sigma + P = 0, \tag{16.105}$$

with S (the trace) and P (the determinant) are given by

$$S = (1 - 2r_0^2),$$

$$P = \frac{1}{4}[(1 - r_0^2)(1 - 3r_0^2) + \beta^2]. \tag{16.106}$$

We have used the fixed point equations in order to substitute $\cos(\theta_0)$ and $\sin(\theta_0)$ by r_0. Figure 16.26 summarizes the stability of the fixed point solutions. Note that $S > 0$ (meaning that at least one eigenvalue has a positive real part, signaling an instability) below the line $r_0^2 = 1/2$, that $P > 0$ outside the ellipse defined by $9(r_0^2 - 2/3) + 3\beta^2 = 1$, and that $S^2 = 4P$ on the lines $r_0^4 = \beta^2$. We usually distinguish between a stable (or unstable) focus (the plural is foci), node and saddle by looking at the eigenvalues σ_1 and σ_2. Each eigenvalue has an associated an eigenvector: if both σ_1 and σ_2 are real we have a node or a saddle. More precisely, if both σ_1 and σ_2 are real and negative, the fixed point is stable. This means that in phase space (here of dimension two) any initial condition will tend towards the fixed point, which will thus be called a stable node. Conversely, if both σ_1 and σ_2 are real and positive, any small deviation around the fixed point will amplify over time and the fixed point is thus called an unstable node. If σ_1 and σ_2 are both real but one is positive and one negative, the fixed point will attract trajectories along the stable direction and repel along the unstable direction, and we have thus a saddle point. If σ_1 and σ_2 are complex (a pair of complex conjugate) we do not have a monotonous convergence or divergence of trajectories, they are instead oscillatory (or spiraling, in phase space). If the real parts of both σ_1 and σ_2 are negative, the trajectory in phase space will spiral towards the fixed point, which is thus a stable focus, whereas if the real parts of both σ_1 and σ_2 are positive the trajectory will spiral out from the fixed point, corresponding to an unstable focus. This will allow you to understand the stability diagram shown in figure 16.26.

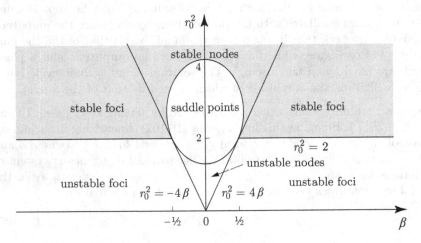

Figure 16.26 – A sketch of the nature of fixed points and their stabilities in the (β, r_0^2) plane.

8.1

1. The map $f(x) = 2x/(1 + x^2)$ has an extrema at $x = \pm 1$ with maximal value $f(1) = 1$ and a minimal value $f(-1) = -1$. Thus f maps $[-\infty, +\infty]$ onto $[-1, 1]$.

2. The fixed points obey $x = 2x/(1 + x^2)$. This equation has three solutions $x = 0$ and $x = \pm 1$. The stability is determined by the value of the slope of f. $f'(0) > 1$, thus $x = 0$ is an unstable fixed point. $f'(\pm 1) = 0 < 1$ so $x = \pm 1$ are stable fixed points.

3. $x_1 = 2x_0/(1 + x_0^2)$ and since the right-hand side is a function which varies from -1 to $+1$ it is always possible to find a real z_1 such that $x_1 = \tanh(z_1)$. Indeed $\tanh(z_1)$ is a function that varies from -1 to $+1$.

4. We start with x_2:

$$x_2 = 2x_1^2/(1 + x_1^2) = 2\tanh(z_1)/(1 + \tanh^2(z_1))$$
$$= 2\sinh(z_1)\cosh(z_1)/(\sinh^2(z_1) + \cosh^2(z_1))$$
$$= \sinh(2z_1)/\cosh(2z_1) = \tanh(2z_1).$$

From here it is easy to complete the proof by a recursive relation. Suppose that $x_{n-1} = \tanh(2^{n-2}z_1)$. We then use

$$x_n = 2x_{n-1}^2/(1 + x_{n-1}^2) = 2\tanh(2^{n-2}z_1)/(1 + \tanh^2(2^{n-2}z_1))$$
$$= 2\sinh(2^{n-2}z_1)\cosh(2^{n-2}z_1)/(\sinh^2(2^{n-2}z_1) + \cosh^2(2^{n-2}z_1))$$
$$= \sinh(2 \times 2^{n-2}z_1)/\cosh(2 \times 2^{n-2}z_1) = \tanh(2^{n-1}z_1).$$

As we have seen above there are two stable fixed points, $x\pm1$. Starting from any initial condition $x_0 > 0$, after the first iteration we have $x_1 = \tanh(z_1)$. After n iterations z_1 is multiplied by $2^{n-1}z_1$ producing larger and larger values as n increases, and $\tanh(2^{n-1}z_1)$ (which is a monotonously increasing function) increases until it reaches (after an infinitely large number of iterations) $+1$, one of the stable fixed points. Thus the basin of attraction of $+1$ is $[0, \infty]$. If the initial value $x_0 < 0$, then x_1 and $z1$ are < 0. n iterations will produce larger values of $2^{n-1}z_1$ which are also more and more negative, until reaching $-\infty$, so that $\tanh(2^{n-1}z_1)$ tends to -1. Thus the bassin of attraction of -1 is $[-\infty, 0]$.

8.2

1. The fixed point equation is $u = a\sin(u)$ with $u = \pi x$. $u = 0$ is a trivial fixed point. The line u/a has a slope $1/a$ and if a is large enough the line u/a intersects the function $\sin(u)$. At $u = 0$, $\sin(u)$ has the largest slope (equal to one). When $a < 1$ (but close enough to 1) the line can intersect $\sin(x)$ at two symmetric points as shown on figure 16.27.

2. Figure 16.27 shows some of the possible fixed points.

3. $x = 0$ is stable if $-1 < a < 1$ and unstable otherwise.

4. The other fixed point u_1 obeys $u_1 = a\sin(u_1)$. The stability is investigated by taking the derivative of f, $\partial f/\partial x = a\cos(\pi x_1) = a\cos(u_1)$. The fixed point is stable if the slope is smaller than 1 (in absolute value) and unstable otherwise. Thus the marginal stability is obtained when $a_* \cos(u_1) = 1$, where a_* is the marginal value of a for the border between stability and instability. Using the fact that $u_1 = a_1\sin(u_1)$ and taking the ratio of $a_1\cos(u_1) = -1$ over $u_1 = a_1\sin(u_1)$, one finds that u_1 obeys $\tan(u) + u = 0$. For small a the fixed point $x = 0$ is stable for $a < 1$, and at $a = 1$ the fixed point looses its stability in favor of two fixed points. This leads to a pitchfork bifurcation. The non-zero fixed points $(x \neq 0)$ are stable if $a < a_1 = u_1/\sin(u_1)$, where $u_1 + \tan(u_1) = 0$

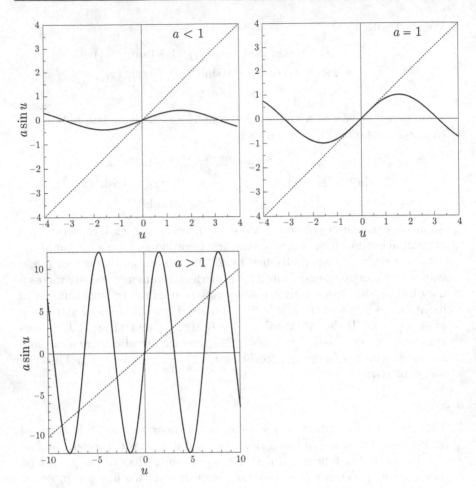

Figure 16.27 – A sketch of fixed points. From left to right: $a = 0.4$, $a = 1$, $a = 12$.

8.3

1. Since $a < 0$, ae^x is a decreasing function while x is an increasing function. Thus the two functions can intersect only once and thus the fixed point is unique. Let us call that fixed point $X(a)$ (satisfying $X = ae^X$). The derivative at the fixed point has the value ae^X and is always negative. X is stable if $ae^X > -1$ (meaning a slope with absolute value smaller than one), and unstable otherwise. Using the fact that $X = ae^X$, one easily deduces that a fixed point such that $X < -1$ is stable, whereas a fixed point such that $X > -1$ is unstable. Using the relation $X = ae^X$ yields the stability conditions for a: stable for $-e < a < 0$ for and unstable for $a < -e$.

2. When $a \to 0$, X is small, and we easily see that $X \sim a$. Taking the logarithm of the fixed point equation $X = ae^X$, one obtains $\ln|X| - X = \ln|a| = 1$, so that $X \to -1$ is a solution of this equation. A Taylor expansion around $X = -1$ of $X = ae^X$ to leading order yields $X \sim -1 + (a+e)/(2e)$. For $a \to -\infty$ one sees from $\ln|X| - X = \ln|a| = 1$ that if $a \to -\infty$, X should be large and in this case X dominates over $\ln(X)$, so that we have $X \sim -\ln|a|$.

3. $g = f(f(x)) = ae^{ae^x}$ so $g_x = a^2 e^x e^{ae^x} = fg$ and $g_{xx} = f_x g + f g_x = fg + fgf = (1+f)fg$. By chain rule we easily obtain that $g_{xxx} = fg(1+3f+f^2)$.

4. Figure 16.28 shows the intersection of $g(x)$ with the line $y = x$. We see that for $a > -e$ we have a single intersection and for $a < -e$ we have three intersections. We have seen above that for $a < -e$ the fixed point of f becomes unstable. Since g has an intersection this is a fixed point for g, which means a bi-periodic (period 2) limit cycle (remember a fixed point of f corresponds to a limit cycle of period 1). We can also use the above results to arrive to these conclusions. For $a = -e$ and $X = -1$ (the corresponding fixed point), we have $g = -1$ (intersection with the line $y = x$ at $X = -1$), which is thus a fixed point of g. In addition, we have $g_x = 1$, $g_{xx} = 0$, and $g_{xxx} = 1$. This means that g has three intersections with line $y = x$ when $a < -e$ (three degenerate solutions when $a = -e$).

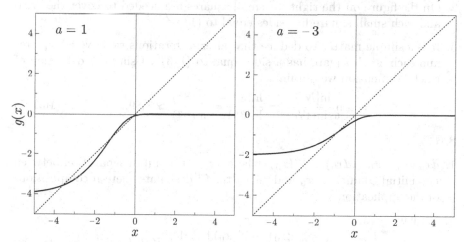

Figure 16.28 – The intersection of x (broken line) and $g(x)$ (continuous line) for $a = 1$ and $a = -3$.

5. One fixed point, belonging to f (denoted as X), always exists and becomes unstable at $a = -e$. The other two are fixed points of g, and are roots of $(x - g)/(x - X) = 0$. Close to the bifurcation where x is close to X, we expand g in Taylor series to get $g = X + (x-X)X^2 + \frac{1}{2}(x-X)^2 X^2(1+X) + \frac{1}{6}(x-X)^3 X^2(1 + 3X + X^2) + \cdots$. Since we have seen that close to the

bifurcation $g = 1$, we can deduce the desired result with a second order equation for $x - X$. Using the expansion $X \sim -1 + (a + e)/(2e)$ (valid for $a \sim -e$; see previous results) we easily find the two other fixed points. We have a pitchfork bifurcation at $a = -e$.

8.4

1. $f = 0$ for $x = 0$ and has a maximum at $x = 1/\sqrt{3}$ where $f = 2a/(3\sqrt{3})$. Thus f maps $[0, 1]$ onto itself provided that $0 \leqslant a \leqslant 3\sqrt{3}/2$.

2. The fixed point equation $x = ax(1 - x^2)$ is solved by $x = 0$ and $x = \pm\sqrt{(a - 1)/a}$.

3. For stability, evaluate the derivatives at the fixed points and impose a slope with absolute value less than 1. The results follow easily.

4. This is a straightforward calculation that follows that of the logistic map treated in this chapter.

8.5

1. Each small square on the left side figure has a side equal to $1/3$. We see that it takes $N = 8$ small squares to fill the area.

2. On the figure on the right, 8^2 small squares are needed to cover the area, and each small square has sides equal to $(1/3)^2$.

3. It is a simple matter to deduce that after n iterations we have 8^n squares and each small square has a side equal to $(1/3)^n$. Using the definition of fractal dimension we obtain

$$D = \lim_{\epsilon \to 0} \frac{\ln[N(\epsilon)]}{\ln(1/\epsilon)} = \frac{\ln(8^n)}{\ln(3^n)} = \frac{\ln(8^n)}{\ln(3^n)} \simeq 1.89... \qquad (16.107)$$

8.6

1. $dx_n = f^n(x_0 + dx_0) - f^n(x_0)$, where dx_0 is the initial separation between two initial conditions x_0 and $x_0 + dx_0$. f^n designates here n compositions of the application f.

2. We have

$$\nu = \frac{1}{n} \ln\left[\frac{f^n(x_0 + dx_0) - f^n(x_0)}{dx_0} \right] = \frac{1}{n} \ln\left[(f^n)'(x_0) \right]. \qquad (16.108)$$

3. For $n = 2$ we have $(f^2)'(x_0) = [f(f(x_0))]' = [f(x_1)]'$ where $x_1 = f(x_0)$. Using the chain rule we can write $[f(x_1)]' = x_1' f'(x_1) = f'(x_0) f'(x_1)$ (where the prime is derivative with respect to the argument). If we take $n = 3$ one easily find $(f^3)'(x_0) = f'(x_0) f'(x_1) f'(x_2)$, and so on. By induction, one proves the results.

4. The result is found using the properties of the ln function.

8.7

1. We have $f' = \pm a$ so that

$$\nu = \lim_{n \to \infty} \frac{1}{n} \sum_{i=0}^{n-1} \ln|f'(x_i)| = \ln(a). \qquad (16.109)$$

2. For $a > 1$ we have chaos, because $\ln(a) > 0$ (we have a positive Lyapunov exponent). The results (8.32) comply with this since there we have $a = 2$.

9.1

1. A stationary and homogeneous solution exists with $u_0 = a$ and $v_0 = b/a$. We refer now to the general necessary Turing condition (9.13) to find

$$f_u = b - 1, \quad f_v = a^2 > 0, \quad g_u = -b < 0, \quad g_v = -a^2 < 0. \qquad (16.110)$$

We know the sign of all but f_u. Turning to the Turing condition (9.13) we see that we must have $b > 1$ for a Turing instability. When this holds, we see that u is an activator ($f_u > 0$) and v is an inhibitor ($g_v < 0$).

2. The necessary and sufficient condition for the Turing instability is

$$b \geqslant \left(1 + a\sqrt{\frac{D_u}{D_v}}\right)^2 \equiv b_c. \qquad (16.111)$$

We also know that for $q = 0$ we must have a negative eigenvalue to ensure that $b < 1 + a^2$, which in turn means that we must have $D_v > D_u$.

3. The critical value for the Turing instability is b_c.

4. We have

$$q_c^2 = \frac{-a^2 D_u + (b_c - 1)D_v}{2D_u D_v}. \qquad (16.112)$$

The fact that this must be positive sets a constraint $D_v/D_u > a^2/(b_c - 1)$.

9.2

The general dispersion relation (equation (9.4)) reads

$$\omega^2 - S\omega + P = 0, \qquad (16.113)$$

where S is the trace (sum of the eigenvalues) and P the determinant (product of eigenvalues) of the stability matrix. Their expressions in terms of the model parameters are given by equation (9.5). The fastest growing mode corresponds to the maximum of ω with respect to q. Let us denote by q_* the corresponding wavenumber. Taking the derivative with respect to q and setting $\omega' = 0$, one easily obtains

$$q_*^2 = \frac{D_u g_v + D_v f_u - \omega(q_*)(D_u + D_v)}{2D_u D_v}. \qquad (16.114)$$

Above the threshold $\omega \neq 0$ (the maximal value of ω), so $q_* \neq q_c$.

9.3

1. The simplest stationary solution is $u = v = 0$.

2. The linear stability analysis yield the following dispersion relation

$$\omega = a - q^2 \pm ibq^2. \tag{16.115}$$

The real part $a - q^2$ is negative for all q when $a < 0$. The bifurcation (or instability) occurs at $a = 0$ for which the critical wavenumber $q_c = 0$, and the imaginary part is zero. For $a > 0$ there is a band of wavevectors $\Delta q = \sqrt{a}$ corresponding to unstable modes with a non-zero imaginary component. This kind of instability is sometimes called type III instability.

3. The Turing condition is not met since there is no critical condition for which $q_c \neq 0$ with a null imaginary component.

9.4

1. The dispersion relation reads $\omega = f'(u_0) - Dq^2$ (prime designating a derivative with respect to the argument). The fastest growing mode satisfies $\omega'(q) = -2Dq = 0$, implying $q = 0$.

2. At $q = 0$, we have $\omega = f'(u_0)$ and we must have $f'(u_0) > 0$ for instability.

3. If u depends on x and y, the eigenmode is proportional to $e^{i(q_x x + q_y y)}$ where q_x and q_y are the components of the wavevector $\mathbf{q}$. Because the equation is isotropic (rotationally invariant) q_x and q_y play the same role and the dispersion relation reads $\omega = f'(u_0) - D(q_x^2 + q_y^2) = f'(u_0) - Dq^2$.

9.5

1. Linearization yields

$$\frac{\partial u}{\partial t} = \left[\epsilon - \left(1 + \frac{\partial^2}{\partial_x^2} \right)^2 \right] u, \tag{16.116}$$

with boundary conditions $u = u_x = 0$.

2. Setting $u = f e^{\omega t}$ yields

$$\epsilon f = \left(1 + \frac{\partial^2}{\partial_x^2} \right)^2 f, \tag{16.117}$$

at the instability threshold ($\omega = 0$).

3. Substituting $f = a e^{iqx}$ into the above equation provides us with $q_\pm = (1 \pm \sqrt{\epsilon})^{1/2}$ (and two other solutions given by $-q_\pm$, but the corresponding solutions can be absorbed in the complex conjugate part of the solution).

4. The general solution is $f = a e^{iq_+ x} + b e^{iq_- x} + c.c. = 2a \cos(iq_+ x) + 2b \cos(iq_- x)$ where we have taken a and b real to ensure symmetry in

the solution at $x = 0$ (thus boundary conditions on one side automatically imply that the boundary conditions on the other side are satisfied).

5. Imposing $f = f_x = 0$ at $x = L/2$ yields two algebraic and homogeneous equations for a and b. Imposing a zero determinant yields

$$q_+ \tan (Lq_+/2) = q_- \tan (Lq_-/2). \tag{16.118}$$

6. First we will show that $\epsilon_c = O(L^{-2})$. From equation (9.94), we obtain that ϵ_c is the lowest positive root of the function

$$f(\epsilon) = q_+ \tan (Lq_+/2) - q_- \tan (Lq_-/2). \tag{16.119}$$

It is easy to check that $f(0) = 0$. A more tedious calculation shows that $f'(\epsilon) > 0$ for any $\epsilon > 0$ such that $f(\epsilon)$ is defined. Let us write $L = 2\pi k + \alpha$, where k is any integer and $\alpha \in [0, 2\pi[$. The first two points for which the function $f(\epsilon)$ satisfies the above condition are given by the following equations:

$$\sqrt{1 + \sqrt{\epsilon_+}} = 2\pi(k + 1)/L, \quad \sqrt{1 - \sqrt{\epsilon_-}} = 2\pi k/L. \tag{16.120}$$

We can see that

$$\lim_{\epsilon \to \epsilon_\pm^-} f(\epsilon) = - \lim_{\epsilon \to \epsilon_\pm^+} f(\epsilon) = +\infty, \tag{16.121}$$

where superscript $-$ and $+$ refer to the left and right one-sided limits, respectively. Therefore, the function $f(\epsilon)$ is continuous and strictly increasing from $-\infty$ to ∞ on the interval between the points $\epsilon_\pm$. From this, we conclude that the function $f(\epsilon)$ has exactly one root between ϵ_+ and ϵ_-. Since $f(\epsilon)$ is positive for $0 < \epsilon < \min\{\epsilon_+, \epsilon_-\}$, we conclude that the root of $f(\epsilon)$ located between ϵ_+ and ϵ_- is indeed ϵ_c. Solving equations (16.120) for $\epsilon_\pm$, gives the desired scaling $\epsilon_c = O(L^{-2})$.

We can now find a more precise expression for ϵ_c. Assuming $\epsilon \ll 1$, we can write $q_\pm = 1 \pm \sqrt{\epsilon}/2 + O(\epsilon)$. Using the Taylor expansion of the tangent function, we get

$$q_\pm \tan (Lq_\pm/2)$$
$$= \tan \left[(Lq_\pm/2) + (q_\pm - 1) \sin (Lq_\pm/2) \cos (Lq_\pm/2) + O(L^{-2}) \right]. \tag{16.122}$$

Here we have brought the expansion inside the tangent function in order to avoid potentially diverging coefficients. It can easily be checked that the Taylor expansion of the right-hand side term is equivalent to the left-hand side term. We can now apply the arctangent function to equation (9.94) after the transformation (16.122). We get

$$\frac{L(q_+ - q_-)}{2} = l\pi + O(L^{-1}), \tag{16.123}$$

where l is any integer. Here we ignore the terms of order $O(L^{-1})$ because $q_+ - q_- = O(L^{-1})$ so both the left-hand side and the right-hand side are of order $O(1)$. We can now write

$$L\sqrt{\epsilon} = 2l\pi + O(L^{-1}). \tag{16.124}$$

Even though its limit is equal to zero, the left-hand side of equation (16.124) is nonetheless positive for a given value of L. Therefore, the lowest value of l in equation (16.124) can be 1 for $L \gg 1$. This gives us the desired approximation (9.95). It is easy to check that ϵ_c indeed lies between the points ϵ_+ and ϵ_-.

10.1

1. The next order term (after the cubic one) in the amplitude equation is $|A|^4 A$. Indeed the fourth order term is forbidden by translational symmetry, exactly as with the second order term (see section 10.2.1). We will see that other type of terms will enter the equation, like $i|A|^2 A_x$ and $iA^2 A_x^*$. This is a consequence of inversion symmetry for a steady bifurcation (see exercises of chapter 14 for a proof based on symmetry).

2. Let us write the steady version order by order. We find to order $\sqrt{\epsilon}$, ϵ, $\epsilon^{3/2}$, ϵ^2 and $\epsilon^{5/2}$ respectively

$$
\begin{aligned}
O(\sqrt{\epsilon}): \quad & L\phi_1 = 0 \\
O(\epsilon): \quad & L\phi_2 = -4\partial_{xX}(1 + \partial_x^2)\phi_1 \\
O(\epsilon^{3/2}): \quad & L\phi_3 = -4\partial_{xX}(1 + \partial_x^2)\phi_2 - 4\partial_{xxXX}\phi_1 - 2\partial_{XX}(1 + \partial_x^2)\phi_1 + b_3\phi_1^3 \\
O(\epsilon^2): \quad & L\phi_4 = -4\partial_{xX}(1 + \partial_x^2)\phi_3 - 4\partial_{xxXX}\phi_2 - 2\partial_{XX}(1 + \partial_x^2)\phi_2 \\
& \qquad - 4\partial_{xXXX}\phi_1 + 3b_3\phi_1^2\phi_2 \\
O(\epsilon^{5/2}): \quad & L\phi_5 = -4\partial_{xX}(1 + \partial_x^2)\phi_4 - 4\partial_{xxXX}\phi_3 - 2\partial_{XX}(1 + \partial_x^2)\phi_3 \\
& \qquad - 4\partial_{xXXX}\phi_2 - \partial_X^4\phi_1 + 3b_3(\phi_1\phi_2^2 + \phi_1^2\phi_3) \tag{16.125}
\end{aligned}
$$

with $L = (1 + \partial_x^2)^2$. The first two orders are solved by

$$\phi_1 = A_1(X)e^{ix} + c.c., \quad \phi_2 = A_2(X)e^{ix} + c.c., \tag{16.126}$$

where $A_{1,2}$ are complex amplitudes which are not yet determined. The third order equation is solved by (a homogeneous plus a particular solution)

$$\phi_3 = A_3(X)e^{ix} + C_3(X)e^{3ix} + c.c., \tag{16.127}$$

where by substitution we obtain

$$4A_1'' = -3b_3|A_1|^2 A_1, \quad C_3 = \frac{b_3}{64}A_1^3. \tag{16.128}$$

The prime designates the derivative with respect to X. In contrast to the situation where we were interested in the leading order equation, here we needed to determine C_3 for use in the next order solutions.

The fourth order solution is given by

$$\phi_4 = A_4(X)e^{ix} + C_4(X)e^{3ix} + c.c. \tag{16.129}$$

Here we will not need to know C_4. The solvability condition yields

$$4A_2'' = 4iA_1''' - 3b_3(2|A_1|^2A_2 + A_1^2A_2^*). \tag{16.130}$$

The fifth order solution reads

$$\phi_5 = A_5(X)e^{ix} + C_5(X)e^{3ix} + D_5e^{5ix} + c.c., \tag{16.131}$$

leading to

$$4A_3'' = 4iA_2''' + A_1'''' + 3b_3(2|A_2|^2A_1 + A_1^*A_2^2 + 2|A_1|^2A_3 + A_1^2A_3^*)$$
$$- \frac{3b_3^2}{64}A_1|A_1|^4. \tag{16.132}$$

We have three equations for A_1, A_2 and A_3 (equations (16.128), (16.130) and (16.132)). We can regroup them into a single equation by setting $Z = A_1 + \sqrt{\epsilon}A_2 + \epsilon A_3$. The trick consists in adding the three equations by multiplying them by 1, $\sqrt{\epsilon}$ and ϵ respectively, and we find to leading order

$$4Z'' = -3b_3|Z|^2Z + 4i\sqrt{\epsilon}Z''' + \epsilon\left(Z'''' - \frac{3b_3^2}{64}Z|Z|^4\right) + O(\epsilon^{3/2}). \tag{16.133}$$

Eliminating the higher order terms in an iterative way (for example by taking the derivative of the above equation, so that on the left-hand side we can express Z''' in terms of first order derivatives, starting from $-3b_3|Z|^2Z$, and so on), we finally obtain

$$4Z'' = -3b_3|Z|^2Z - 3ib_3\sqrt{\epsilon}(2Z'|Z|^2 + Z^2(Z^*)')$$
$$+ \epsilon\left\{\frac{9b_3}{2}(2Z|Z'|^2 + Z'^2Z^*) - \frac{327b_3^2}{64}|Z|^4Z\right\} + O(\epsilon^{3/2}). \tag{16.134}$$

10.2

1. Setting $A_0 = Re^{iqx}$ and inserting into the amplitude equation we obtain

$$R_\pm^2 = \frac{1 \pm \sqrt{1 + 4(\epsilon - q^2)}}{2}. \tag{16.135}$$

2. R_+ exists for $\epsilon \geqslant -1/4 + q^2$, while R_- exists for $1/4 \leqslant \epsilon < q^2 \leqslant 0$.

3. Setting $A = \rho e^{i\psi}$ we find that ρ and ψ obey the following set of equations:

$$\rho_T = \epsilon\rho + \rho_{XX} - \rho\psi_X^2 + \rho^3 - \rho^5,$$
$$\rho\psi_T = \rho\psi_{XX} + 2\rho_X\psi_X. \tag{16.136}$$

The steady solution corresponds to $A_0 = Re^{i\psi_0}$ with $\psi_0 = qX$.

4. Linearizing the above set around R and ψ_0 directly yields

$$\rho_{1T} = \epsilon\rho_1 + \rho_{1XX} - (2Rq\psi_{1X} + q^2\rho_1) + 3R^2\rho_1 - 5R^4\rho_1,$$
$$R\psi_{1T} = R\psi_{1XX} + 2q\rho_{1X}. \tag{16.137}$$

5. Imposing a zero determinant we obtain the following dispersion relation

$$\Omega^2 + 2\Omega(Q^2 + g) + Q^2(Q^2 + f) = 0$$
$$g = 2\epsilon - 2q^2 + R^2, \quad f = 4R^4 - 2R^2 - 4q^2 \tag{16.138}$$

where we have used the fact that R obeys $\epsilon - q^2 + R^2 - R^4 = 0$. The two eigenvalues are given by

$$\Omega_\pm = -(Q^2 + g) \pm \sqrt{(Q^2 + g)^2 - Q^2(Q^2 + f)} \tag{16.139}$$

Using the fact that $2g - f = 4q^2$, the term under the square root simplifies to $g^2 + 4q^2Q^2$. It is easy to see that $Q = 0$ is a marginal mode, which is a consequence of translational invariance. For $Q = 0$, $\Omega_+ = 0$ provided that $g > 0$. We can see by expansion around $Q = 0$ (since this is a dangerous mode) that $\Omega_+ \sim -(f/2g)Q^2$ and thus it is positive if $f < 0$ (instability). It can be checked from the system of algebraic equations (as we did for the Eckhaus instability) that for $\Omega = \Omega_+$ and small Q that the phase is large as compared to the amplitude, so that the phase mode dominates (as in the study of the Eckhaus instability).

6. Using the fact that the amplitude is slaved to the phase and is equal to its steady-state value (16.135), and substituting into equation (16.136) one easily finds

$$\psi_T = D\psi_{XX}, \quad D = \frac{f}{2R^2}\sqrt{1 + 4(\epsilon - q^2)}. \tag{16.140}$$

One sees that $D < 0$ (phase instability) if $f < 0$, in agreement with the previous analysis.

10.3

We have A, A^2A^*, A^3A^{*2}, A^4A^{*3}. These are the only terms that satisfy translational invariance, since if we change x into $x + x_0$, the terms must transform like $A \to Ae^{ix_0}$ to guarantee the translational invariance symmetry of the amplitude equation.

10.4

1. This equation can be written in a variational form whose functional reads

$$V = \int dx \left\{ -\frac{\epsilon \psi^2}{2} - \frac{\alpha \psi^3}{3} + \frac{\psi^4}{4} + \frac{1}{2} \left[(\partial_{x^2}^2 + 1)\psi \right]^2 \right\} \qquad (16.141)$$

2. This equation can not be written as a variational equation. The nonlinear term u_x^2 can not be written as a functional derivative.

10.5

We follow step-by-step the method developed in this chapter for the Swift-Hohenberg equation. The nonlinear term scales as $\epsilon^{3/2}$. By eliminating secular terms we obtain the following amplitude equation

$$A_T = A + 4A_{XX} - |A|^2 A. \qquad (16.142)$$

10.6

1. From equation (10.52) we obtain

$$\rho_t = \rho - \rho^3 + \rho_{xx} - \rho \psi_x^2,$$
$$\rho \psi_t = \rho \psi_{xx} + 2\rho_x \psi_x. \qquad (16.143)$$

The right-hand side of the second equation reads $(\rho^2 \psi_x)_x / \rho$, so that for a stationary solution we have $\rho^2 \psi_x = I_1$ which is a constant. Susbtituting this into the first equation we get

$$\rho_t = \rho - \rho^3 + \rho_{xx} - \frac{I_1^2}{\rho^3}. \qquad (16.144)$$

Multiplying this equation by ρ_x we can rewrite it as

$$\rho_x \rho_t = \partial_x \left[\frac{\rho^2}{2} - \frac{\rho^4}{4} + \frac{\rho_x^2}{2} + \frac{I_1^2}{2\rho^2} \right]. \qquad (16.145)$$

For a steady-state solution we have the following invariant

$$\frac{\rho^2}{2} - \frac{\rho^4}{4} + \frac{\rho_x^2}{2} + \frac{I_1^2}{2\rho^2} = I_2, \qquad (16.146)$$

where I_2 is a constant.

2. If, at a point, $\rho = 0$ then $I_1 = 0$ as well. Since I_1 is a constant it should thus equal zero at every x, meaning (since ρ is generally nonzero, except at isolated points) that $\psi_x = 0$. Since ψ_x measures (as seen in this chapter) the deviation from the critical wavenumber, this implies that the total pattern wavenumber is equal to q_c.

10.7

1. We take the amplitude equation as a starting point (from which we have studied the Eckhaus instability). This equation reads

$$A_t = \nu A + A_{xx} - |A|^2 A, \tag{16.147}$$

where ν is the small parameter that measures the distance from threshold (which we have often denoted by ϵ; we need to introduce another small parameter below which will be denoted by ϵ). We have seen in this chapter that $A_0 = \sqrt{\nu - q_0^2}e^{iq_0 x}$ is a steady solution. If we set $A = \rho(x,t)e^{i\phi(x,t)}$ we get from equation (16.147) the following system

$$\rho_t = \nu\rho - \rho^3 + \rho_{xx} - \rho\phi_x^2,$$
$$\rho\phi_t = \rho\phi_{xx} + 2\rho_x\phi_x, \tag{16.148}$$

where the steady solution corresponds to $\rho_0 = \sqrt{\nu - q_0^2}$ and $\phi_0 = q_0 x$.

2. When the system is perturbed the phase and amplitude deviate from their steady values $q_0 x$ and $\sqrt{\nu - q_0^2}$. We have seen in a linear stability analysis that the Eckhaus instability takes place if $\nu - 3q_0^2 < 0$. We will study the nonlinear behavior of the instability by focusing on the vicinity of the threshold. For that purpose we set $\epsilon^2 = \nu - 3q_0^2$ (ϵ is a small parameter). We have seen in section 10.3 that the growth rate of the phase instability behaves as $\sigma \sim (\nu - 3q_0^2)Q^2 - aQ^4 \sim \epsilon^2 Q^2 - aQ^4$, where Q is the wave-number of the instability and a is a quantity of order one (note that in that section $\nu = 1$, but if we leave ν arbitrary, then the factor $1 - 3q_0^2$ becomes $(\nu - 3q_0^2)$).

3. This means that the band of active modes (modes having a positive growth rate) extends from zero to about ϵ. This implies that the growth rate $\sigma \sim \epsilon^4$. In a multiscale spirit we introduce the following slow variables

$$X = \epsilon x, \quad T = \epsilon^4 t \tag{16.149}$$

4. The set of equations is easily written.

5. The deviation from the steady-state solution is expanded in power series of ϵ. We set $\phi = q_0 x + \epsilon\psi$ (with ψ of order one) to rewrite the above system (16.148) in terms of the slow variables

$$\epsilon^4\rho_T = \nu\rho - \rho^3 + \epsilon^2\rho_{XX} - \rho[q_0^2 + 2q_0\epsilon^2\psi_X + \epsilon^4\psi_X^2]$$
$$\rho\epsilon^5\psi_T = \epsilon^3\rho\psi_{xx} + 2\epsilon\rho_X(q_0 + \epsilon^2\psi_X) \tag{16.150}$$

6. We then look for solutions in the form of power series

$$\rho = \rho_0 + \epsilon\rho_1 + \epsilon^2\rho_2 + \cdots, \quad \psi = \psi_0 + \epsilon\psi_1 + \epsilon^2\psi_2 + \cdots \tag{16.151}$$

To order ϵ^0 we obtain

$$\rho_0 = 0, \quad \rho_0 = \pm\sqrt{\nu - q_0^2}. \tag{16.152}$$

Only the nonzero solution is of interest.

7. To order ϵ we obtain

$$(\nu - q_0^2)\rho_1 - 3\rho_0\rho_1 = 0, \tag{16.153}$$

implying $\rho_1 = 0$

8. To order ϵ^2 we obtain

$$\rho_2 = -\frac{q_0\psi_{0X}}{\rho_0}. \tag{16.154}$$

To order ϵ^3 we obtain

$$\rho_3 = -\frac{q_0\psi_{1X}}{\rho_0}. \tag{16.155}$$

At first sight the phase equation also provides an equation at this order which reads

$$\epsilon^3(\rho_0\psi_{0XX} + 2\rho_{2X}q_0) = 0. \tag{16.156}$$

However, close inspection shows that the above expression is of order ϵ^5. Indeed using equation (16.154) we find that equation 16.157 has a small prefactor, and can be rewritten as

$$\epsilon^3\frac{(\nu - 3q_0^2)}{\rho_0}\psi_{0XX} = 0. \tag{16.157}$$

Since $\nu - 3q_0^2 = \epsilon^2$, the left-hand side is of order ϵ^5.

To order ϵ^4 we obtain

$$\rho_4 = -\frac{q_0\rho_2\psi_{0X}}{\rho_0^2} - \frac{\psi_{0X}^2}{2\rho_0} + \frac{\rho_{2XX}}{2\rho_0^2} - \frac{q_0\psi_{2X}}{\rho_0} - \frac{3\rho_2^2}{2\rho_0}. \tag{16.158}$$

9. To order ϵ^5 we obtain (after using the results of previous orders)

$$\psi_{0T} = D\psi_{0XX} - \alpha\psi_{0XXXX} + \beta\psi_{0X}\psi_{0XX}. \tag{16.159}$$

10. We have

$$D = \frac{\nu - 3q_0^2}{\rho_0^2}, \quad \alpha = \frac{q_0^2}{\rho_0^4}, \quad \beta = \frac{q_0}{\rho_0^4}(3q_0^2 - 5\nu). \tag{16.160}$$

11. Setting $x = X\sqrt{D/\alpha}, t = TD^2/\alpha$ and $\phi = \psi\beta/\alpha$ transforms the equation into a canonical form.

12. By virtue of translational invariance we can change the phase by a contant term meaning that only derivatives of ψ enter the equation. This means, for example, that no term like ψ^2 could enter the equation. In addition, since the field is a fast variation times an amplitude, it reads as $Ae^{iq_c x} + A^* e^{iq_c x} = \rho e^{i\psi} e^{iq_c x} + \rho e^{-i\psi} e^{-iq_c x}$ (where ρ and ψ depend on slow variables X and T). Thus if we change x into $-x$ and ψ into $-\psi$ we have the same information. The evolution equation should thus exhibit this invariance. The first nonlinear term we could think of is ψ_X^2, but then the nonlinear equation would not be invariant under the simultaneous transformation $x \to -x$ and $\psi \to -\psi$. The next possibility allowed by symmetry is thus $\psi_X \psi_{XX}$. Note that the reflection symmetry evoked here is not anymore valid if we have a Hopf bifurcation (see chapter 12). As a consequence, the phase equation in the case of a Hopf bifurcation contains ψ_X^2 as a leading term (see an explicit derivation of the phase equation in section 12.5.1).

11.1

1. Since v is real (while q in general is not) the real part of the argument is $\Re(\omega) + v\Im(q)$. Imposing the stationarity condition one easily gets

$$v^* = \frac{\Re(\omega)}{\Im(q)}\bigg|_{q=q^*}. \tag{16.161}$$

2. The linear dispersion relation is given by $\omega = \epsilon - q^2$.

3. Using equations (11.27) and (16.161) we easily obtain $v^* = -2iq^*$ and $v^* = i(\epsilon - q^{*2})/q^*$. Since v^* is real, q^* is imaginary.

4. Setting $q^* = iq_i$ with q_i real, and confronting the two expressions of v^* we easily obtain $q^* = \pm i\sqrt{\epsilon}$ and $v = \pm 2\sqrt{\epsilon}$, giving us the same result as we found by other means earlier in this chapter.

5. For $v > v^*$ the real component $\Re(\omega) - v\Im(q)$ is negative and thus the solution is stable, while for $v < v^*$ we have the opposite, and the solution is unstable.

6. Since for $v = v^*$ the real component is zero, we are left with a purely imaginary argument, which reads $\Im(\omega) + v^*\Re(q^*)$. This is a pure oscillation. If the propagation speed is v^*, the relation between wavenumber and frequency is simply $q_s = \Omega/v^*$.

7. Since q^* is purely imaginary and ω purely real for that equation, we easily deduce that $q_s = 0$.

11.2

1. By using (11.27) we obtain $v^* = 4iq^*(1 - q^{*2})$, and by imposing that v^* must be real we obtain the following relation between the real and

imaginary part of q: $3q_i^2 = q_r^2 - 1$. Then using (11.27) and (16.161) we obtain $v^* = 8q_i(1 + 4q_i^2)$ and $v^* = (\epsilon + 8q_i^4 + 4q_i^2)/q_i$. Using the two relations we can write the values of q_i and q_r as functions of ϵ. Then substituting either relation in for v^*, we obtain the velocity in terms of ϵ.

2. For small ϵ we have $v^* = 4\sqrt{\epsilon}$, which is the same as that obtained from the amplitude equation. We also see that $q_r = 1 + O(\epsilon)$ and that $q_i \simeq \sqrt{\epsilon}/2$. The fact that here we have $q_r = 1$ is a result of the fact that the field ϕ in the Swift-Hohenberg equation is a product of the fast oscillation e^{ix} and amplitude A. So q_r corresponds the fast oscillations.

11.3

1. From (11.27) we obtain, by linearizing the equation, the following result:

$$v^* = (2i - 4iq^{*2} + 3bq^*)q^*. \tag{16.162}$$

As we have discussed above, if a fixed observer sees the instability developing at some given point, we have absolute instability. It is thus sufficient to analyze the result for $v = 0$. We find from the above equation three solutions

$$q^* = 0, \quad q_\pm^* = \frac{1}{8}[3ib \pm (32 - 9b^2)^{1/2}] \tag{16.163}$$

As seen above the growth rate of the packet is determined by $\Re(\omega) - v\Im(q)$. Since we have set $v = 0$, the growth rate is determined by $\Re(\omega)$. For q_+^* we have

$$\Re(\omega) = \frac{256 + 54b^4 - 288b^2}{1024}. \tag{16.164}$$

$\Re(\omega) > 0$ for $|b| < (9 - 3\sqrt{3})/16$ and for $|b| > (9 + 3\sqrt{3})/16$. These intervals of b correspond to absolute instability. Outside these intervals, the instability is convective. For further discussion about the spatiotemporal dynamics of the Benney equation, see reference [72].

12.1

1. Because of translational invariance, the equation for A contains odd powers of A and even powers of B. A similar argument applies for the equation for B. Furthermore, since A and B can be interchanged without affecting the physics, the coefficient in front of $|A|^2A$ is identical to that in front of $|B|^2B$, and the coefficient in front of $|A|^2B$ is identical to that in front of $|B|^2A$.

2. For the travelling wave $A \neq 0$ and $B = 0$, and in this case we have

$$A_t = \epsilon A - (1 + i\alpha)|A|^2A. \tag{16.165}$$

This equation admits a travelling solution of the form $A_0(t) = ae^{i\Omega t}$. Plugging into the above equation yields

$$a^2 = \epsilon, \quad \Omega = -\alpha a^2, \tag{16.166}$$

with the condition that $\epsilon > 0$. A linear stability analysis is performed by looking for solutions in the form

$$A = A_0(1 + A_1), \quad B = B_1, \tag{16.167}$$

where A_1 and B_1 are small amplitudes. Carrying these into the original equations for A and B we obtain, by keeping only first order terms in A_1 and B_1, the following system:

$$A_{1t} + i\Omega A_1 = \epsilon A_1 - (1 + i\alpha)a^2(A_1^* + 2A_1), \tag{16.168}$$
$$B_{1t} = \epsilon B_1 - \beta(1 + i\gamma)a^2 B_1. \tag{16.169}$$

We then decompose the amplitudes into their real and imaginary components, by setting $A_1 = X + iY$ and $B_1 = U + iV$. Substituting into the above system (note that the two equations for A_1 and B_1 are decoupled) and looking for solutions in the form $e^{\sigma t}$, we can then set the determinant to zero to obtain the following dispersion relations

$$\sigma(\sigma + 2\epsilon) = 0, \quad \sigma = \epsilon(1 - \beta) \pm i\sqrt{\beta\epsilon\gamma}. \tag{16.170}$$

The first relation shows that $\sigma = 0$ and $\sigma = -2\epsilon < 0$ (stability). However, the second relation shows that the real part is positive if $\beta < 1$, signaling instability. In conclusion, the traveling waves are stable for $\beta > 1$ and unstable otherwise.

3. For standing waves we have $A = B$ and the stationary solution $A_0(t) = ae^{i\Omega t}$, with

$$a^2 = \frac{\epsilon}{1 + \beta}, \quad \Omega = -(\alpha + \beta\gamma)a^2, \tag{16.171}$$

under the condition $\beta > -1$. A linear stability analysis is performed by looking for solutions of the form

$$A = A_0(1 + A_1), \quad B = A_0(1 + B_1) \tag{16.172}$$

Substituting into the original equations we obtain two coupled equations for A_1 and B_1. Decomposing each into a real and imaginary part, we obtain four algebraic and linear equations (after seeking solution in the form $e^{\sigma t}$). Finally, imposing a zero determinant, we obtain

$$\sigma^2 = 0, \quad \sigma = -2\epsilon, \quad \sigma = 2\epsilon\frac{\beta - 1}{\beta + 1}. \tag{16.173}$$

There is instability for $\beta > 1$, and stability for $\beta < 1$ (recall that $\beta > -1$ in order to guarantee existence of a standing wave solution).

12.2

The field of interest takes the form

$$h(x, X, T) = A(X, T)e^{iq_c x} + A^*(X, T)e^{-iq_c x}, \tag{16.174}$$

for a steady bifurcation and

$$h(x, t, X, T) = A(X, T)e^{i(q_c x + \omega_i t)} + A^*(X, T)e^{-i(q_c x + \omega_i t)}, \tag{16.175}$$

for a Hopf bifurcation. A is the complex amplitude that can be written as $A = \rho e^{i\psi}$ where ρ is the amplitude and ψ the phase. We note (as has been discussed before) a major difference between the two cases. For a steady bifurcation, if x is changed into $-x$, this is equivalent to changing A into A^* (meaning changing ψ into $-\psi$). This implies that the phase equation should enjoy the following simultaneous symmetry operations: $x \to -x$ and $\psi \to -\psi$. Thus, when seeking an equation for the phase of the form $\psi_T = F(\psi)$ (where F is some function of ψ and its spatial derivatives), we must keep only terms which are consistent with symmetry. On one hand, translational invariance imposes that only spatial derivatives of ψ enter the equation. Moreover, owing to the above inversion symmetry, the only linear terms which are allowed are those with even derivatives, like ψ_{XX}, ψ_{XXXX}. The first nonlinear term compatible with the above symmetries is $\psi_X \psi_{XX}$.

For a Hopf bifurcation, if x is changed into $-x$, there is no equivalence with the transformation $A \to A^*$ (meaning that there is no equivalence when changing ψ into $-\psi$). Due to the absence of this symmetry constraint, the first nonlinear term in its phase equation is ψ_X^2.

12.3

The next order terms in the equation for A are $|A|^4 A$, $|B|^4 A$, $|A|^2|B|^2 A$. These are the only nonlinear terms respecting translational invariance; terms like $|A||B|A$ respect this invariance, but not analyticity. Indeed, in order to obtain in the final equation $|B|$, the only possibility is to have a product $\sqrt{B}\sqrt{B^*}$, meaning that in the nonlinear expansion of terms in the amplitude we should have terms like $\sqrt{B}$, which are non-analytic. The same reasoning applies for the equation for B by interchanging A and B.

12.4

1. The dispersion relation reads $\omega^2 = 1 + q^2$. There is no instability since there is no exponential growth. We are in the presence of a nonlinear wave. The group velocity is $U = d\omega/dq = q/\sqrt{1 + q^2}$, which depends on q (the so-called dispersive medium). This means, since the general solution is a superposition of all possible waves with different wavenumbers, that each mode will have different velocity, and in general we have a wavepacket peaked around a specific wavenumber q_0.

2. A wavepacket is in general decomposed in Fourier space as

$$u(x,t) = \int_{-\infty}^{\infty} a(q) e^{i[\omega(q) - qx]} dq. \tag{16.176}$$

Note when we studied the nonlinear evolution in the presence of instability (chapter 10, section 6.7), we developed the solution in a similar way (we could as well use an integral instead of a sum without changing our final result). We can rewrite equation (16.176) as

$$u(x,t) = e^{i(\omega(q_0)t - q_0 x)} \int_{-\infty}^{\infty} a(q) e^{i\{[\omega(q) - \omega(q_0)]t - (q - q_0)x\}} dq,$$

$$= e^{i(\omega(q_0)t - q_0 x)} \int_{-\infty}^{\infty} a(q_0 + Q) e^{i[\Omega(Q)t - Qx]} dQ,$$

$$\simeq A(x,t) e^{i[\omega(q_0)t - q_0 x]}, \tag{16.177}$$

where we have set $Q = q - q_0$ and $\Omega(Q) = \omega(Q + q_0) - \omega(q_0)$. We have in mind a wavepacket which has an intrinsic oscillation with wavenumber q_0 and with an envelope $A(x,t)$, represented by the integral above. Since we assume the wavepacket is peaked around $q = q_0$, it is legitimate to expand $\Omega(Q)$ around $Q = 0$ to obtain

$$\Omega(Q) = U_0 Q + \frac{\omega_0'' Q^2}{2} + \cdots, \tag{16.178}$$

where $U_0 = U(q_0)$ and $\omega_0'' = (d^2\omega/dk^2)_{q=q_0}$.

3. Since Q is small we set $Q = \epsilon \bar{Q}$ (with $\bar{Q}$ of order unity), so that the envelope can be written as

$$A(x,t) = \int_{-\infty}^{\infty} \epsilon d\bar{Q} a(q_0 + \epsilon \bar{Q}) e^{i[U_0 \bar{Q} \epsilon t + \omega_0''(\bar{Q}^2/2)\epsilon^2 t - \bar{Q}\epsilon x]}. \tag{16.179}$$

It follows that the amplitude A depends on t and x via the combinations $\epsilon t \equiv T_1$, $\epsilon^2 t \equiv T_2$ and $\epsilon x \equiv X$, so that we can write $A(X, T_1, T_2)$. This means that the total field u is a function of five variables, the fast ones (x,t) and the slow ones (X, T_1, T_2), $u(x, t, X, T_1, T_2)$.

4. According to the multiscale method we must make the following substitutions

$$\partial_x \to \partial_x + \epsilon \partial_X,$$
$$\partial_t \to \partial_t + \epsilon \partial_{T_1} + \epsilon^2 \partial_{T_2}. \tag{16.180}$$

5. The Sine-Gordon equation now reads in terms of the new variables

$$(\partial_t + \epsilon \partial_{T_1} + \epsilon^2 \partial_{T_2})^2 u - (\partial_x + \epsilon \partial_X)^2 u + \sin(u) = 0. \tag{16.181}$$

6. Looking for solutions in the form $u = \epsilon u_1 + \epsilon^2 u_2 + \cdots$, we find to order ϵ

$$\partial_{tt} u_1 - \partial_{xx} u_1 + u_1 \equiv L u_1 = 0, \qquad (16.182)$$

where L is the left-hand side operator acting on u_1. This equation is solved by $e^{i[\omega(q_0)t - q_0 x]}$, where the integration constant depends on the slow variables, since the above equation is a partial differential equation and not a differential equation. We have thus for the solution

$$u_1 = A(X, T_1, T_2) e^{i(\omega(q_0)t - q_0 x)} + c.c. \qquad (16.183)$$

7. To next order we obtain

$$L u_2 = -2i(\omega_0 \partial_{T_1} A + q_0 \partial_X A) e^{i(\omega(q_0)t - q_0 x)} + c.c. \qquad (16.184)$$

Since the right-hand side is a null eigenfunction of the operator L, we must set the prefactor on the right-hand side to zero. This is the Fredholm alternative theorem (or solvability condition) which we have seen several times in this book. This yields

$$\partial_{T_1} A = -U_0 \partial_X A, \qquad (16.185)$$

where $U_0 q_0 / \omega_0$ is the group velocity at q_0. U_0 is the group velocity of the leading mode having wavenumber q_0. The general solution of this equation is $f(X - U_0 T_1, T_2)$ (with f any arbitrary function which is fixed by initial conditions). This means that to leading order the envelope moves in a shape-preserving manner with speed U_0; higher order terms will modify this conclusion. Note also that since the right-hand side of equation (16.184) is zero (as a result of solvability condition), the solution for u_2 is similar to that of u_1, so it can be absorbed in a redefinition of A (as we have seen in chapter 10), and thus we can set $u_2 = 0$.

8. To order ϵ^3 we obtain

$$L u_3 = -\partial_{T_1 T_1} u_1 + \partial_{XX} u_1 - 2\partial_{t T_2} u_1 + \frac{u_1^3}{6}. \qquad (16.186)$$

Eliminating secular terms (to fulfill the solvability condition), yields

$$\partial_{T_2} A = i\alpha \partial_{XX} A - i\beta |A|^2 A, \qquad (16.187)$$

where we have obtained $\alpha = (U_0^2 - 1)/(2\omega_0) = \omega_0''$ and $\beta = 1/(4\omega_0)$. This equation is known as the nonlinear Schrödinger equation. Among many interesting properties of this equation, we can cite the existence of soliton solution.

13.1

1. By neglecting nonlinear terms and setting $\phi = ae^{i(q_x x + q_y y) + \omega t}$, we can then substitute into the equation to obtain $\omega = \epsilon - (1 - q^2)^2$, with $q^2 = q_x^2 + q_y^2$.

2. The critical condition is fulfilled when $\omega = 0$ and $d\omega/dq = 0$, which gives us $q_c = 1$ and $\epsilon_c = 0$. For $\epsilon < 0$ all modes have a negative ω, while for $\epsilon > 0$ there is a band of q with positive ω (instability). The band of active modes is obtained by setting $\omega = 0$ which yields $\Delta q = q_+ - q_-$ with $q_\pm = \sqrt{1 \pm \sqrt{\epsilon}}$. Since we have a two-dimensional system, the critical condition $q = 1$ corresponds to a circle of radius one in the (q_x, q_y) plane. ω is positive for $q_- < q < q_+$, which represents a band between two circles of radii q_- and q_+.

3. Because we assume a stripe (or a band) structure along Ox, this means that $q_x = q_c$ at the bifurcation, and since there is no modulation along y, we have $q_y = 0$.

4. Setting $q = \sqrt{q_x^2 + q_y^2}$, $q_x = 1 + \Delta q_x$ and $q_y = \Delta q_y$, we easily obtain the desired result thanks to a Taylor expansion.

5. Reporting into the dispersion relation we obtain to leading order

$$\omega \simeq \epsilon - 4\left(\Delta q_x^2 + \Delta q_x \Delta q_y^2 + \frac{\Delta q_y^4}{4}\right), \tag{16.188}$$

while the other terms (like Δq_x^3) omitted, since they are of higher order, as it will appear clear below. We have thus $a = b = 4$ and $c = 1$. Suppose we focus on what happens along the q_x axis, $\Delta q_y = 0$, and we see that the band of active modes $\Delta q_x \sim \epsilon^{1/2}$. If now we set $\Delta q_x = 0$ in the above dispersion relation, we find that the active band scales as $\Delta q_y \sim \epsilon^{1/4}$. Had we retained Δq_x^3, we would then have obtained a correction of higher order in ϵ. Due to the scaling of Δq_x and Δq_y, we see that $\omega \sim \epsilon$.

6. The above scalings suggest the introduction of slow variables $X = \epsilon^{1/2}$, $Y = \epsilon^{1/4}$ and $T = \epsilon t$. Since the emerging bifurcating state is by definition a stripe, that state depends only on x (with $q_x = 1$ as the modulation wave-vector). Therefore any dependance on y will appear only in the envelope. The operators have to be changed as follows (in a multiscale spirit)

$$\partial_x \to \partial_x + \epsilon^{1/2}\partial_X,$$
$$\partial_y \to \epsilon^{1/4}\partial_Y,$$
$$\partial_t \to \epsilon\partial_T. \tag{16.189}$$

This has to be substituted then into the original equation.

7. We seek solutions in the form $\phi = \epsilon^{1/2}\phi_1 + \epsilon\phi_2 + \cdots$. We find to order $\epsilon^{1/2}$

$$L\phi_1 \equiv (1 + \partial_{xx})^2 \phi_1 = 0. \tag{16.190}$$

The solution of is given by (the integration factor depending on slow variables)
$$\phi_1 = A(X, Y, T)e^{ix} + c.c. \tag{16.191}$$
To order ϵ we find $L\phi_2 = 0$. Thus ϕ_2 has the same form as ϕ_1 and can be absorbed into ϕ_1 upon a redefinition of A. The interesting result emerges for order $\epsilon^{3/2}$. We find

$$L\phi_3 = \left\{ A - A_T - 3|A|^2 A - [-4A_{XX} + A_{YYYY} + 4iA_{XYY}] \right\} e^{ix} + \cdots, \tag{16.192}$$

where we have kept only the resonant terms on the right-hand side, since e^{ix} is a solution of L with zero eigenvalue (see chapter 10 where we discuss this at length). The solvability condition amounts to setting the coefficient of e^{ix} to zero, yielding

$$A_T = A - 3|A|^2 A + (2\partial_X - i\partial_{YY})^2 A, \tag{16.193}$$

which is the desired amplitude equation. By defining new variables, $\bar{X} = X/2$, $\bar{Y} = Y/\sqrt{2}$ and $\bar{A} = \sqrt{3}A$, we can easily obtain the canonical form.

13.2

The functional is given by

$$V[A] = \int dX dY \left[-|A|^2 + \frac{|A|^4}{2} + \left| \left(\partial_X + \frac{i}{2} \partial_{YY} \right) A \right|^2 \right] \tag{16.194}$$

13.3

1. By substituting into the equation we obtain $a_{q_0} = \sqrt{1 - q_0^2}$ with $-1 < q_0 < 1$.

2. Following the same strategy as with the Eckhaus instability studied in chapter 10 we obtain a set of two linear and coupled equations for ρ_1 and ψ_1. Imposing a null determinant for the system of equations for a and b, we obtain a relation between Ω and (Q_X, Q_Y) (dispersion relation).

3. The whole band with $q_0 < 0$ is unstable. However, we can check that we still have an Eckhaus instability (to do so, set $Q_Y = 0$). In conclusion, the band solution is stable for $0 < q_0 < 1/\sqrt{3}$. Recall that in one dimension we had a larger stable band $-1/\sqrt{3} < q_0 < 1/\sqrt{3}$. The present instability is called zig-zag. The band is limited on the right part of the wavenumber axis ($q_0 > 0$) by the Eckhaus instability and on the left part ($q_0 < 0$) by the zig-zag instability.

14.1

1. As we have seen in chapter 10 inversion symmetry corresponds to changing A into A^*. In other words, the equation obeyed by A^* and that obeyed by A

must be equivalent after having exchanged x into $-x$. Taking the conjugate of the equation gives the equation obeyed by A^*. Then changing x into $-x$ we obtain another equation. Since both equations are equivalent, the coefficients must be identical. This leads to $\alpha = \alpha^*$ and $\beta = \beta^*$, meaning that α and β are real, so all coefficients of the new terms (next order terms) are purely imaginary.

2. $\rho^2 = (\epsilon - q^2)/(1 - \alpha q)$.

3. We have already encountered the Euler-Lagrange (or functional) derivative. The following functional derivative

$$\frac{\delta}{\delta A^*} = \frac{\partial}{\partial A^*} - \frac{d}{dx}\frac{\partial}{\partial A_x^*} \tag{16.195}$$

applied to the functional $V[A]$ yields the following:

$$\frac{\partial V}{\partial A^*} = \epsilon A - |A|^2 A - \frac{i\alpha}{4}(2A_x|A|^2 - A_x^* A^2),$$

$$\frac{d}{dx}\frac{\partial V}{\partial A_x^*} = \frac{d}{dx}\left(-A_x + \frac{i\alpha}{4}A^2 A^*\right)$$

$$= -A_{xx} + \frac{i\alpha}{4}\left(A_x^* A^2 + 2|A|^2 A_x\right). \tag{16.196}$$

Regrouping the terms we find

$$\frac{\delta V}{\delta A^*} = \epsilon A - |A|^2 A + A_{xx} - i\alpha A_x|A|^2 = 0. \tag{16.197}$$

Thus the equation corresponds to a stationary solution of V (zero functional derivative). Note also that V is a real quantity and is thus a Lyapunov functional.

4. We have seen that this solution corresponds to a steady solution, which corresponds (according to the above result) to a functional derivative of V equal to zero. Thus this solution is an optimum of V.

5. V is optimal for $\rho^2 = (\epsilon - q^2)/(1 - \alpha q)$; in this case the potential per unit length is $(\epsilon - q^2)^2/[2(1 - \alpha q)]$. We write $A_t = -\delta V/\delta A^*$ and take $-V$ (instead of V) in order to obtain the dynamical equation. Therefore, the optimum corresponds to a minimum of $-V$ (and maximum of V). The minimum is obtained by differentiating with respect to q $(\epsilon - q^2)^2/[2(1 - \alpha q)]$ which tells us that the minimum of $-V$ occurs for $q = [2 - \sqrt{4 - 3\alpha^2\epsilon}]/3 \simeq \alpha\epsilon/4$, where in the last expression we have exploited the smallness of ϵ. We have seen in this chapter that in the leading order equation the optimum is obtained for $q = 0$. Thus, taking into account the higher order terms in the amplitude equation modifies the conclusion.

6. In the presence of the β-term, the additional terms in the functional could be $\sim A_x^* A^* A^2 - c.c.$ However, application of the functional derivative operators produces extra terms of the form $\beta |A|^2 A_x$, which are already accounted for in the α-term.

7. By substituting into the equation we easily obtain

$$\rho = \left[\frac{\epsilon - q^2}{1 + (\beta - \alpha)q} \right]^{1/2}. \tag{16.198}$$

8. The local wavenumber is a derivative of a phase, because q is defined related to ϵ which is a space-dependent quantity.

9. If we substitute into the equation we obtain

$$(\epsilon - \psi_x^2)\rho - \rho^3[1 + (\beta - \alpha)\psi_x]$$
$$= -\rho_{xx} - 2i\rho_x\psi_x + i\rho\psi_{xx} + i\rho_x\rho^2(\alpha + \beta). \tag{16.199}$$

We see that if ρ and $q = \psi_x$ are independent of x, then the right-hand side vanishes and we obtain the above discussed solution.

10. If ϵ depends on x the assumption of constant ρ and ψ_x ceases to be valid and the right-hand side is nonzero.

11. The ramp variation is slow, so we define $X = \nu x$ with ν a small parameter. We expect the solution to depend on x and X separately. The operators must thus be rewritten in a multiscale spirit as

$$\partial_x \to \partial_x + \nu\partial_X$$
$$\partial_{xx} \to \partial_{xx} + 2\nu\partial_{xX} + \nu^2\partial_{XX}. \tag{16.200}$$

12. To order ν^0 we obtain

$$\epsilon f + f_{xx} - |f|^2 f = i(\alpha f_x |f|^2 + \beta f_x^* f^2), \tag{16.201}$$

which is solved by $f = \rho e^{i\psi}$, with $q = \psi_x$.

$$\rho = \left[\frac{\epsilon(X) - q(X)^2}{1 + (\beta - \alpha)q(X)} \right]^{1/2}. \tag{16.202}$$

13. By expanding we easily obtain this equation.

14. Taking the derivative of equation (14.23) with respect to x, we find that the resulting operator acting on f_x is exactly that appearing on the left-hand side of equation (14.24). In other words, the homogeneous part of equation (14.24) is solved by the translational mode f_x.

15. Applying the solvability condition we obtain easily the result (since f_x is nonzero in general). Note that ρ depends on X only via q and ϵ, so that we can write $\rho_X = f_q q' + \epsilon' f_\epsilon$ where differentiation is subscripted and prime designates a derivative with respect to X. The result of solvability condition reads

$$q'[\rho + \rho_q(2q - (\alpha + \beta)\rho^2)] = \epsilon' \rho_\epsilon [(\alpha + \beta)\rho^2 - 2q]. \qquad (16.203)$$

16. Truncating equation (16.202) to the leading order yields $\rho \simeq \sqrt{\epsilon}$. Taking the derivative with respect to X and substituting into (14.25) provides us with

$$\epsilon'[2q - (\alpha + \beta)\epsilon] + 2q'\epsilon = 0, \qquad (16.204)$$

17. This can easily be integrated to yield

$$\frac{d\phi}{dx} = 0, \quad \phi = 2q\epsilon - (\alpha + \beta)\frac{\epsilon^2}{2}. \qquad (16.205)$$

Thus ϕ is a constant (reminiscent of adiabatic invariants in mechanics).

18. Since for $\epsilon = 0$ we have $q = 0$, this implies that the constant is zero. Setting $\phi = 0$, the result follows. Had we considered the leading order solution one would have obtained that the ramp problem gives $q = 0$ as the selected wavenumber, like with the variational principle. We have seen above that if we have a ramp, then higher order terms must be taken into account for consistency reasons, and this leads to a different wavenumber.

This problem is extracted from an article by Pomeau and Zaleski [86].

Bibliography

[1] M. Abramowitz, I.A. Stegun (eds.), *Handbook of Mathematical Functions* (Dover Publications, New York, 1970)

[2] I.S. Aranson, L. Kramer, The world of the complex Ginzburg–Landau equation. Rev. Mod. Phys. **74**(1), 99 (2002)

[3] D.G. Aronson, H.F. Weinberger, Multidimensional nonlinear diffusion arising in population genetics. Adv. Math. **30**, 33–76 (1978)

[4] R. Asai, E. Taguchi, Y. Kume, M. Saito, S. Kondo, Zebrafish leopard gene as a component of the putative reaction-diffusion system. Mech. Dev. **89**, 87–92 (1999)

[5] P. Ball, *Branches, Shapes, Flow: Nature's Patterns: a Tapestry in Three Parts* (Oxford University Press, 2009)

[6] Y. Bar-Yam, *Dynamics of Complex Systems*, The Advanced Book Studies in Nonlinearity Series (Westview Press, 2005)

[7] N. Bekki, K. Nozaki, Formations of spatial patterns and holes in the generalized Ginzburg–Landau equation. Phys. Lett. **110A**(3), 133–135 (1985)

[8] T.B. Benjamin, J.E. Feir, The disintegration of wave trains in deep water. J. Fluid Mech. **27**(3) (1967)

[9] P. Bergé, Y. Pomeau, C. Vidal, *Order within Chaos* (Hermann, 1988)

[10] A. Bernheim-Groswasser, S. Wiesner, R.M. Golsteyn, M.F. Carlier, C. Sykes, The dynamics of actin-based motility depend on surface parameters. Nature **417**(6886), 308–311 (2002)

[11] A. Boettigera, B. Ermentroutb, G. Oster, Solitary modes and the Eckhaus instability in directional solidification. PNAS **106**(22), 6837–6842 (2009)

[12] J. Boissonade, V. Castets, P. De Kepper, E. Dulos, Les structures de Turing. Pour la Science **157**, 10–12 (1990)

[13] A.J. Bray, Theory of phase-ordering kinetics. Adv. Phys. **43**, 357–459 (1994)

[14] P. Brunet, J.-M. Flesselles, L. Limat, Parity breaking in a one-dimensional pattern: a quantitative study with controlled wavelength. Europhys. Lett. **56**, 221–226 (2001)

[15] V. Castets, E. Dulos, J. Boissonade, P. De Kepper, Experimental evidence of a sustained standing Turing-type nonequilibrium chemical pattern. Phys. Rev. Lett. **64**(24), 2953–2956 (1990)

[16] S. Chandrasekhar, *Hydrodynamic and Hydromagnetic Stability*, International Series of Monographs on Physics (Clarendon, Oxford, 1961)

[17] F. Charru, *Hydrodynamic Instabilities* (Cambridge University Press, 2011)

[18] H. Chate, Spatiotemporal intermittency regimes of the one-dimensional complex Ginzburg–Landau equation. Nonlinearity **7**(1) (1994)

[19] Y. Chen, A.F. Schier, The zebrafish nodal signal squint functions as a morphogen. Nature **411**, 607–610 (2001)

[20] S. Ciliberto, P. Coullet, J. Lega, E. Pampaloni, C. Perez-Garcia, Defects in roll-hexagon competition. Phys. Rev. Lett. **65**(19), 2370–2373 (1990)

[21] P. Coullet, R.E. Goldstein, G.H. Gunaratne, Parity-breaking transitions of modulated patterns in hydrodynamic systems. Phys. Rev. Lett. **63**(18), 1954–1957 (1989)

[22] P. Coullet, G. Iooss, Instabilities of one-dimensional cellular patterns. Phys. Rev. Lett. **64**(8), 866–869 (1990)

[23] M. Cross, H. Greenside, *Pattern Formation and Dynamics in Nonequilibrium Systems* (Cambridge University Press, 2009)

[24] M.C. Cross, P.C. Hohenberg, Pattern formation outside of equilibrium. Rev. Mod. Phys. **65**(3), 851–1112 (1993)

[25] Z. Csahok, C. Misbah, A. Valance, A class of nonlinear equations based on geometry and conservation. Physica D **128**, 87–100 (2000)

[26] G. Dee, J.S. Langer, Propagating pattern selection. Phys. Rev. Lett. **50**, 383–386 (1983)

[27] M. Demazure, *Bifurcations and Catastrophes: Geometry of Solutions to Nonlinear Problems*, Universitext (Springer, 2000)

[28] W.L. Ditto, S.N. Rauseo, L.L. Spano, Experimental control of chaos. Phys. Rev. Lett. **65**(26), 3211–3214 (1990)

[29] P.G. Drazin, *Nonlinear Systems* (Cambridge University Press, 2002)

[30] W. Eckhaus, *Studies in Non-linear Stability Theory*. Springer Tracts in Natural Philosophy vol. 6 (Springer-Verlag, Berlin, 1965)

[31] G. Faivre, S. de Cheveigne, C. Guthmann, P. Kurowski, Solitary tilt waves in thin lamellar eutectics. Europhys. Lett. **9**, 779–784 (1989)

[32] S. Fauve, in *Hydrodynamics and Nonlinear Instabilities*. Collection Aléa-Saclay: Monographs and Texts in Statistical Physics (Cambridge University Press, 1998)

[33] M.J. Feigenbaum, Quantitative universality for a class of non-linear transformations. Journal of Statistical Physics **19**, 25–52 (1978)

[34] T. Galla, E. Moro, Defect formation in the Swift–Hohenberg equation. Phys. Rev. E **67**, 035101 (2003)

[35] A. Garfinkel, M. Spano, W.L. Ditto, J. Weiss, Controlling cardiac chaos. Science **257**(5074), 1230–1235 (1992)

[36] Z. Gills, C. Iwat, R. Roy, I.B. Schwartz, I. Triandaf, Tracking unstable steady states: Extending the stability regime of a multimode laser system. Phys. Rev. Lett. **69**(22), 3169–3172 (1992)

[37] P. Glendinning, *Stability, Instability and Chaos: an Introduction to the Theory of Nonlinear Differential Equations* (Cambridge University Press, 1994)

[38] C. Godrèche (ed.), *Lasers* (Wiley, New York, 1988)

[39] R.J. Goldstein, D.J. Graham, Stability of a horizontal fluid layer with zero-shear boundaries. Phys. Fluids **12**, 1133 (1969)

[40] J. Guckenheimer, P. Holmes, *Nonlinear Oscillations, Dynamical Systems, and Bifurcations of Vector Fields* (Springer, New York, 1983)

[41] E. Guyon, J.-P. Hulin, L. Petit, *Ce que disent les fluides* (Belin, 2005)

[42] E. Guyon, J.-P. Hulin, L. Petit, *Hydrodynamique physique*. Savoirs Actuels Series (EDP Sciences, Les Ulis, 2012)

[43] M.P. Hassell, *The Dynamics of Arthropod Predator-Prey Systems* (Princeton University Press, 1978)

[44] E. Hernández-Garcia, M. San Miguel, R. Toral, Noise and pattern selection in the one-dimensional Swift–Hohenberg equation. Physica D **61**, 159–165 (1992)

[45] P.C. Hohenberg, J.B. Swift, Effects of additive noise at the onset of Rayleigh–Bénard convection. Phys. Rev. A **46**, 4773–4785 (1992)

[46] R.B. Hoyle, *Pattern Formation: an Introduction to Methods* (Cambridge University Press, 2006)

[47] B. Janiaud, A. Pumir, D. Bensimon, V. Croquette, H. Richter, L. Kramer, The Eckhaus instability for traveling waves. Physica D **55**, 269–286 (1992)

[48] K. John, P. Peyla, K. Kassner, J. Prost, C. Misbah, Nonlinear study of symmetry breaking in actin gels: Implications for cellular motility. Phys. Rev. Lett. **100**(6), 068101 (2008)

[49] B. Kaoui, G. Biros, C. Misbah, Why do red blood cells have asymmetric shapes even in a symmetric flow? Phys. Rev. Lett. **103**(18), 188101 (2009)

[50] K. Kassner, C. Misbah, Parity breaking in eutectic growth. Phys. Rev. Lett. **65**(12), 1458–1461 (1990)

[51] K. Kassner, C. Misbah, Amplitude equations for systems with long-range interactions. Phys. Rev. E **66**(2), 026102 (2002)

[52] K. Kassner, C. Misbah, Non-linear evolution of a uniaxially stressed solid: a route to fracture? Europhys. Lett. **28**(4), 245 (1994)

[53] K. Kassner, C. Misbah, H. Müller-Krumbhaar, Transition to chaos in directional solidification. Phys. Rev. Lett. **67**(12), 1551–1554 (1991)

[54] K. Kassner, C. Misbah, H. Müller-Krumbhaar, A. Valance, Directional solidification at high speed. II. Transition to chaos. Phys. Rev. E **49**(6), 5495–5516 (1994)

[55] D.M Kern, C.H. Kim, Iodine catalysis in chloride-iodine reaction. J. Am. Chem. Soc. **87**, 5309–5313 (1965)

[56] L. Kramer, E. Ben-Jacob, H. Brand, M.C. Cross, Wavelength selection in systems far from equilibrium. Phys. Rev. Lett. **49**, 1891–1894 (1982)

[57] Y. Kuramoto, T. Tsuzuki, On the formation of dissipative structures in reaction-diffusion systems. Prog. Theor. Phys. **54**, 687 (1975)

[58] L.D Landau, E.M Lifshitz, *Mechanics* (Elsevier Butterworth–Heinemann, Oxford, 1976)

[59] J. Lega, Traveling hole solutions of the complex Ginzburg–Landau equation: a review. Physica D **152**, 269–287 (2001)

[60] I. Lengyel, I.R. Epstein, Modeling of Turing structures in the chlorite–iodide–malonic acid-starch reaction system. Science **251**, 650–652 (1991)

[61] H. Levine, W.J. Rappel, Numerical study for traveling waves in directional solidification. Phys. Rev. A **42**(12), 7475–7478 (1990)

[62] R.T. Liu, S.S. Liaw, P.K. Maini, Two-stage Turing model for generating pigment patterns on the leopard and the jaguar. Physical Review E **74**(1), 011914 (2006)

[63] T.P. Loisel, R. Boujemaa, D. Pantaloni, M.F. Carlier, Reconstitution of actin-based motility of listeria and shigella using pure proteins. Nature **401**, 613–616 (1999)

[64] A.J. Lotka, *Elements of Physical Biology* (Williams and Wilkins Co, Baltimore, 1925)

[65] D.J. Ludwig, C.S. Holling, Qualitative analysis of insect outbreak systems: the spruce budworm and forest. J. Anim. Ecol. **47**, 315 (1978)

[66] M.-A. Mader, V. Vitkova, M. Abkarian, A. Viallat, T. Podgorski, Dynamics of viscous vesicles in shear flow. Eur. Phys. J. E **19**, 389–397 (2006)

[67] P. Manneville, *Instabilities, Chaos and Turbulence* (Imperial College Press, 2009)

[68] J. De Meeus, J. Sigala, Cinétique et mécanisme de la réduction du chlorite par l'iodure. J. Chim. Phys. Biol. **63**, 453–459 (1966)

[69] H. Meinhardt, *The Algorithmic Beauty of Sea Shells* (Springer-Verlag, Berlin, 2003)

[70] H. Meinhardt, *Models of Biological Pattern Formation* (Academic Press, London, 1982)

[71] C. Misbah, Vacillating breathing and tumbling of vesicles under shear flow. Phys. Rev. Lett. **96**, 028104 (2006)

[72] C. Misbah, O. Pierre-Louis, Pulses and disorder in a continuum version of step-bunching dynamics. Phys. Rev. E **53**(5), R4318–R4321 (1996)

[73] C. Misbah, O. Pierre-Louis, Y. Saito, Crystal surfaces in and out of equilibrium: a modern view. Rev. Mod. Phys. **82**(1), 981–1040 (2010)

[74] C. Misbah, A. Valance, Secondary instabilities in the stabilized Kuramoto–Sivashinsky equation. Phys. Rev. E **49**(1), 166–183 (1994)

[75] F.C. Moon, *Chaotic and Fractal Dynamics: an Introduction for Applied Scientists and Engineers* (Wiley, 1992)

[76] J. Murray, *Mathematical Biology* (Spinger-Verlag, Berlin, 1993)

[77] A.H. Nayfeh, *Introduction to Perturbation Techniques* (Wiley Inter-Science, New York, 1981)

[78] F.J. Nedelec, T. Surrey, A.C. Maggs, S. Leibler, Self-organization of microtubules and motors. Nature **389**(22), 305–308 (1997)

[79] G. Odell, *Qualitative Theory of Systems of Ordinary Differential Equations, Including Phase Plane Analysis of the Hopf Bifurcation Theorem* (Cambridge University Press, 1980)

[80] E. Ott, *Chaos in Dynamical Systems* (Cambridge University Press, 1993)

[81] Q. Ouyang, H.L. Swinney, Transition from a uniform steady state to hexagonal and striped Turing patterns. Nature **352**, 610–612 (1991)

[82] V. Petrov, V. Gaspar, J. Masere, K. Showalter, Controlling chaos in the Belousov–Zhabotinsky reaction. Nature **361**, 240–243 (1993)

[83] P. Politi, C. Misbah, When does coarsening occur in the dynamics of one-dimensional fronts? Phys. Rev. Lett. **92**(9), 090601 (2004)

[84] Y. Pomeau, P. Manneville, Intermittent transition to turbulence in dissipative dynamical systems. Comm. Math. Phys **74**(2), 189–197 (1980)

[85] Y. Pomeau, P. Manneville, Stability and fluctuations of a spatially periodic convective flow. J. Phys. **40**(23), 609–612 (1979)

[86] Y. Pomeau, S. Zaleski, Pattern selection in a slowly varying environment. J. Phys. Lett. **44**, L135 (1983)

[87] T. Poston, I. Stewart, *Catastrophe Theory and its Applications* (Courier Corporation, 2014)

[88] M. Rabaud, S. Michalland, Y. Couder, Dynamical regimes of directional viscous fingering: Spatiotemporal chaos and wave propagation. Phys. Rev. Lett. **64**(2), 184–187 (1990)

[89] F. Rioual, T. Biben, C. Misbah, Analytical theory of vesicle tumbling. Phys. Rev. E **69**, 061914 (2004)

[90] D. Ruelle, F. Takens, On the nature of turbulence. Comm. Math. Phys. **20**(2), 167–191 (1971)

[91] P.T. Saunders, *An Introduction to Catastrophe Theory* (Cambridge University Press, 1986)

[92] J. Schnackenberg, Simple chemical reaction systems with limit cycle behaviour. J. Theor. Biol. **81**, 389–400 (1979)

[93] M. Schroeder, *Fractals, Chaos, and Power Laws: Minutes from an Infinite Paradise* (Freeman, 1991)

[94] A.A. Seyraniana, A.P. Seyranian, The stability of an inverted pendulum with a vibrating suspension point. *J. App. Math. Mech.* **70**, 754–761 (2006)

[95] B.I. Shraiman, Order, disorder, and phase turbulence. Phys. Rev. Lett. **57**(3), 325–328 (1986)

[96] A.J. Simon, J. Bechhoefer, A. Libchaber, Solitary modes and the Eckhaus instability in directional solidification. Phys. Rev. Lett. **61**(22), 2574–2577 (1988)

[97] G.I. Sivashinsky, Nonlinear analysis of hydrodynamic instability in laminar flames: 1. Derivation of basic equations. Acta Astronautica **4**, 1177–1206 (1977)

[98] H.E. Stanely, *Introduction to Phase Transitions and Critical Phenomena* (Oxford University Press, 1971)

[99] S.H. Strogatz, *Nonlinear Dynamics and Chaos* (Levant Books, Kolkata, India, 2007)

[100] G.I. Taylor, D.H. Michael, On making holes in a sheet of fluid. J. Fluid Mech. **58**, 625–639 (1973)

[101] R. Thom, *Structural Stability and Morphogenesis* (Westview Press, Oxford, 1989)

[102] D'Arcy Thompson, *On Growth and Form* (Cambridge University Press, 1961)

[103] C. Tresser, P. Coullet, Iterations of endomorphisms and renormalization group. Comptes Rendus Acad. Sci. (Paris) **287A**, 577–580 (1978)

[104] A. Turing, The chemical theory of morphogenesis. Phil. Trans. Roy. Soc. B **237**, 32 (1952)

[105] W. van Saarloos, Front propagation into unstable states: Marginal stability as a dynamical mechanism selection. Phys. Rev. A **37**, 211–229 (1988)

[106] W. van Saarloos, P.C. Hohenberg, Fronts, pulses, sources and sinks in generalized complex Ginzburg–Landau equations. Physica D **56**, 303–367 (1992)

[107] A. Varlamov, J. Villain, A. Rigamonti, *Le Kaleidoscope de la Physique* (Belin, 2014)

[108] V. Volterra, Variazioni e fluttuazioni del numero di individui in specie animali conviventi. Mem. R. Accad. Naz. dei Lincei, Ser. VI (1926)

[109] T.A. Witten, L.M. Sander, Diffusion-limited aggregation, a kinetic critical phenomenon. Phys. Rev. Lett. **47**(19), 1400–1403 (1981)

[110] D. Yarar, W. To, A. Abo, M.D. Welch, The Wiskott–Aldrich syndrome protein directs actin-based motility by stimulating actin nucleation with the Arp2/3 complex. Curr. Bio **9**, 555–558 (1999)

Index

© Springer Science+Business Media B.V. 2017
C. Misbah, *Complex Dynamics and Morphogenesis*,
DOI 10.1007/978-94-024-1020-4

Printed in the United States
By Bookmasters